# God Made
# LIFE

Science/Worldview | 6-8 Grade

Copyright © 2021 by Generations. All rights reserved. No part of this book may be reproduced in any form or by any means without permission in writing from the publisher.

Printed in the United States of America.

ISBN: 978-1-954745-26-1

Cover Design: Justin Turley
Interior Design: Sarah Lee Bryant

Cover Image: Baby sucking thumb at 20 weeks. LENNART NILSSON, TT/SCIENCE PHOTO LIBRARY. Used by permission.

Published by:
Generations
19039 Plaza Drive Ste 210
Parker, Colorado 80134
Generations.org

Unless otherwise noted, Scripture taken from the New King James Version®. Copyright © 1982 by Thomas Nelson. Used by permission. All rights reserved.

For more information on this and other titles from Generations, visit Generations.org or call (888) 389-9080.

# TABLE OF CONTENTS

**Chapter 1**
What is True?..................................................................................7

**Chapter 2**
What is Life?..................................................................................27

**Chapter 3**
God Loves Reproduction ..............................................................59

**Chapter 4**
God Made Microscopic Organisms ..............................................83

**Chapter 5**
God Made Plants ........................................................................119

# GOD MADE LIFE

### Chapter 6
God Made Food ................................................................................................. 155

### Chapter 7
God Made Animals ............................................................................................ 187

### Chapter 8
God Made Man .................................................................................................. 221

### Chapter 9
God Sustains Human Life .................................................................................. 257

### Chapter 10
God Restores and Reproduces Life ................................................................... 285

Endnotes ............................................................................................................ 317

Image Credits .................................................................................................... 319

Index .................................................................................................................. 321

# CHAPTER 1
# WHAT IS TRUE?

> Thomas said to Him, "Lord, we do not know where You are going, and how can we know the way?" Jesus said to him, "I am the way, the truth, and the life. No one comes to the Father except through Me." (John 14:5-6)

During the trial of Jesus, Pilate asked Him, "What is truth?" The answer to this question is simple. Jesus already told His disciples in John 14:6: "I am the way, the truth, and the life." Surely, the Son of God Himself is the truth. Jesus also said that God's Word is truth. Above all, we know for certain that all of God's Word is trustworthy. The Bible is absolutely true, and we can depend on it.

Humans want to know the truth. They seek the truth, but many do not look for truth in the right places. They do not look to God for truth. They seek truth, but they do not like the truth when they hear it. There are many lies in the world, and people are often content to believe lies. Those who reject God think they are smart, and they do not trust God to give them truth. They are too proud to submit to God's Word, the very source of truth.

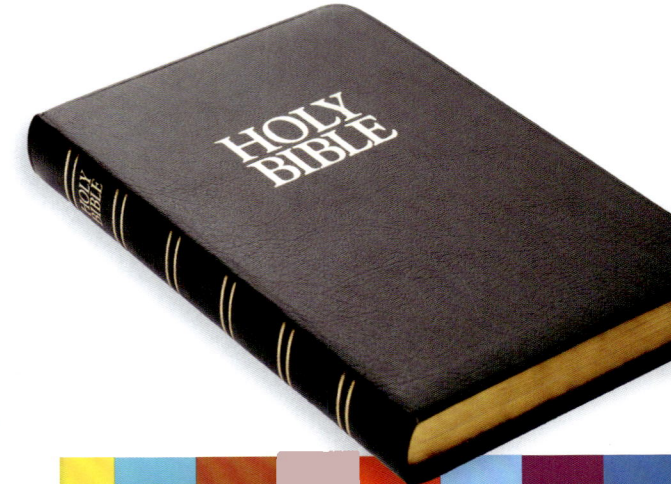

# GOD MADE LIFE

But we know that we are not very smart. We do not know all things, and we know that only God knows everything about everything. So, we cannot live without the Word of God. We cannot know anything for sure unless God speaks to us through His Word.

## Does Science Give Us Knowledge of Truth?

*The words of the LORD are pure words,*
*Like silver tried in a furnace of earth,*
*Purified seven times. (Psalm 12:6)*

As we study science, we want to be careful not to be proud. This is a strong temptation for scientists in our day. There are certain things science cannot do. Science cannot provide truth with total certainty. We can be partially sure of scientific conclusions. We can be totally sure of the truths in God's Word. We can't know *for sure* what is true by scientific studies. Science can give us clues about what is true. And science can help us solve problems. It can help make life easier. It can help to find a cure for diseases.

Science can help us answer certain questions. Some children might think that the moon is made out of cheese. By investigating the moon, we can come to the conclusion that the moon is NOT made out of cheese. In fact, men have explored the moon, and they have made sure that it is not made out of cheese. Science can help us disprove certain theories.

Does lemonade cure cancer?

Consider the following theories, for example:

- *You can cure the disease of cancer by drinking lots of lemonade. Is that true?*
- *Eating three eggs every day will cause you to get a terrible heart disease and make you die sooner than the average person. Is that true?*
- *Leaves on trees turn yellow and brown in the autumn or winter because of the colder temperatures. Is that true?*

Is the moon made out of cheese?

**CHAPTER 1: WHAT IS TRUE?**

1. Just guessing or theorizing is not a good way to do science. A friend may have told you that leaves turn yellow in the fall because of the colder weather. But then, you could test out his opinion on the matter. You could place green leaves in a refrigerator, and you would quickly learn that they do not turn yellow in the cold. So your friend's guess was wrong. Scientists who make guesses or theories can fool a lot of people until their theories are tested.

2. How would you know if lemonade could cure cancer? How much lemonade would it take to cure cancer? You would have to test your theory on a hundred patients or a thousand patients who were willing to try it out. Even if your theory seemed to help a few people, that doesn't necessarily prove the theory. When a medicine is tested on 10 people who are suffering from some disease, the medicine might only help two of the patients a little bit. It might cure one patient, while it doesn't help the remaining seven people at all. At the end of the study, we don't know for sure that the medicine helped any of the patients. The one patient, or 10% of the patients, was cured but might have gotten better for some other reason. Science does not give us certainty. Science can only provide probable solutions or probable answers for our problems.

3. How would you know for sure that people who ate three eggs a day would end up with more cases of heart disease than the average person later in life? There are many theories about diet. What is a good diet? What kind of foods would make for a bad diet? Some people say eggs are good for you, and other people say eggs are bad for you. Of course, you would have to test out your theory. You would have to get 1,000 people to agree to eat three eggs a day for 20 years (or for a very long time). Then, you would have to ask another set of 1,000 people not to eat eggs for 20 years. Suppose that 300 of the egg-eaters succumbed to heart disease and died at an earlier age than the average person. And then, suppose that only 100 of the non egg-eaters succumbed to heart disease and suffered early deaths. Would this prove that eating three eggs a day was bad for you? But what if the egg-eaters were more likely to smoke cigarettes and drink whiskey? Or what if the non

9

## GOD MADE LIFE

egg-eaters tended to exercise more than the egg-eaters during the 20 years that you were conducting the test? Science can give you a hint about truth, but there is always a possibility that the theory is still wrong.

## The Limitations of Science

There are many questions that science cannot answer. Science cannot answer the question of origins. How did this world get here? What is a thought? What is right and wrong? What is a worthwhile goal for human existence? What is the invisible part of the human being? What is the soul? What are emotions? Why do humans feel guilt? What is love?

Scientists invent things. But science cannot tell us whether an invention is good or bad. Science cannot change the heart of somebody who doesn't believe in God. Suppose, for example, that Jesus were to raise somebody from the dead right in front of an unbelieving scientist. What would he say? Would he fall on his knees and repent of his unbelief? No. He wouldn't. Rather, the unbeliever would say, "It's a magic trick. The man was never really dead to begin with." Or he would say, "There must be some natural reason why people can rise from the dead. If we figure this out, we can solve the problem of death, and people can live forever in this world."

This is what Jesus taught in the parable of the rich man and Lazarus:

There was a certain rich man who was clothed in purple and fine linen and fared sumptuously every day. But there was a certain beggar named Lazarus, full of sores, who was laid at his gate, desiring to be fed with the crumbs which fell from the rich man's table. Moreover the dogs came and licked his sores. So it was that the beggar died, and was carried by the angels to Abraham's bosom. The rich man also died and was buried. And being in torments in Hades, he lifted up his eyes and saw Abraham afar off, and Lazarus in his bosom.

## CHAPTER 1: WHAT IS TRUE?

Then he cried and said, "Father Abraham, have mercy on me, and send Lazarus that he may dip the tip of his finger in water and cool my tongue; for I am tormented in this flame." But Abraham said, "Son, remember that in your lifetime you received your good things, and likewise Lazarus evil things; but now he is comforted and you are tormented. And besides all this, between us and you there is a great gulf fixed, so that those who want to pass from here to you cannot, nor can those from there pass to us."
Then he said, "I beg you therefore, father, that you would send him to my father's house, for I have five brothers, that he may testify to them, lest they also come to this place of torment." Abraham said to him, "They have Moses and the prophets; let them hear them." And he said, "No, father Abraham; but if one goes to them from the dead, they will repent." But he said to him, "If they do not hear Moses and the prophets, neither will they be persuaded though one rise from the dead." (Luke 16:19-31)

## God's Designs are Very Complicated

For You formed my inward parts;
You covered me in my mother's womb.
I will praise You, for I am
fearfully and wonderfully made;
Marvelous are Your works,
And that my soul knows very well.
My frame was not hidden from You,
When I was made in secret,
And skillfully wrought in the lowest parts of the earth.
Your eyes saw my substance, being yet unformed.
And in Your book they all were written,
The days fashioned for me,
When as yet there were none of them.
(Psalm 139:13-16)

God has made an extremely complicated world. As we shall see, the human body is the most complex and beautiful of His material creation. Think about how the human body is more complicated than a rock. The human body can move gracefully, even capable of running, jumping, skating on ice, and twirling around. Whereas rocks cannot move or think, the human mind can think logically, make decisions, design buildings, and invent very complex machines. The human eye can distinguish 10,000,000 shades of colors. Thousands of processes are going on in your body at one time, most of which you are not even thinking about. Cells are reproducing. Wounds are healing themselves. You are breathing. Your blood is taking nutrients and oxygen throughout your body. Your immune system is fighting off disease. What an amazing, intricate design made by an all-powerful, all-wise, and all-good God!

Have you not known?
Have you not heard?

## GOD MADE LIFE

**The everlasting God, the LORD,
The Creator of the ends of the earth,
Neither faints nor is weary.
His understanding is unsearchable.
(Isaiah 40:28)**

The body is so complicated that doctors and scientists have only learned a little bit about it over 6,000 years. Some of what we learn also turns out to be wrong. Scientists change their opinions throughout history. For example, doctors used to try to drain blood in order to get rid of a disease. This went on for about 2,500 years until people figured out it didn't really cure the patient. For one thing, they did not check with the Bible which said, "The life of the flesh is in the blood" (Leviticus 17:11).

Scientists and doctors disagree about many things. For example, some say that vaccines prevent disease and reduce the possibility of children dying. Some say that vaccines can cause problems like autism, asthma, or auto-immune disorders, and should be avoided. Remember that science cannot prove these theories beyond all doubt. It is hard to be 100% sure about any of these opinions.

Some traditions never had much proof, and lots of theories turn out to be wrong after many studies. Here are a few examples:

Eating something that has fallen on the floor within five seconds isn't a problem. Actually, germs can contaminate the food within a few milliseconds (hundredths of a second).

Coffee will stunt your growth. Some scientists now say that a tablespoon of milk will

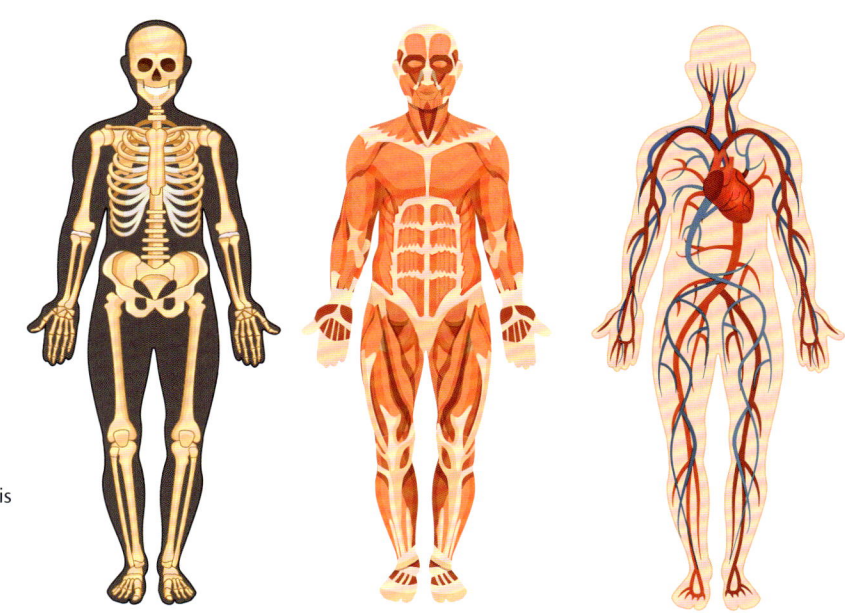

The human as created by God is very complex.

12

CHAPTER 1: WHAT IS TRUE?

compensate for the decrease of calcium produced by drinking a caffeinated drink.

## How to Gain Certainty in Science

So, how do we begin to improve our certainty about the things we know? Have you ever sliced up onions and found that your eyes begin to water? What causes your eyes to water?

1. We must observe. That means we must look at something over a lengthy period of time. Or, we must look at it many times over. We must compare one thing with another, and try to identify the differences between them. Do your eyes water when you cut up carrots, green peppers, or green onions?

2. We must observe cause-and-effect relationships. Is there a physical relationship between the onion and your eyes? Or does everybody just get emotional when they cut up onions? Is there something about onions that makes us all start to tear up? What is the cause which specifically brings about these tears?

Are there liquids splashing out of the onion while it is being cut? Or is it a gaseous vapor that is released from the onion which causes tears? Sometimes, scientists will block other potential factors to isolate the real cause. Do the tears flow because of vapors coming into the nose or into the eyes? So you put a clothespin on your nose, and cut an onion. Then, you remove the clothespin, put on goggles, and cut another onion. When you do this, you will discover that the cause of the tears is not an odor picked up by

Science requires observation.

# GOD MADE LIFE

the nose. Something is emitted from the onion, coming into contact with the eyes.

3. You might use instruments like microscopes or spectrometers to analyze the stuff coming out of the onion. You could study the chemical reactions going on at the surface of the onion when you cut through it.

Finally, you would want to repeat your observations over and over again to improve the certainty of your conclusions.

But most importantly, the godly scientist should pray to God for wisdom and insight to better understand this complicated world. This was Solomon's request of God in 1 Kings 3. And God answered his prayer.

And Solomon said: . . ."O Lord my God, You have made Your servant king instead of my father David, but I am a little child; I do not know how to go out or come in. And Your servant is in the midst of Your people whom You have chosen, a great people, too numerous to be numbered or counted. Therefore give to Your servant an understanding heart to judge Your people, that I may discern between good and evil." (1 Kings 3:6-9)

## Check with the Experts

To get a fast answer for a scientific question, some students will want to look up the answer on the internet. They might check "Wikipedia" or some other encyclopedia. Keep in mind that many sources on the internet can be wrong. Even experts can be wrong. They may pretend to know the answer to your question beyond any doubt. But experts have been shown to be wrong over and over again.

Around AD 2002, scientists discovered the causes of the onion tears. They say that a chemical (or enzyme) called **Lachrymatory-factor synthase** is released when we cut an onion. This converts the amino acid sulfoxides into "Syn-propanethial-S-oxide gas" which floats up into the air and comes in contact with your eyes, causing them to tear up. As you can see, this explanation is very complicated, and it took scientists 6,000 years to figure it out. But still, this explanation could be wrong.

## The Scientific Method

If you come up with an idea that might be helpful for your health, for animal health, or

CHAPTER 1: WHAT IS TRUE?

for plant health, you should use the scientific method. This is a disciplined approach to using science effectively.

Let us say that you want to grow healthy tomato plants and yield the most tomatoes possible. The world-record yielding tomato plant yielded 32,194 tomatoes (or 1,152 pounds/522 kg of tomatoes) during one season. But you are hoping to get the most tomatoes from your plants. Here is how you would use the scientific method.

1. Establish the goal or state the problem you are addressing. In this case, you want to yield more tomatoes per plant in your garden.

2. Form your hypothesis. This is usually an educated guess based on some limited experience you have had with the subject (in this case, gardening). You pick the best fertilizer you can find. You believe that Fertilizer A is better than Fertilizers B and C. You believe that one-half gallon of water per plant, per day, will produce the best tomatoes.

3. Experiment with the hypothesis. Use Fertilizer A for 10 plants, and then use Fertilizers B and C for 10 plants each. Provide one-half gallon of water for each plant.

    For a second set of 30 plants, use Fertilizer A exclusively. Then, provide half a gallon a day for 10 plants, one-quarter gallon a day for 10 plants, and three-quarters gallon a day for the other 10 plants.

It is important to keep everything in a greenhouse so rainfall doesn't mess up your experiment. This is called blocking confounding factors.

4. Harvest your crop and record your data by weighing all the tomatoes from each plant. Then analyze the data. Average the total weight from each set of plants.

5. Decide which was the best fertilizer. Taking a look at this data, we would conclude that the plants which did the best used Fertilizer A and received three-quarters gallon of water per day.

Fertilizer

# GOD MADE LIFE

6. Verify your conclusions. During the next planting season, you should experiment again. This time, you might try seven-eighths gallon of water per day, one gallon of water per day, and 1 and 1/8 gallons of water per day. There are confounding problems too. For example, too much water might invite bugs into the greenhouse during the second year of experimenting, which would decrease your yield.

Tomatoes

### First Set: Average Harvest

- Fertilizer A—104 pounds (47 kg)
- Fertilizer B—40 pounds (18 kg)
- Fertilizer C—62 pounds (28 kg)

### Second Set: Average Harvest

- One-quarter gallon per day—32 pounds
- One-half gallon per day—85 pounds
- Three-quarters gallon per day—112 pounds

7. Use your conclusions to predict future outcomes. If you increase your planting to 100 plants, you could yield as much as 11,200 pounds (5080 kg) of tomatoes. That would be a great harvest!

## Making Discoveries in Science

Yet, there have been many helpful scientific discoveries throughout history. Many of these breakthroughs come by "accident." But really, there are no accidents. All inventions that have helped mankind come by God's blessing when He provides a special insight to a humble scientist. In the history of medical science, the following were some of the most helpful breakthroughs:

- **1720**—Some terrible pandemics such as smallpox, tuberculosis, and cholera could kill 5% to 50% of the population. Pastor Cotton Mather of Boston, Massachusetts, was the first to discover the smallpox vaccine in an effort to make it widely available in America.

- **1846**—Anesthesia was discovered by several dentists in America. The use of certain chemicals would put a patient to sleep while doctors performed surgery.

- **1861**—A French microbiologist named

CHAPTER 1: WHAT IS TRUE?

Louis Pasteur identified germs as the cause of infectious diseases.

- **1895**—The X-ray was accidentally discovered by German physicist Wilhelm Conrad Röntgen.

- **1928**—The first antibiotic was discovered by a Christian researcher, Alexander Fleming. His discovery came about accidentally when he forgot to clean a petri dish in his laboratory. He returned after two weeks and found a mold growing, preventing the bad germ from multiplying. Later he wrote, "One sometimes finds what one is not looking for. When I woke up just after dawn on Sept. 28, 1928, I certainly didn't plan to revolutionize all medicine by discovering the world's first antibiotic, or bacteria killer. But I guess that was exactly what I did."

- **1970**—A Christian inventor named Raymond Damadian developed Magnetic Resonance Imaging. The MRI machine can be used to find diseases like cancer in the body without cutting it open to look at it. This technology has saved countless lives throughout the years.

### Alexander Fleming

The Christian who discovered antibiotics, Alexander Fleming, once wrote, "My greatest discovery was that I needed God, and that I was nothing without Him and that He loved me and showed His love by sending Jesus to save me."

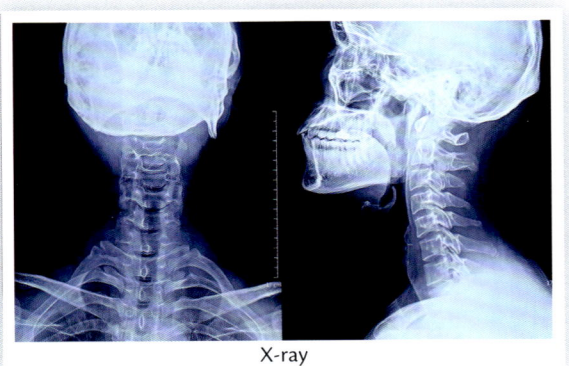

X-ray

## History is Not Science

History is not the same subject as science. History is the records others have made of certain events that they have witnessed in

# GOD MADE LIFE

### Dr. Raymond Damadian

Dr. Raymond Damadian, the inventor of the MRI machine, attributed all his discoveries to God. He wrote, "How could a scientist achieve his goal of discovering the absolute truths that govern the natural world without the blessing of the Author of those truths? For me now the true thrill of science is the search to understand a small corner of God's grand design, and to lay the glory for such discoveries at the Grand Designer's feet."[3]

the past. History does not require repetition and verification because you cannot repeat the very same historical event. Those who signed the U.S. Declaration of Independence in 1776 only signed the document once. It never happened again. Science will verify a cause-and-effect relationship in the natural universe. This is done by attempting the same experiment under the same conditions over and over again. That is how scientists can be more certain of the truth of their findings.

Towards the end of the 1700s and on into the 1800s, a false kind of science developed. One scientist, Charles Lyell (1797-1875), looked at geological changes on the earth's surface, and he tried to guess what happened 1,000 year ago or 1,000,000 years ago. Of course, he assumed the world was very old. Lyell dated the strata of rock layers by their location, presuming they had been laid down over a very long period of time. He rejected the idea of Noah's flood as recorded in Scripture. Lyell assumed that the conditions on the surface of the earth had been the same, and the changes occurring by earthquakes and volcanoes remained constant through the years. But, how can we know for sure that this is the case? Science cannot possibly prove this. Suppose that somebody had recorded every meteorite that hit the earth, and every flood, every earthquake, and every volcanic eruption happening over thousands of years (or even millions of years). We would have to rely on the accuracy of single observations. We would have to rely on an expert, who could very well make a mistake in his observations and recordings. This is not science. This is history. These are the records taken down by somebody who was observing what happened. This cannot possibly provide a high level of reliability for these "truth claims."

About the same time that Lyell came out with his imaginative claims, Charles Darwin (1809-1882) also delivered a very weak hypothesis. He tried to explain how animals and humans appeared on the earth by rejecting God's revelation in Genesis 1 and 2. Darwin suggested that humans

CHAPTER 1: WHAT IS TRUE?

evolved from ape-like animals, and that all animals somehow evolved out of lower life forms. He said this may have happened by a process of "natural selection" or by chance. He thought that somehow complex animals came from simple animals. And somehow, living organisms came from non-life. He thought that animals could change through the generations by **genetic mutation**. These changes would come about because certain mutated animals could adapt better to their environment. Thus, the animals which had adapted to their environment would live longer and have more babies and healthier babies. This was the "hypothesis." But there was a big problem with Darwin's ideas. These radical changes, where one kind of animal changes into another kind of animal, has never been observed in a laboratory. Darwin hoped the fossil layer would show lots of varieties of animals developing over time. There was nothing like this to be found in the fossil layers. Even if there were millions of varieties of animals of varying complexities (that have gone extinct),

Charles Lyell

Charles Darwin

this still would not prove that one variety of animal adapted from another. Unless scientists observe the mothers giving birth to mutated animals over millions of years, then the field of evolution can't be considered true science.

Occasionally, you might find a fossilized animal in the mountains. Maybe the animal had gone extinct, which means there are none of these to be found anywhere in the world today. Does this prove Darwin's hypothesis? Of course not. All we can say is that some mighty flood must have swept over the mountain and encased this animal (and many of his friends) in the rock. With the fossil in your hand, you have evidence that this animal died. That is all you can say. You do not know for certain when he died. You don't know why all of his friends died, or why this animal is extinct. You don't know for certain why he died. He could have died in a local flood, or it could have happened during the worldwide flood. Humble scientists are careful not to say too much about these fossils.

# GOD MADE LIFE

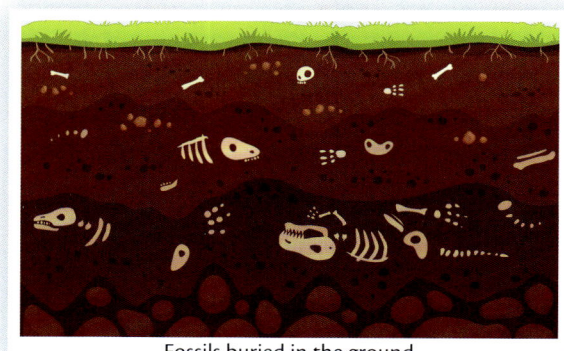
Fossils buried in the ground

Proud scientists pretend that these fossils prove a weak hypothesis concerning the origins of life on the earth.

Some unbelieving scientists try to date the age of a fossil or a rock by what is known as Carbon 14 dating. When a living organism dies, part of its Carbon content breaks down by radioactive decay. The Carbon 14 decays, but the Carbon 12 stays the same. So, scientists monitor the ratio of Carbon 14 to Carbon 12. From this they try to guess the age of the fossil. But they assume that this ratio has remained the same in the atmosphere of the earth from the time the fossil died until now. We don't know what the ratio of Carbon 14 to Carbon 12 was 4,000 years ago, or 6,000 years ago. For one thing, we know the world was very different before the worldwide flood. So we can't trust this dating method.

Also, when it comes to dating methods, the scientists make assumptions about the starting conditions. They assume uniform conditions throughout history. Suppose you were to come upon a bicycle tire that was almost flat, but not quite. You measure how fast the air is escaping, and you conclude that the tire was filled with air exactly 8 hours ago. What is the problem with this assumption? Of course, the problem is that you do not know when the tire was punctured. You also do not know whether the size of the hole got bigger hours after the tire was punctured.

## Accuracy in History

We cannot completely trust human records concerning what happened in history. There are few human eyewitnesses in the very ancient records. In fact, the oldest writing known to man is a papyrus record from about 2100 BC. This would have been about 400 years after the worldwide flood. It is the diary of a man named Merer who was supervising the building of the Great Pyramid of Giza. We may have an accurate account of what Merer was doing 4,000 years ago, organizing his crew on the construction project. To understand for sure what happened in history we must go back to God's Word.

We are absolutely certain of God's revelation concerning these historical events. There was a worldwide flood, which took place around 2518 BC (+/- 200 years). It was a worldwide catastrophe, and it impacted the whole globe. Beyond this global event, we don't have reliable records of any other large floods, volcanoes, or earthquakes. But, according to biblical record, this impacted the face of the whole earth. And a lot of animals died in the floodwaters.

**CHAPTER 1: WHAT IS TRUE?**

Then the LORD said to Noah, "Come into the ark, you and all your household, because I have seen that you are righteous before Me in this generation. You shall take with you seven each of every clean animal, a male and his female; two each of animals that are unclean, a male and his female; also seven each of birds of the air, male and female, to keep the species alive on the face of all the earth. For after seven more days I will cause it to rain on the earth forty days and forty nights, and I will destroy from the face of the earth all living things that I have made. . . ."
The waters prevailed and greatly increased on the earth, and the ark moved about on the surface of the waters. And the waters prevailed exceedingly on the earth, and all the high hills under the whole heaven were covered. The waters prevailed fifteen cubits upward, and the mountains were covered. And all flesh

Great Pyramid of Giza

Dinosaur fossil

## GOD MADE LIFE

died that moved on the earth: birds and cattle and beasts and every creeping thing that creeps on the earth, and every man. All in whose nostrils was the breath of the spirit of life, all that was on the dry land, died. So He destroyed all living things which were on the face of the ground: both man and cattle, creeping thing and bird of the air. They were destroyed from the earth. Only Noah and those who were with him in the ark remained alive. And the waters prevailed on the earth one hundred and fifty days. (Genesis 7:1-4, 18-24)

## The Most Important Event in History

The most important historical event took place about 2,000 years ago, when Jesus Christ, the Son of God, died on the cross for our sins. On the third day, He rose again from the dead according to the Scriptures. Though the world has been terribly affected by sin and death, the Lord Jesus Christ came to fix all of this. If we believe in the coming of Christ in history as the Scriptures tell us, and we believe that He died on the cross for our sins and rose again, we will be saved. God said this happened in His Word. He cannot lie. This is truth, and we must believe it.

The most important historical event: The death and resurrection of Jesus Christ

For I delivered to you first of all that which I also received: that Christ died for our sins according to the Scriptures, and that He was buried, and that He rose again the third day according to the Scriptures, and that He was seen by Cephas, then by the twelve. After that He was seen by over five hundred brethren at once, of whom the greater part remain to the present, but some have fallen asleep. After that He was seen by James, then by all the apostles. Then last of all He was seen by me also, as by one born out of due time. (1 Corinthians 15:3-8)

## Pray

- Take a moment and praise God for His very complicated creation of the human body, the human eye, and the human brain.
- Thank God for communicating His truth in the Word. Thank Him that we can know the truth for sure as we read His Word.
- Pray for humility as we explore His creation and gain some knowledge of the world.
- Pray for special wisdom and understanding. Pray that God might help you to discover something new about His world.
- Ask for God's forgiveness for overlooking God's wonderful creation or for taking it for granted.

## Sing

We have thought about God's truth and God's wisdom to make this world. We have studied a little bit of science and a little bit of history. The appropriate response must be worship and praise to God. If the student is unfamiliar with the hymn or psalm, some version of it is available on the internet, and may be accessed (with supervision) for singing along.

*How Great Thou Art*
O Lord my God, when I in awesome wonder
Consider all the worlds Thy hands have made,
I see the stars, I hear the rolling thunder,
Thy power throughout the universe displayed:

*Refrain:*
Then sings my soul, my Savior God, to Thee:
How great Thou art! How great Thou art!
Then sings my soul, my Savior God, to Thee:
How great Thou art! How great Thou art!

When through the woods and forest glades I wander
And hear the birds sing sweetly in the trees,
When I look down from lofty mountain grandeur,
And hear the brook and feel the gentle breeze:

# CHAPTER 1: WHAT IS TRUE?

And when I think that God, His Son not sparing,
Sent Him to die, I scarce can take it in,
That on the cross, my burden gladly bearing,
He bled and died to take away my sin.

When Christ shall come with shout of acclamation
And take me home, what joy shall fill my heart!
Then I shall bow in humble adoration,
And there proclaim, "My God, how great Thou art!"

## Do

Choose at least one of the following activities and apply the lessons you learned in this chapter.
1. Obtain a historical artifact or antique in your house, obtain one from a friend, or research an antique in a shop. Try to figure out when it was made. How certain are you of the date in which it was made? What are you trusting in to be sure of that date? Is it possible that you could be wrong? Suppose you found a fossil in the back yard. How certain could you be of the fossil's age? Would you be more certain of the date the antique was made than of the date the fossilized animal was alive?
2. Organize your own experiment in the garden, in the kitchen, or with your household chores. Are there ways you could improve your efficiency? How might you get the job done more quickly and do a better job? Or, how might you find the best baking results for pies, cakes, or cookies? Choose your household experiment. Plan out each step of the experiment. Conduct the experiment, and draw a conclusion.
3. Make a list of the problems with an evolutionary explanation for how this world came about. Do a little more research using materials from Answers in Genesis (*aig.org*) or *creation.org*. Make your own short list and memorize these so you can discuss them with your friends.

## Watch

To watch the recommended videos for this chapter, go to **generations.org/GodMadeLife** and scroll down until you find the video links for Chapter 1. Our editors have been careful to avoid films with references to evolution; however, we would still encourage parents or teachers to provide oversight for all internet usage. The producers of these videos may not themselves give God the glory for His amazing creative work, but we encourage the student and parent/teacher to respond with prayer and praise.

Butterfly and chrysalis

# CHAPTER 2
# WHAT IS LIFE?

> Let heaven and earth praise Him,
> The seas and everything that moves in them. (Psalm 69:34)

Science must never be boring. If science is boring, then God is boring, and this is never the case. As Christians, we believe that God made the world and everything in it. This is His handiwork. So, the first and most important reason why we must study plants, animals, and humans is to discover more reasons to praise God. We learn about God and we worship God while gazing into a microscope in the laboratory, surveying the gorgeous landscapes from the mountain tops, and swimming through the coral reefs in the South Seas. We love God. We love God's world, and we are excited to learn more about it. Beyond this, to ignore God as we study His work is to insult Him.

## The Purpose of This Study—To Glorify God

We look at a great work of art and admire the artist. We read a great work of literature and admire the writer. We watch a rocket ship soar into the sky, and we consider the genius of the scientists who designed the craft. But the greatest work of all is that of the Creator of the human brain. We want to know more of God who made us with outstanding capabilities to do these things. We want to know more of the Creator of life. God is the Genius who created all of the life around us—plant, animal, and human life. And so, with every page of this book the student is called to wonder and praise. After studying this creation of life, we should love God

# GOD MADE LIFE

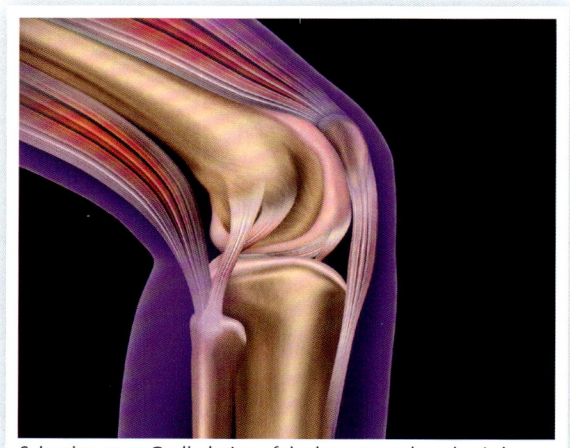
Scientists copy God's design of the knee to make robotic knees.

mechanisms God used in the human joint. They wanted the same compactness, the same strength, and the same mechanical advantage for the robotic knee as the original design for the human knee. Also, they wanted the same kind of stiffness in the locked position when the robot stood upright.[1]

Velcro

more. This is the major purpose for gaining more knowledge in this field of science.

Also, we study the body to properly oversee the creation, for the Creator has assigned us to this. Mosquitoes bite and spread disease. Bacteria and viruses introduce diseases to the human body. We need to understand how this happens and how we might help the body fight off these diseases.

## Science Copies God's Designs

You will discover in this study that God is the original Genius. We are created in His image, and we think His thoughts after Him. Therefore, scientists will copy God's design for their own purposes. George de Mestral invented Velcro® in the 1950s as he studied burdock seeds that would cling to his coat and get tangled in his dog's fur. Also, robotic limbs have been designed to mimic the human knee joint. Scientists used the four bar-and-cam

## What is Life as Defined by God?

There is a big difference between a rock and a bird. Immediately, every child can see the differences. The bird can move, grow, reproduce, eat food, and run away from predators. Rocks don't move, grow, reproduce, eat food, or run away from predators. So, the first thing to notice here is the amazing complexity of life. Of course, none of us could make a rock out of nothing. This is something only God can do. But, birds and man are much more complicated than a rock.

God's Word defines life by two characteristics:

28

# CHAPTER 2: WHAT IS LIFE?

### 1. Breath of Life

And all flesh died that moved on the earth: birds and cattle and beasts and every creeping thing that creeps on the earth, and every man. All in whose nostrils was the breath of the spirit of life, all that was on the dry land, died. So the LORD destroyed all living things which were on the face of the ground: both man and cattle, creeping thing and bird of the air. (Genesis 7:21-23)

### 2. Blood

"If the place where the LORD your God chooses to put His name is too far from you, then you may slaughter from your herd and from your flock which the LORD has given you, just as I have commanded you, and you may eat within your gates as much as your heart desires. . . . Only be sure that you do not eat the blood, for the blood is the life; you may not eat the life with the meat. You shall not eat it; you shall pour it on the earth like water." (Deuteronomy 12:21, 23-24)

From these words, we understand that all creatures with the breath of life and blood are **higher life forms**. This includes creeping things on the earth, cattle, and man. This does not include plants and trees. Those animals without the breath of life and blood like jelly fish, sea anemones, and invertebrates (insects) are **lower life forms**. Because insects have **hemolymph** instead of blood, the Scriptures would not include insects under the category of animals that have the blood of life and the breath of life. Keeping this in mind, the biological sciences would still include plants and insects under the category of biological life. For this course, we will separate lower life forms from higher life forms.

Yet, in the order of His creation, God has

Sea anemone

29

## GOD MADE LIFE

established various levels of dignity upon the creatures. This is important for the way we think about the world. For example, we cannot treat a whale with the same level of dignity we would apply to humans.

Often, unbelievers will treat apes, whales, and bald eagles like they treat humans. That's because they believe humans descended from apes. They do not believe in a Creator God, who placed varying levels of honor on His creatures. Psalm 8 provides a clear scale of honor as laid down by God at creation.

O LORD, our Lord,
How excellent is Your name in all the earth,
Who have set Your glory above the heavens! . . .
What is man that You are mindful of him,
And the son of man that You visit him?
For You have made him a little lower than the angels,
And You have crowned him with glory and honor.
 You have made him to have
  dominion over the works of
   Your hands;
  You have put
   all things under his feet,
 All sheep and oxen—
 Even the beasts of the field,
 The birds of the air,
 And the fish of the sea
That pass through the paths of the seas.
(Psalm 8:1,4-8)

### The Scale of Honor Established by the Creator

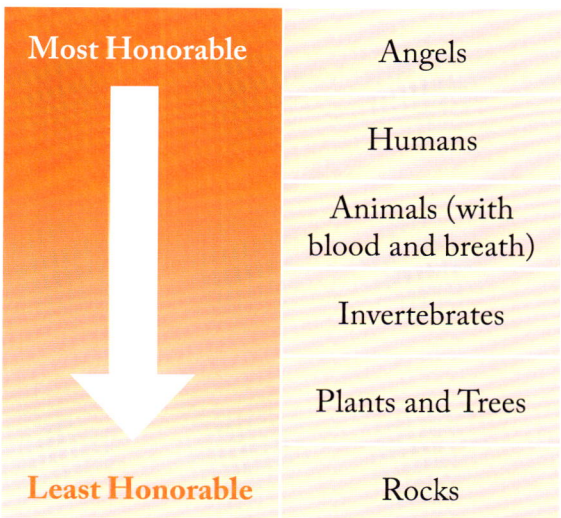

| Most Honorable ↓ Least Honorable | |
|---|---|
| | Angels |
| | Humans |
| | Animals (with blood and breath) |
| | Invertebrates |
| | Plants and Trees |
| | Rocks |

Man and monkey

# CHAPTER 2: WHAT IS LIFE?

## Why Humans are Different

*And the LORD God formed man of the dust of the ground, and breathed into his nostrils the breath of life; and man became a living being. (Genesis 2:7)*

Human life is different from animal life because God breathed into Adam's nostrils "the breath of life." Man has the breath of God within him. This is not said of any of the other creatures. Scripture goes on to speak of three kinds of life—physical life, spiritual life, and eternal life.

When Adam fell into sin in the garden, all mankind entered into a state of spiritual death. Man was now "dead in trespasses and sins" (Eph. 2:1). Adam brought physical death, spiritual death, and eternal death to us by the fall. Though Adam was still physically alive, he was the "walking dead" as it were. Only by the second Man, that is Jesus Christ, did new life come. Jesus brought us spiritual life. This was an eternal life. By Christ's resurrection came complete victory over death. We will be resurrected too, because we believe in Jesus. There will be no more physical or spiritual death for all those who put their trust in Him.

Adam was also created in the image of God in knowledge and righteousness (moral capacity). Animals do not have a sense of morality. Black widows will eat their mates. Hyena siblings kill each other for food. Dogs will fight over their food, but that is only because they have an instinct to eat and to survive. Dogs and cats do not protest moral problems in the world. Sometimes female dogs will eat their puppies, and they have no conscience about it. They don't feel bad about doing this afterward.

Man was also given the responsibility of taking dominion over the creation. He was supposed to be God's caretaker over the created world. This was not assigned to dogs and cats, or even to the angels. Man is very special, and he is very different from the rest of the material creation.

Animals are not moral creatures. They don't protest for moral causes.

## The Source of Life

*Jesus said to her, "I am the resurrection and the life. He who believes in Me, though he may die, he shall live. And whoever lives and believes in Me shall never die. Do you believe this?" (John 11:25-26)*

# GOD MADE LIFE

Life is a mystery to scientists. Where did this very complicated creation come from? The Bible tells us clearly this life comes from God. But the modern world does not believe this. Governments have spent several trillions of dollars investigating other planets, searching for life. Unbelieving scientists do not want to believe that this miracle of life comes from an intelligent Personality. They want to believe that life can come from non-life. They imagine that somehow a rock can start to move and then reproduce baby rocks.

Some people believed in "spontaneous generation" in the 1800s. They thought life came out of non-life and that maggots formed spontaneously on dead meat until Louis Pasteur disproved it. Now, scientists have a new name for it—**abiogenesis**. They tell us that water is pretty much the same as a life form. Or, they tell us that an organic compound like **isopropyl alcohol** (rubbing alcohol) is pretty much the same as life. These scientists are living in a make-believe world.

Where did life come from? There is a very simple answer to this. Jesus is the source of life. In John 6, He told everybody He came down from heaven to "give life to the world."

## What This Study is All About

With this basic introduction, we want to study a little bit about zoology, botany, anatomy, and physiology. Let's define these terms.

Louis Pasteur

| | |
|---|---|
| **Biology** | The study of all lifeforms |
| **Microbiology** | The study of micro-organisms |
| **Zoology** | The study of animals |
| **Botany** | The study of plants |
| **Anatomy** | The study of the human body and how all the parts fit together |
| **Physiology** | The study of the functions of the parts in the human body |

# CHAPTER 2: WHAT IS LIFE?

## Cell-Based Organisms

The biblical definition of life is all those creatures with breath and blood. For this study, we will consider everything made of cells. Any part of God's creation made of one or more cells is referred to as an organism.

These cell-based, created entities have seven characteristics:

1. **Organisms are made of cells.** A cell is an amazing creation of God made of a wall or membrane on the outside and a **cytoplasm** (SYE-toe-plaz-uhm) on the inside. The cytoplasm contains water and carbon compounds—starches, sugars, fats, and proteins.

2. **Rocks and dirt don't die.** But, all organisms will wither and pass away. Flowers and trees grow for a while, then they wither, fade, and are no more. Man and animals all experience the sad reality of death.

**Cell Structure**

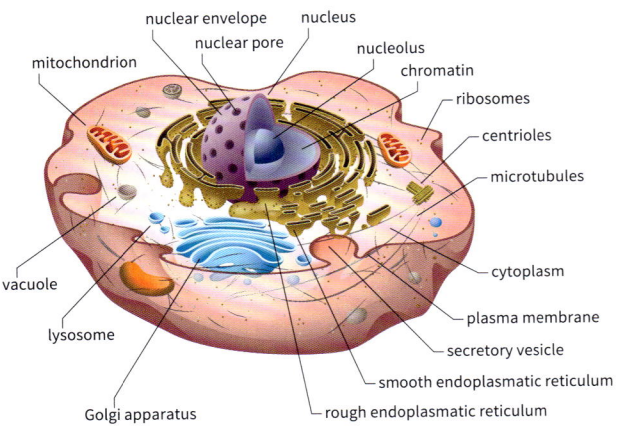

I said in my heart, "Concerning the condition of the sons of men, God tests them, that they may see that they themselves are like animals." For what happens to the sons of men also happens to animals; one thing befalls them: as one dies, so dies the other. Surely, they all have one breath; man has no advantage over animals, for all is vanity. (Ecclesiastes 3:18-19)

"The grass withers, the flower fades,
Because the breath of the LORD blows upon it;
Surely the people are grass.
The grass withers, the flower fades,
But the word of our God stands forever."
(Isaiah 40:7-8)

3. **Organisms reproduce.** At the beginning, God created plants and animals to reproduce "according to their kind." Cats will never give birth to puppies, and pine trees will never reproduce oak trees. The creation was never designed such that one life form would develop into higher levels of complexity, or vastly different kinds of creature.

Then God said, "Let the earth bring forth grass, the herb that yields seed, and the fruit tree that yields fruit according to its kind, whose seed is in itself, on the earth"; and it was so. (Genesis 1:11)

33

# GOD MADE LIFE

Then God said, "Let the earth bring forth the living creature according to its kind: cattle and creeping thing and beast of the earth, each according to its kind"; and it was so. (Genesis 1:24)

4. **Organisms need food for energy to exist.** You do not need to feed rocks to keep them from withering away. But plants will wither and pass away if they do not receive sunlight (energy from the sun) for a long period of time. If a human doesn't get food, at some point, he won't have the energy to breathe. He will die without food.

5. **Also, cell-based organisms move.** Obviously, rocks don't move around on their own unless they are pushed along by something else. Animals and humans will walk around without anything pushing on them. There is always some kind of movement going on in every living organism, even in the case of plants and trees. Single-cell organisms like **amoebas** are moving stuff inside the cell.

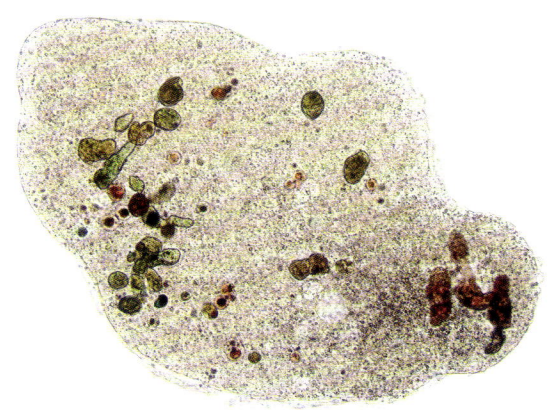
Amoeba (a single-celled organism)

6. **Organisms start out very small as a single cell or a seed, and they grow.** God has designed organisms to take in food and turn it into more cells to replenish the body. Parts of the organism will also wear out over time, so they need to be replaced throughout the organism's cycle of existence. Growth happens when the body takes in more mass than it is using up.

7. **Organisms react to outside stimuli.** A stimulus is a change or force that interacts with the organism. The organism can sense this force and responds to it. A rock cannot tell when the wind is blowing. A cat can sense when the wind blows or when the temperatures drop, and he will fluff up his hair to keep himself warmer in the cold. A grizzly bear can smell a

Food is essential for life.

34

Sunflowers

dead carcass at 18 miles away, and then travel the distance for a meal. The stimulus is the smell, and the grizzly picks it up with his nose. He reacts by lumbering over to the meat and consuming it. Sunflowers will respond to the movement of the sun. Throughout the day, the young plants will rotate their yellow blossoms, tracking the sun's movement. During the nighttime, they will reset their faces towards the east. Rocks don't do this.

## Energy Conversion

It takes energy to move stuff around, and since rocks don't move anywhere, they don't need to eat or convert energy. However, living things always move, and they need energy to do it. They must get their energy from food or convert energy from other forms. Also, it may not appear that plants are moving. If you looked at the cells that make up the plants under a microscope, you would see many moving parts.

When you eat a sandwich, you are consuming matter. Somehow this matter must be transferred into energy if it will help you to move, talk, run, and jump. When your body takes in food, it is processed for building new cells and providing energy for your systems. The body uses a process called **anabolism** (uh-NA-bowl-izm) to transfer the food into fat, carbs, and protein. Also, the body can create new cells out of the food using this same process. Another process called **catabolism** (cuh-TA-bowl-izm) will break down the body tissues, fats, and carbohydrates. This provides energy that can be used to heat the body or contract muscles for movement, organ function, and cell mobility. **Enzymes** help in this process, but more on this later in this study.

This is an amazing capability created by

# GOD MADE LIFE

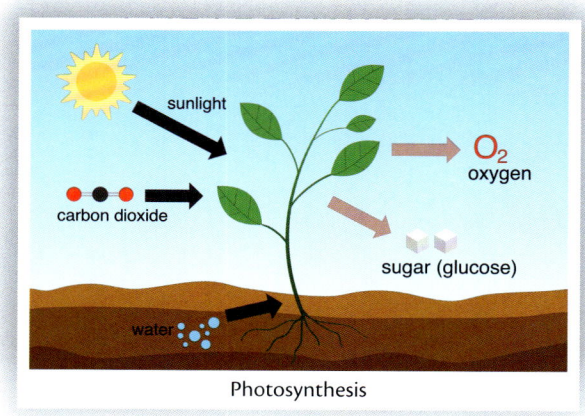
Photosynthesis

God for the functioning of living bodies and organisms in this world. Every living thing can process food for energy and store food or matter to be used later for energy. Plant organisms will store sugars in their roots. Plants and trees also take in water from the roots under the ground. The leaves and stems get carbon dioxide and sunlight from the atmosphere. They use that energy and carbon dioxide to make sugar food for themselves by **photosynthesis**.

The energy that keeps life going on earth comes from the sun. What a powerful energy source God provided the world! He designed this huge nuclear reactor to burn for 5-10 billion years. This sun keeps the plants growing, feeding man and animal year after year.

There are three kinds of organisms, each with its own way of getting energy for life and movement.

1. **Producers** make their own food using sunlight. This includes plants.

2. **Consumers** eat the producers and other consumers. This includes carnivores (tigers), omnivores (humans), and herbivores (sheep). Carnivores eat meat. Herbivores eat plants. And omnivores eat both meat and plants.

3. **Decomposers** are the little organisms like bacteria and fungus which God designed to break down other dead organisms. This makes good fertilizer for new plants growing in the soil.

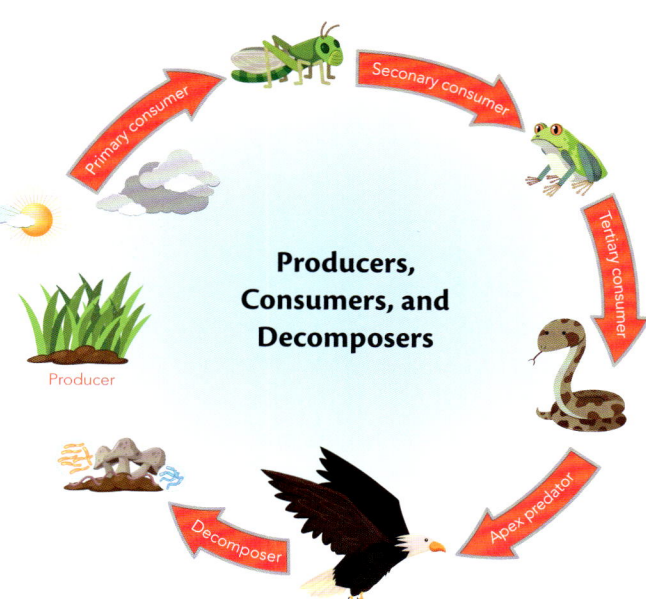

The decomposers turn the dead organisms into plant food. The plants grow by producing more food for themselves. The consumers eat the plants, and the cycle just keeps going, day by day, and year by year.

This is how God provides a perfect **ecosystem** to sustain life and continue repro-

36

ducing new plants, trees, and animals. An ecosystem is all the living and nonliving parts of a community. All these parts work together perfectly. Everything gets to grow and reproduce. Every creature gets something to eat. And, altogether, they trade or cycle food and energy.

In this ecosystem, green plants make food energy from sunlight and nutrients in the soil. The cow eats the plants to get energy. The fox hopes to eat the cow to get energy. Tiny organisms in the soil eat animal droppings and dead animals for energy. Their waste is used by plants as nutrients to grow and be eaten by animals again.

Venus flytrap

gory very well.

A plant called the Venus flytrap (Dionaea muscipula) eats bugs and small frogs. It will digest the frog or insect and translate the food into energy. Interestingly, the Venus flytrap will not close its "mouth" over a dead insect. Somehow, the plant senses movement, and this signals the plant to crunch down on a live bug. The flytrap secretes digestive juices and over 5-12 days dissolves the soft innards of the bug. Only the hard exoskeleton or the bones remain. Is this a producer or a consumer? Is this a plant or an animal? It doesn't really matter, because this organism wasn't created with the "breath of life."

Then He will give the rain for your seed
With which you sow the ground,
And bread of the increase of the earth;
It will be fat and plentiful.
In that day your cattle will feed
In large pastures.
Likewise the oxen and the young donkeys that work the ground
Will eat cured fodder,
Which has been winnowed with the shovel and fan. (Isaiah 30:23-24)

## Bug-Eating Plants

Some plants eat frogs and bugs. Although they are not animals, and they do have the breath of life, God has created these organisms to surprise us. As scientists try to categorize God's creation, they often run into strange creatures that don't fit into any cate-

But God has chosen the foolish things of the world to put to shame the wise, and God has chosen the weak things of the world to put to shame the things which are mighty. (1 Corinthians 1:27)

GOD MADE LIFE

## The Basic Building Block of All God's Created Organisms

Buildings are made of bricks and boards. Sentences are made of letters and words. In a similar manner, all of God's created organisms and life forms are made of cells. The cell is the most basic building block for plants, animals, and the human body.

Robert Hooke (1635-1703) discovered these building blocks in 1665. Examining a wooden cork under a microscope, he observed a honeycomb of "dead" cells. In his words, "These pores, or **cells**, were not very deep, but constituted of a great many little Boxes, separated out of one continued long pore, by certain Diaphragms...." He was the first to name the cell. That was what he called it.

The cell is a very busy place with lots of things going on at the same time. The closest analogy we have to a cell is a city. But it operates at a microscopic level. The cell has machines, factories, highways, and communication systems.

Your skin, bones, and organs are made of tissues, which are made out of cells. And, God made two kinds of cells:

The **prokaryotic** cell is the simpler design, with no organelles floating around inside. All that is swimming around in this cell is the DNA and ribosomes, which we will look at later. The prokaryotic cells are used for unicellular (single-celled organisms) like bacteria. However, these cells can team up in chains and do the same job together.

### Robert Hooke

Robert Hooke, the Christian scientist who first identified the cell, was overwhelmed by the sheer complexity and the variety of designs in nature when he wrote: "So prodigiously various are the works of the Creator, and so All-sufficient is he to perform what to man would seem impossible, they being both alike easy to him, even as one day, and a thousand years are to him as one and the same time.[2]

The **eukaryotic** cell is the more complex design, with distinct little organs (organelles) floating around inside of it. Every eukaryotic cell has a nucleus containing the DNA (or

## CHAPTER 2: WHAT IS LIFE?

the genetic material). Organelles are small parts that float around in the cell, all of which have a job to do. There are also different kinds of eukaryotic cells with different jobs to do. Your blood cells, nerve cells, and muscle cells are all doing different jobs, but they are teaming up to run your body. What a wondrous design the Lord has built into the body! Imagine designing an organism with 200 different kinds of cells, making up 37 trillion cells all functioning as a unit! God did that in the creation of your body! Give Him all the credit, all the honor, and all the praise for this wonderful creation.

Since the human body is made of eukaryotic cells, we will spend most of our time studying these amazing little factories. There is a lot going on in the cell. Picture an automobile assembly plant with all the activity going on inside. All the workers know what to do as they assemble cars. They are all very busy, and they are all working for a common goal. It is an intelligent process, planned out by engineers and plant managers.

God has made the cell even more complex and more impressive than man's creations. The seven basic processes of a cell are:

- Movement
- Reproduction
- Response to external stimuli
- Excretion (or getting rid) of unneeded material
- Provision of nutrition
- Respiration or converting food into useful energy using a chemical reaction
- Growth

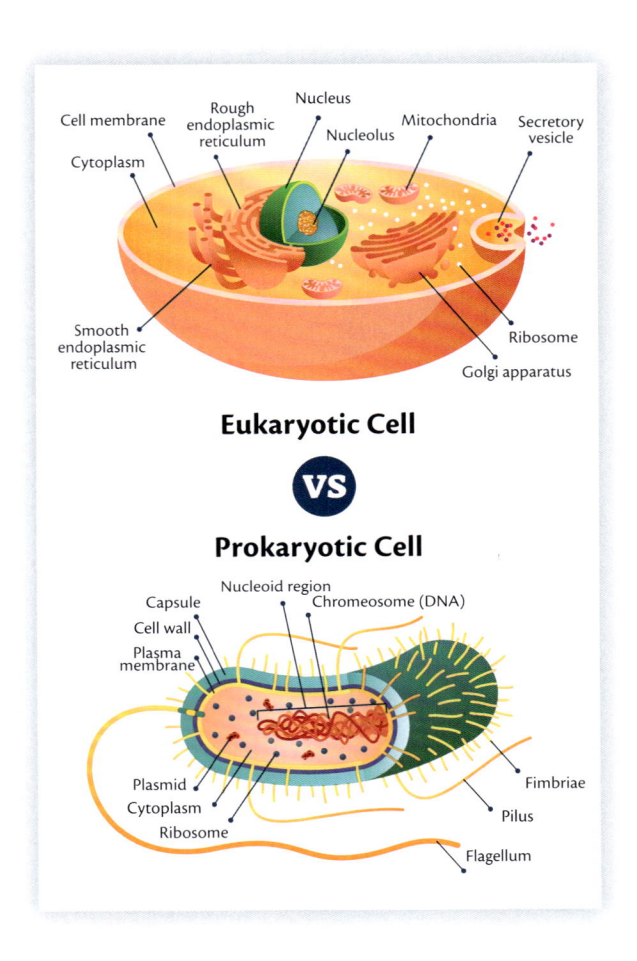

## The Raw Materials of the Cell

Cells are variously sized. The yoke of a bird's egg is a single cell. In fact, the largest cell on earth would have to be the yoke of an ostrich egg. If the ostrich egg is 6 inches long (15 cm), the yoke would be about 1 inch (2.5 cm) in diameter. One nerve cell can be very thin and long (3.3 feet or 1 meter). You could fit 10,000 human cells on the head of a pin.

Yet, the cell is not the smallest piece of

# GOD MADE LIFE

matter. As far as scientists are currently aware, the smallest piece of matter is the quark. Breaking down a cell into its smallest division, here is the order from smallest to largest:

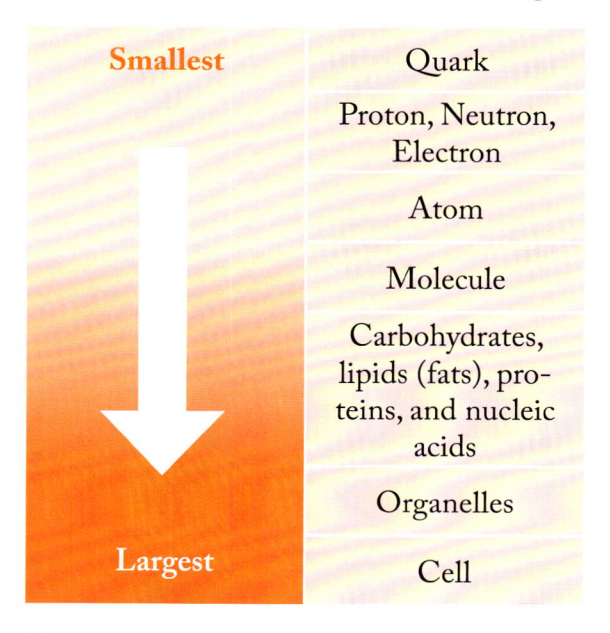

| Smallest → Largest | |
|---|---|
| Smallest | Quark |
| | Proton, Neutron, Electron |
| | Atom |
| | Molecule |
| | Carbohydrates, lipids (fats), proteins, and nucleic acids |
| | Organelles |
| Largest | Cell |

Houses are mostly made of wood, bricks, and drywall. Our amazing Creator used the raw materials of water, carbohydrates, lipids, proteins, and nucleic acids to make the human cell and the human body. A human body is made up of the following materials:

| | |
|---|---|
| Water | 60-65% |
| Fat | 14-20% |
| Protein | 14-20% |
| Minerals | 6% |
| Carbohydrates | 1% |

Carbohydrates are familiar to us because we eat bread, sugar, fruit, rice, potatoes, and corn. Glucose (also a carbohydrate) is the cell's basic food. It is stored as a **polymer** in plants, and **glycogen globules** in humans.

Lipids or fats can store energy, but they are also used to make the cell membrane. Also, lipid molecules can also serve as little taxis carrying other molecules over the cell boundaries.

However, the most impressive and the most exciting part of God's creation of life and plants has to be the proteins. There are between 80,000-400,000 different proteins in the human body, all of which have different functions. Proteins are extremely versatile, generating every kind of function imaginable for the human body, plants, and animals.

Some proteins can even make light like little flashlights. Some proteins can detect light. Some can communicate signals, or even detect a signal. Some can function like little motors. They can tie stuff together and untie other things. They can get chemical reactions going. They can check to see if something is right or wrong. They can make hard turtle shells as well as feathers for a bird. Some proteins can combine with other molecules to make complicated machines that replicate DNA, transmit signals to other parts of the cell, and perform other functions. Some proteins are rigid, but others can bend and act like springs or hinges. Proteins are like a super high-tech Lego® set that can do just about anything at a microscopic level.

40

## CHAPTER 2: WHAT IS LIFE?

**Human Body Composition**

- 62% Water
- 16% Protein
- 16% Fat
- 6% Minerals
- 1% Carbohydrate

Elemental Composition — The Human Body:
- Oxygen 65%
- Carbon 18%
- Hydrogen 9.5%
- Nitrogen 3.2%
- Calcium 1.5%
- Phosphorus 1.2%
- Potassium 0.4%
- Sulfur 0.2%
- Sodium 0.2%
- Chlorine 0.2%
- Magnesium 0.1%
- Other >1%

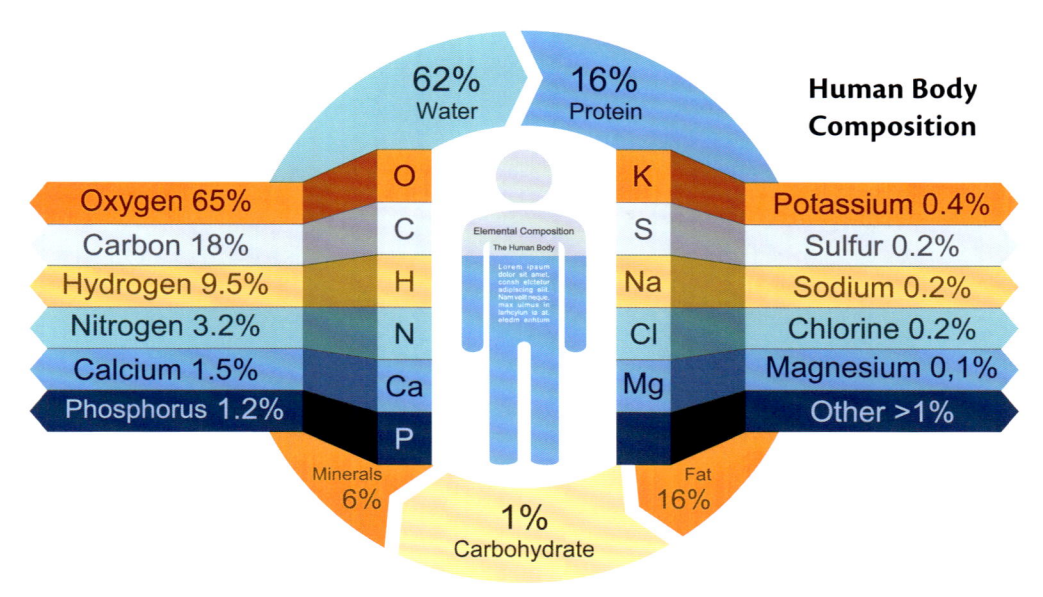

Elemental composition

Certainly, Psalm 139:15 applies to the Lord's extraordinary design of proteins!

*My frame was not hidden from You,
When I was made in secret,
And skillfully wrought in the lowest parts of the earth.*

## What are Proteins Made Of?

Proteins are made of little molecules called **amino acids**. There are 20 essential amino acids used for proteins, and they come together like a complicated 3D jigsaw puzzle. Not all 20 amino acids will be found in every protein. Amino acids are called **organic compounds** because they are made of **carbon,** and they are the building blocks of life. Amino acids always include the chemical building blocks of an amino and a carboxylic acid. These molecules are described by the chemical formulas $NH_2$ and $COOH$. But, how does a cell know how to build the jigsaw puzzle, or how does a certain protein know how to perform a certain function? That's the role of the DNA.

## Special Proteins—Enzymes

The body produces thousands of special proteins called **enzymes**. God made enzymes to serve as a **catalyst** for certain chemical reactions. Catalysts are used to trigger a chemical reaction. The reaction would not happen if there was no catalyst like an enzyme in the mix. Each type of enzyme has a specific job to do. Enzymes will either break down a

41

# GOD MADE LIFE

**Amino Acids**

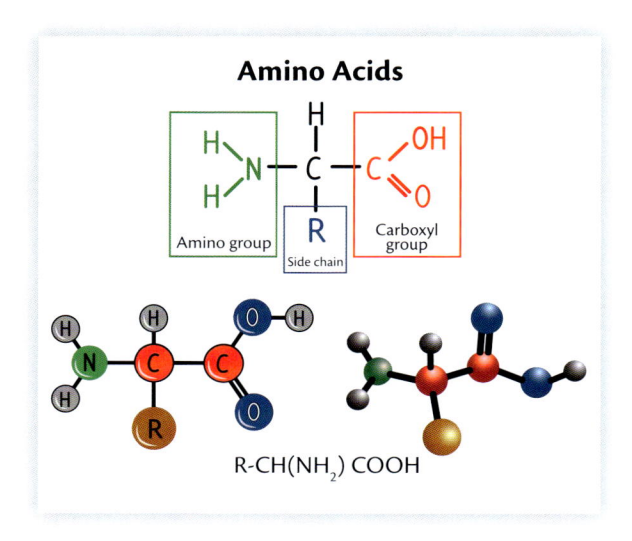

R-CH(NH$_2$) COOH

substance or will combine two substances—always to make something new.

An army surgeon in the 1800s experimented with **fly larvae** or maggots. He packed some wounds with larvae but left other wounds alone. Those wounds containing the larvae healed quicker. Later, scientists learned that the maggots were producing enzymes which slowly digested the dead skin, speeding up the healing process.

There are still other uses for enzymes. For example, enzymes help the stomach to digest food. They can break down fat and protein stains on clothing. Plumbers use enzymes to dissolve hair clumps and other organic garbage clogging up drains.

## Nucleic Acids

The wisest and most intricate parts of God's wonderful design of life are the **nucleic acids**. These are made up of smaller molecules called **nucleotides**. There are two kinds of nucleic acids:

**DNA (Deoxyribonucleic acid)**: The order of the nucleotides in the DNA chain provides a code or a blueprint for producing proteins. Just like a contractor uses a blueprint to figure out how to build a particular house, the DNA is the blueprint for a protein. The DNA is a double strand, shaped like two worms twisted around each other (in the pattern of a candy cane). The DNA

Maggots

Atoms usually have the same number of negatively-charged electronics and positively-charged protons. Ions are atoms with more electrons than protons, or more protons than electrons. If an atom has more electrons than protons, it is called a **negative ion**. If there are more protons than electrons, it is called a **positive ion**.

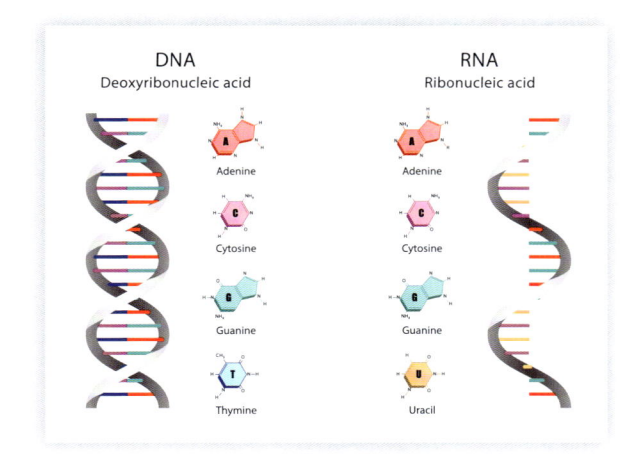

stays inside of the nucleus of the cell.

**RNA (Ribonucleic acid)**: The RNA helps build the protein using the instructions from the DNA. The RNA is a single strand, capable of carrying the information outside of the nucleus into other parts of the cell.

## Organelles and Cytoplasm

A cell is like your body. Your body has lots of distinct parts and organs, each tasked with a certain job. Similarly, a cell has organelles, each with its own job to do. Our good and wise Creator has assigned quite a few organelles to work together as a team in the cell.

The organelles swim around in a jelly-like substance called cytoplasm. This jelly is made up of water, fats, proteins, carbohydrates, ions, and small organic (carbon-based) molecules.

## Cytoplasmic Streaming

Remember, there is a lot of movement going on inside a living cell. God has designed the cell such that it can move stuff around inside of it through the cytoplasm. He uses a protein called **Myosin** to get things moving in the cytoplasm.

## Jobs for the Organelles

On the next page is a short list of some of the organelles floating around in the cytoplasm of the cell, along with the jobs the Creator gave to each of them. At the end of the chapter, you will find the longer list.

In addition, all cells have a **cytoskeleton** that acts sort of like the bones and muscles of a cell. It provides structure and helps move things around the cell. Some cells have tails called **flagella**. Other cells have finger-like **cilia** on the outside.

You can see that God has designed the tiny cell with more complexity than an automobile. A car runs down the road because the fuel pump provides fuel, the carburetor or fuel injectors feed the gas, the spark plugs light up the gas, the pistons push up and down, the driveshaft turns, and the wheels roll down the road. Somebody has to steer the car and push the brakes from time to time. And that's pretty much it. But, a human cell is 1,000 times more complicated than that!

There are still lots of things scientists do not understand about the cell. They don't know how cells vary from each other. They haven't figured out all the different functions performed by different kinds of cells. Scientists have a hard time determining the state

GOD MADE LIFE

## A Short List of the Organelles in the Cell

| | |
|---|---|
| **Nucleus** | • Control center of the cell<br>• Tells the rest of the cell what to do<br>• Contains DNA |
| **Rough Endoplasmic Reticulum (R.E.R.)** | • Makes proteins with the help of Ribosomes |
| **Smooth Endoplasmic Reticulum (S.E.R.)** | • Helps make fats (or lipids)<br>• Lipids will be used to construct new cell membranes<br>• Detoxifies the cell (cleans out the bad stuff) |
| **Golgi Apparatus** | • Modifies the proteins it receives from the R.E.R.<br>• Sends just the right proteins to other organelles in the cell or to the cell membrane<br>• Makes Lysosomes out of enzyme proteins from the R.E.R. |
| **Mitochondria** | • The digestive system of the cell, converting food into energy<br>• Known as "The Power Plant of the cell" |
| **Chloroplast (in plant cells)** | • Makes food for plants by converting sunlight energy to sugar |
| **Lysosome** | • The stomach of the cell; digests food |
| **Ribosomes** | • Protein factories, using instructions from RNA |

of a cell. How old is the cell? Is it getting ready to "die?" They don't know when a cell finds out what kind of cell it is supposed to be as it forms. Scientists still don't know what most of the DNA is supposed to do, or how it determines human and animal traits. And these cells are functioning all over the place! There are only 1.4 billion cars in the world, but there are 30,000 times more cells in your body than cars in the world. Now, consider all the people, animals, trees, and plants in the world. Think of all the cells that make up all of these organisms. All of these things are busy growing, eating, reproducing, moving, or responding to stimuli all over the place, all the time. Do you see how God's creation is much more expansive, more complicated, and more impressive than what man has produced? Praise God for His infinite wisdom and power—far beyond all of man's collective knowledge and abilities!

Also, think about this: Cars don't reproduce themselves. You don't find little baby cars squeezing out of the tail pipes of cars.

# CHAPTER 2: WHAT IS LIFE?

**Cilia and Flagella**

Cilia

Flagella

For thus says the LORD,
Who created the heavens,
Who is God,
Who formed the earth and made it,
Who has established it,
Who did not create it in vain,
Who formed it to be inhabited:
"I am the LORD, and there is no other."
(Isaiah 45:18)

## God Makes Cells with Irreducible Complexity

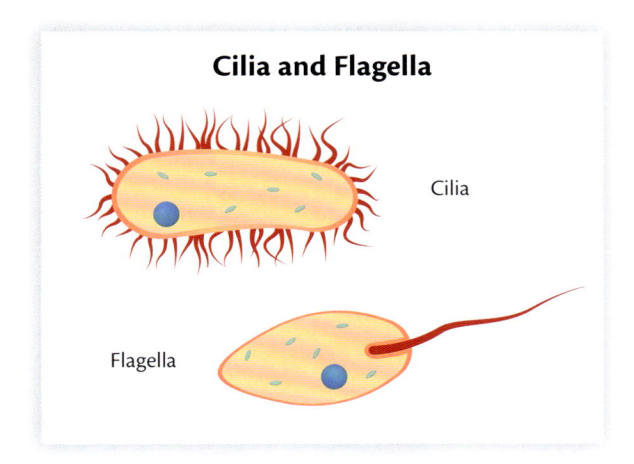

Cars can't repair themselves either.

Of the 400,000 different proteins in the human body, one protein is especially active. He's the repair guy, assigned to take care of the DNA molecule. The busy little **53BP1 protein** will fix a single DNA molecule as many as 100,000 times a day. There are over 30 trillion cells in your body, so there are a lot of DNA molecules to work on.

Think about this for a moment. To learn how to fix cars, automotive repair guys usually have to study the ins and outs of cars for six months to two years. Automotive engineers go to school for four to six years. Yet, biologists and doctors have been studying the human cell for 200 years, and they still don't understand the inner workings of it very well. There are thousands of diseases and problems they have no idea how to fix. The human cell is far more complex than all of man's designs.

Evolutionists are baffled. Some cells swim about using a tail called a flagellum. How could something as complicated as a flagellum appear on protozoa or bacteria? For one thing, these tails are so tiny you could fit 100 of them in a human hair. But each of them is propelled by a very complex little motor. How difficult would it be to build a motor that small? How would you assemble it if you couldn't even see it?

Scientists have come up with a term called **irreducible complexity**. The concept is not hard to understand. What needs to work for a car to run down the road? At the very least, you need the engine to work, along with the fuel injectors, the fuel pump, the spark plugs, the drive shaft, and the wheels. If the fuel pump doesn't work, the engine will never run. If the engine runs, but there are no wheels, the car will make a lot of noise, but it won't go anywhere. All of the

# GOD MADE LIFE

A car needs all these parts to run.

> Now to Him who is able to establish you according to my gospel and the preaching of Jesus Christ, according to the revelation of the mystery kept secret since the world began . . . to God, alone wise, be glory through Jesus Christ forever. Amen. (Romans 16:25, 27)

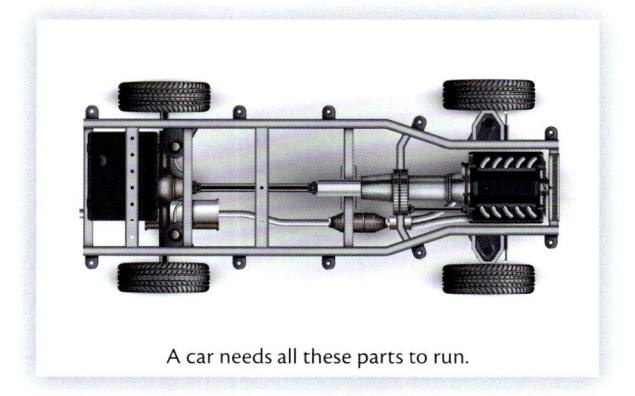

critical parts must work at the same time if the car is going to run. This is called irreducible complexity.

The same principle applies to the flagellum. God designed a complicated motor to propel this little tail. It has a universal joint, bushings, a stator, a rotor, and a driveshaft. Little clamps hold the motor to the cell membrane. The contraption also has a clutch and brakes to slow it down. Some of these tiny motors can spin at 100,000 revolutions a minute. They can also switch directions in a flash—all designed by the Genius of all geniuses. Now, what if no clamps held the motor to the cell membrane? Would the motor propel the little flagellum around? Or what if there was no driveshaft or rotor to spin? Would the little motor still work? Of course not. It is **irreducibly complex**. The Designer had to produce the whole motor at the same time, and of course, this could never have come about by chance.

## Why Evolutionists Are Befuddled

Evolutionists reject the Creator. They do not believe that God made the motor for the flagellum. The famous evolutionist Charles Darwin thought that somehow a cell developed its little motor over many generations. He thought that everything came about gradually. He observed that some birds have developed slightly longer beaks over generations. When a mother bird has babies, the chick with the longest beak might be healthier, live longer, and produce more baby birds. So sometimes we do see gradual change through the generations. But, this flagellum motor has many parts that must work together. This is something very different from a little beak which grows bigger over a thousand years, through many generations. Regardless of the size of the beak, it is always functional for the birds. They can still peck around for food. Now, what about the flagellum motor? Did it come about by gradual processes over a long period of time? Suppose that just the motor clamps and the driveshaft appeared on the flagellum. Without the rest of the parts, the motor clamps and driveshaft would be useless. Besides, how could such a complicated device appear by chance, unless there was some very intelligent Designer who laid out a blueprint in the DNA for building it? These are the things that befuddle the evolutionist who does not believe in God. This is how "smart" evolutionists become foolish in their thinking. According to Paul in Romans 1, "although they knew God, they did not glorify Him as God, nor were thankful, but became futile in their thoughts, and their foolish hearts were darkened. Professing to be wise, they became fools. . ." (Romans 1:21-22).

## Healthy Cells

A healthy cell works as God designed it to function. It turns food into energy. It repairs wounds. It provides for movement. And, it should provide for its own demise (or death) when it has finished its job. However, when cells do not complete their tasks or end their natural life, they can "go rogue." These old cells begin to divide over and over again, creating tumors in the body. This is called **cancer**. They don't pay any attention to the signals sent to them by the Ribosomes. They shouldn't reproduce like this. Instead, these "rebel colonies" get surrounding cells to send more nutrients into the colony to feed it and help it to grow.

Cancer usually forms in a "primary tumor." If it doesn't spread, or if it is cut out or reduced by some other treatment, then the patient probably won't die. However, if the cancerous cells make it into the blood-

**The Flagellum Motor: A Complicated Design**

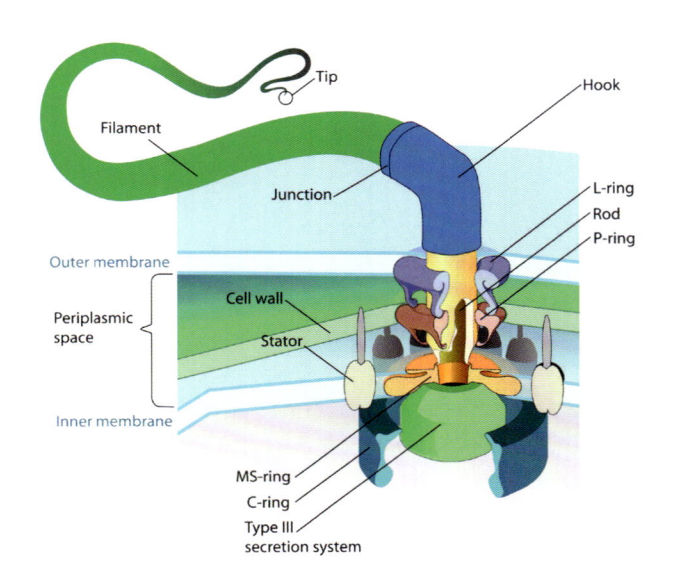

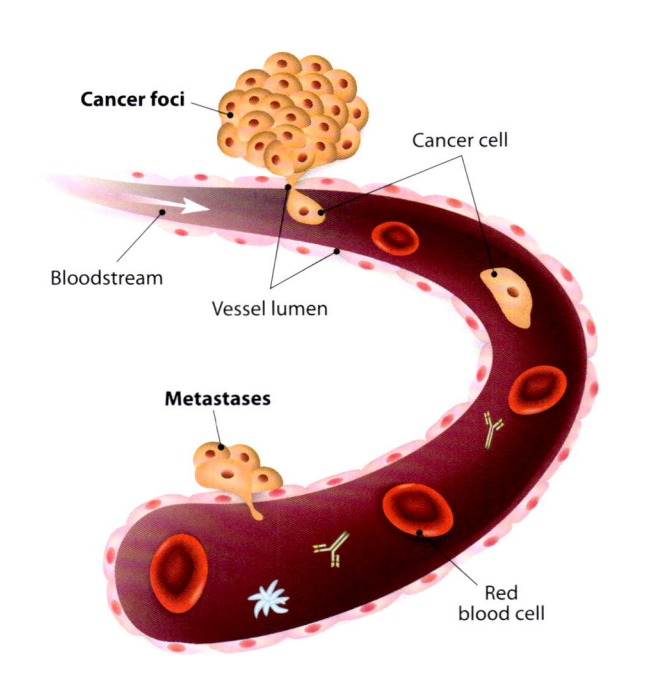

# GOD MADE LIFE

Doctor examining x-ray film

stream and spread throughout the body, this is bad news. It is called **metastasis**. Most cancer deaths are due to this problem.

Cancer is the second leading cause of death in the world. About 22% of deaths are caused by cancer. But the good news is that the death rate has dropped 27% over the last 25 years. This is mostly due to the reduction in smoking and the increase in testing. The sooner you can detect cancer the more likely you are to prevent it from spreading and killing the patient. The death rate for skin cancer has dropped the most recently because of a new treatment called **immunotherapy**. This treatment gets the body's immune system to work better, and that takes care of the cancer problem.

We can't always control the things that

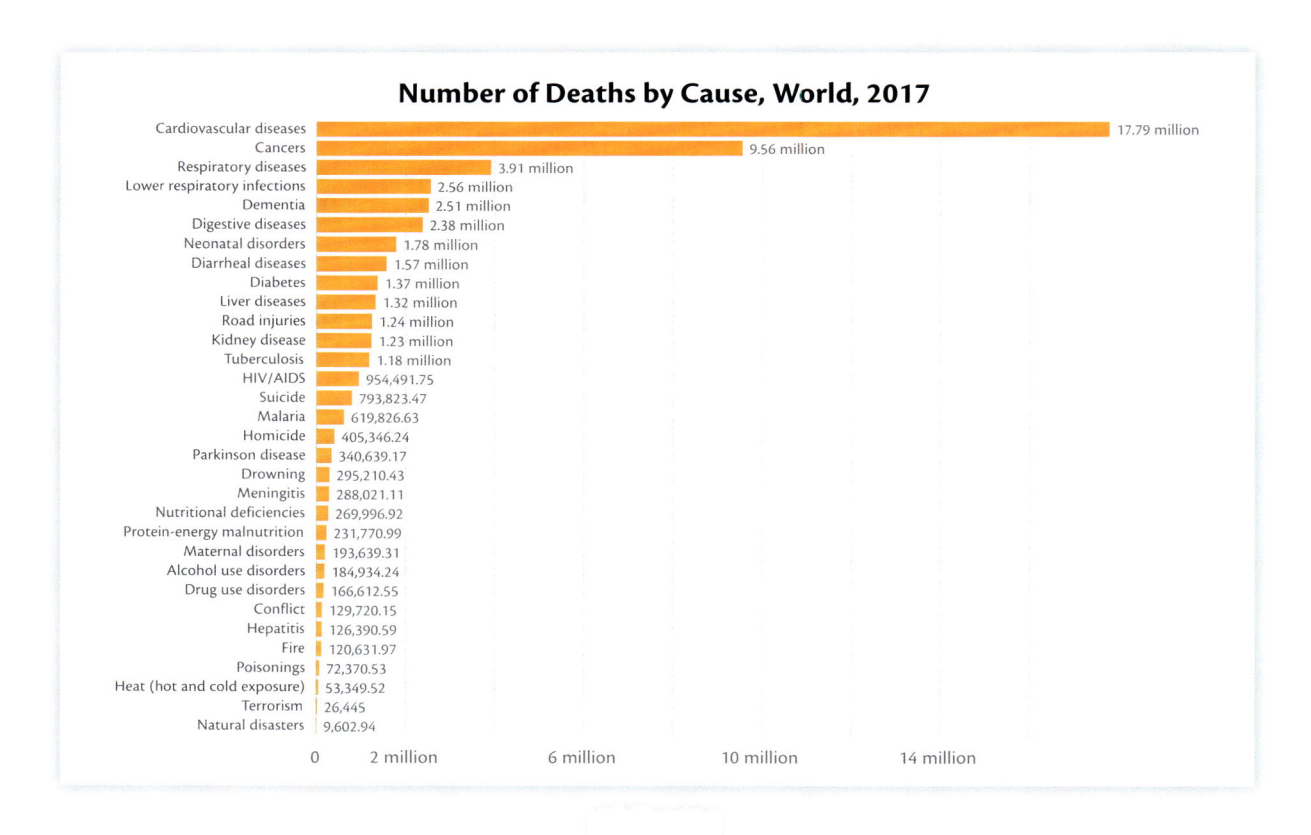

## Number of Deaths by Cause, World, 2017

| Cause | Deaths |
|---|---|
| Cardiovascular diseases | 17.79 million |
| Cancers | 9.56 million |
| Respiratory diseases | 3.91 million |
| Lower respiratory infections | 2.56 million |
| Dementia | 2.51 million |
| Digestive diseases | 2.38 million |
| Neonatal disorders | 1.78 million |
| Diarrheal diseases | 1.57 million |
| Diabetes | 1.37 million |
| Liver diseases | 1.32 million |
| Road injuries | 1.24 million |
| Kidney disease | 1.23 million |
| Tuberculosis | 1.18 million |
| HIV/AIDS | 954,491.75 |
| Suicide | 793,823.47 |
| Malaria | 619,826.63 |
| Homicide | 405,346.24 |
| Parkinson disease | 340,639.17 |
| Drowning | 295,210.43 |
| Meningitis | 288,021.11 |
| Nutritional deficiencies | 269,996.92 |
| Protein-energy malnutrition | 231,770.99 |
| Maternal disorders | 193,639.31 |
| Alcohol use disorders | 184,934.24 |
| Drug use disorders | 166,612.55 |
| Conflict | 129,720.15 |
| Hepatitis | 126,390.59 |
| Fire | 120,631.97 |
| Poisonings | 72,370.53 |
| Heat (hot and cold exposure) | 53,349.52 |
| Terrorism | 26,445 |
| Natural disasters | 9,602.94 |

48

## CHAPTER 2: WHAT IS LIFE?

go wrong in our bodies. Doctors try hard to get rid of the cancer, but they are not always successful. For example, about 30% of people with stomach cancer will survive. Only 18% of people with lung cancer survive. And 68% of people with bone cancer survive. Yet, we must always remember that God is always in control of everything going on in our bodies. He counts every hair on our heads. And He can miraculously heal men, women, and children with cancer. If you know somebody with cancer, remember to pray for them.

> "Are not five sparrows sold for two copper coins? And not one of them is forgotten before God. But the very hairs of your head are all numbered. Do not fear therefore; you are of more value than many sparrows." (Luke 12:6-7)

## How to Keep Your Body Healthy

> Now may the God of peace Himself sanctify you completely; and may your whole spirit, soul, and body be preserved blameless at the coming of our Lord Jesus Christ. (1 Thessalonians 5:23)

Although we will all die someday (until Jesus returns), we are still supposed to take good care of our bodies. Scientists are pretty sure that some environmental conditions can encourage this cancer. Smoking cigarettes and radiation (too much exposure to X-rays) can confuse the body and produce cancer in it. There are many theories on how to maintain healthy cells, but these are generally accepted:

1. Avoid using tobacco.

# GOD MADE LIFE

2. Maintain a healthy diet. Almost universally, doctors say you should eat plenty of fruits and vegetables.
3. Limit the consumption of red meat and processed meats (hot dogs and bacon).
4. Avoid obesity. Eat and drink in moderation.
5. Get lots of exercise.
6. Protect yourself from too much exposure to the sun. Use sunscreen in the summertime.

The foods that pose the highest dietary risks when it comes to cancer are as follows:[3]

1. Too much processed meat
2. Not enough dairy
3. Not enough whole grains
4. Too much red meat or too many sugary beverages

Some doctors and researchers believe certain foods are helpful for maintaining healthy cells. They contain **phytochemicals**. These include berries, broccoli, tomatoes, walnuts, grapes, and other vegetables, fruits and nuts.

Some scientists believe that **antioxidants** like Vitamin C, Vitamin E, and Beta-carotene can prevent cancer, but there is not much evidence confirming this.

Scientists are pretty much agreed that a healthy immune system can also fight off cancer. It is better if your body is doing the work God has designed it to do.

Finally, some studies have found that supplements can help improve the cell's mitochondrial health. These include minerals like Magnesium, Phosphate, and Calcium, some co-factors like CoQ10 and NADH, L-Carnitine transporters, and herbs like Curcumin and Schisandrin.[4]

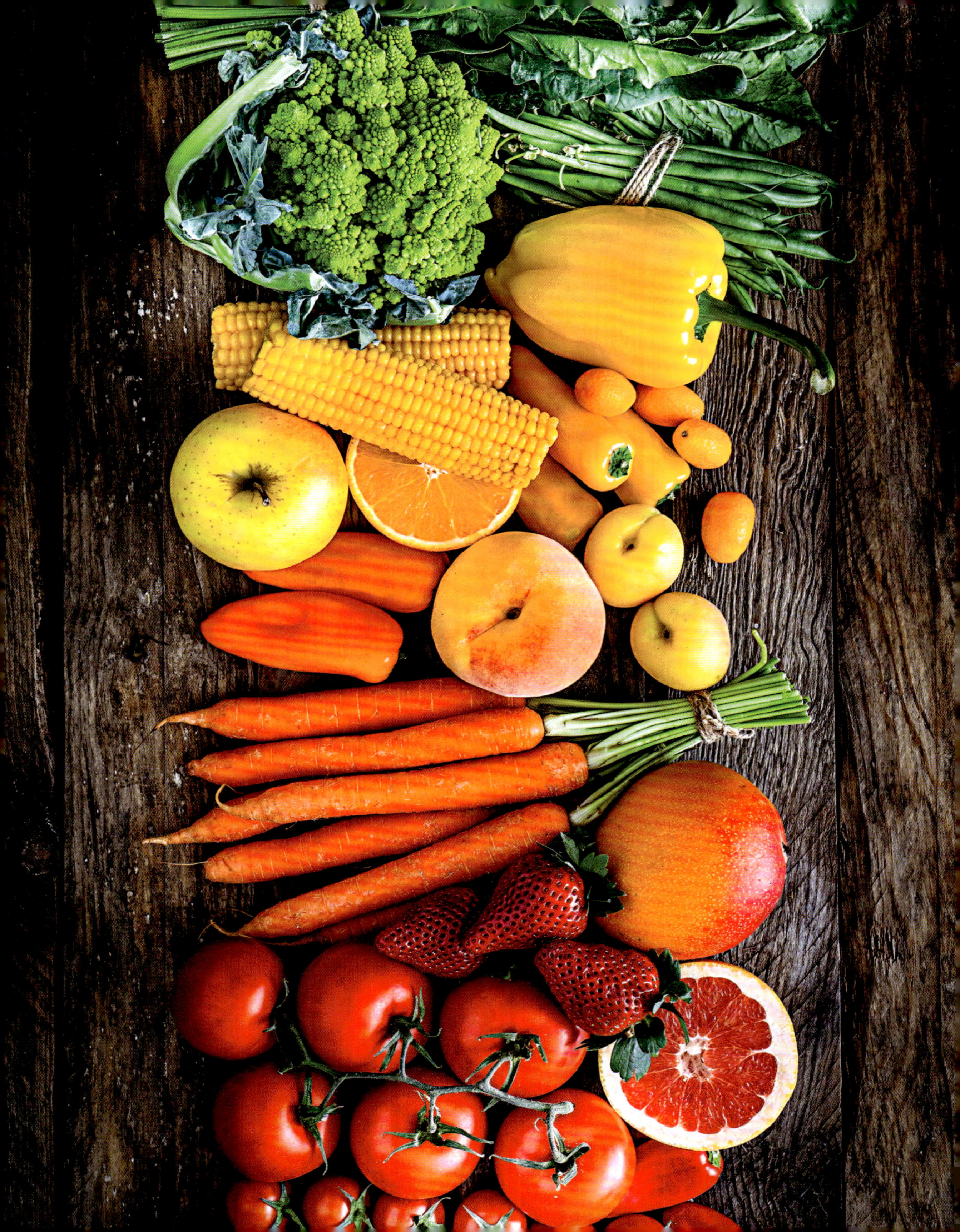

# GOD MADE LIFE

## Sing

Having seen the marvelous work of God in this spectacular creation, an appropriate response is always worship and praise. If the student is unfamiliar with the hymn or psalm, some version of it is available on the internet and may be accessed (with supervision) for singing along.

*Psalm 100*

All people that on earth do dwell,
Sing to the LORD with cheerful voice.
Serve Him with joy, His praises tell,
Come now before Him and rejoice!

Know that the LORD is God indeed;
He formed us all without our aid.
We are the flock He surely feeds,
The sheep who by His hand were made.

O enter then His gates with joy,
Within His courts His praise proclaim!
Let thankful songs your tongues employ.
O bless and magnify His name!

Because the LORD our God is good,
His mercy is forever sure.
His faithfulness at all times stood
And shall from age to age endure.

## Do

Choose at least one of the following activities and apply the lessons you have learned in this chapter.

1. **Weigh yourself.** Calculate your Body Mass Index using the following equation. You may use a calculator if you would like.

   BMI = [Weight (kg) ÷ Stature (cm) ÷ Stature (cm)] x 10,000
   **My BMI** _____

   Refer to these charts provided by the Centers for Disease Control to compare your BMI to the national average.
   - **Boys:** https://www.cdc.gov/growthcharts/data/set1clinical/cj41l023.pdf
   - **Girls:** https://www.cdc.gov/growthcharts/data/set1clinical/cj41l024.pdf

   Where do you lie on this chart by percentile? _____

   Based on the following table, what is your weight status? _____

## CHAPTER 2: WHAT IS LIFE?

## Weight Status for Children

| Weight Status Category | Percentile Range |
|---|---|
| Underweight | Less than the 5th percentile |
| Normal or Healthy Weight | 5th percentile to less than the 85th percentile |
| Overweight | 85th to less than the 95th percentile |
| Obese | 95th percentile or greater |

2. There are many different recommendations for diet, and don't consider any single recommendation as the best for you. However, here is a diet recommended by the Mayo clinic. How does your diet line up with this? You may refer to the Calorie Count Table in Appendix 2 of this book to estimate your daily calorie intake.

|  | **Recommendation** | **How much are you getting?** |
|---|---|---|
| **Calories** | Girls 1,400-2,200<br>Boys 1,600 -2,200<br>*depending on growth and activity level* | _____ |
| **Protein** | Girls 4-6 ounces per day<br>Boys 5-6.5 ounces per day | _____ |
| **Fruits** | Girls 1.5-2 cups per day<br>Boys 1.5-2 cups per day | _____ |
| **Vegetables** | Girls 1.5-3 cups per day<br>Boys 2-3.5 cups per day | _____ |
| **Grains** | Girls 5-7 ounces per day<br>Boys 5-9 ounces per day | _____ |
| **Dairy** | Girls 3 cups per day<br>Boys 3 cups per day | _____ |

## GOD MADE LIFE

Pray with your family about your dietary needs. Are there any areas in which you might adjust your current eating habits?

_____

_____

_____

3. **Investigate enzymes.** Experiment on different kinds of stains using enzyme cleaners. Then, create a chart for the family so they will use the right enzyme on the right stain. Smear a cloth with the following substances:

   a. Fat or grease (fat)

   b. Egg or blood (protein)

   c. Wine or grape juice (carbohydrate)

## Watch

To watch the recommended videos for this chapter, go to **generations.org/GodMadeLife** and scroll down until you find the video links for Chapter 2. Our editors have carefully selected films without references to evolution; however, we would still encourage parents or teachers to provide oversight for all internet usage. The videos may not give God the glory for His amazing creative work in every case. Therefore, the student and parent/teacher should consider responding to these insights with prayer and praise.

## Appendix 1—The Organelles in the Cell

### Membrane-Bound Organelles

| | |
|---|---|
| **Nucleus** | • Control center of the cell<br>• Tells the rest of the cell what to do<br>• Contains DNA |
| **Rough Endoplasmic Reticulum (R.E.R.)** | • Makes proteins with the help of Ribosomes |
| **Smooth Endoplasmic Reticulum (S.E.R.)** | • Helps make fats (or lipids)<br>• Lipids will be used to construct new cell membranes<br>• Detoxifies the cell (cleans out the bad stuff) |
| **Golgi Apparatus** | • Modifies the proteins it receives from the R.E.R.<br>• Sends just the right proteins to other organelles in the cell or to the cell membrane<br>• Makes Lysosomes out of enzyme proteins from the R.E.R. |
| **Mitochondria** | • The digestive system of the cell, converting food into energy<br>• Known as "The Power Plant of the cell," creating ATP<br>• Contains its own DNA (or instructions for operation)<br>• Makes certain steroids<br>• Regulates the metabolism of the cell |
| **Chloroplast (in plant cells)** | • Converts sunlight energy into sugar for plant food |
| **Vacuole (in plant cells)** | • Maintains turgor pressure against the cell wall (in plant cells) to keep the plant standing straight and rigid |
| **Lysosome** | • The stomach of the cell; digests food<br>• If the cell has no food, it will digest parts of the cell itself<br>• The cell's garbage disposal |
| **Peroxisome** | • Similar to Lysosomes; digests fatty acids<br>• In the liver, these guys will breakdown poisons |
| **Secretory Vesicle** | • Carries hormones or neurotransmitters in and out of the cell |

## Non Membrane-Bound Organelles

| | |
|---|---|
| **Ribosomes** | • Protein factories, using instructions from RNA |
| **Cytoskeleton** | • A network of fibers that keeps the cell intact |
| **Microtubules** | • Building blocks for the cytoskeleton<br>• Form cilia and flagella for cells that move using these organelles |
| **Microfilaments** | • Located in the cytoskeleton; contracts muscles<br>• Facilitates cell movement<br>• Serve as tracks on which organelles move around inside the cell |
| **Intermediate Filaments** | • Much thicker than microfilaments; gives the cell structure |
| **Junctions** | • Connect cells to other cells |
| **Centrosomes** | • Help in cell division (reproduction) |
| **Cilia** | • Little hairlike fingers that move a cell, or move something over the surface of a cell |
| **Flagellum** | • The tail of a special cell designed to propel the cell |

# Appendix 2—Calorie Count for Typical Foods

1. Apple, medium: 72
2. Bagel: 289
3. Banana, medium: 105
4. Beer (regular, 12 ounces): 153
5. Bread (one slice, wheat or white): 66
6. Butter (salted, 1 tablespoon): 102
7. Carrots (raw, 1 cup): 52
8. Cheddar cheese (1 slice): 113
9. Chicken breast (boneless, skinless, roasted, 3 ounces): 142
10. Chili with beans (canned, 1 cup): 287
11. Chocolate chip cookie (from packaged dough): 59
12. Coffee (regular, brewed from grounds, black): 2

13. Cola (12 ounces): 136
14. Corn (canned, sweet yellow whole kernel, drained, 1 cup): 180
15. Egg (large, scrambled): 102
16. Graham cracker (plain, honey, or cinnamon): 59
17. Granola bar (chewy, with raisins, 1.5-ounce bar): 193
18. Green beans (canned, drained, 1 cup): 40
19. Ground beef patty (15 percent fat, 4 ounces, pan-broiled): 193
20. Hot dog (beef and pork): 137
21. Ice cream (vanilla, 4 ounces): 145
22. Jelly doughnut: 289
23. Ketchup (1 tablespoon): 15
24. Milk (2 percent milk fat, 8 ounces): 122
25. Mixed nuts (dry-roasted, with peanuts, salted, 1 ounce): 168
26. Mustard, yellow (2 teaspoons): 6
27. Oatmeal (plain, cooked in water without salt, 1 cup): 147
28. Orange juice (frozen concentrate, made with water, 8 ounces): 112
29. Peanut butter (creamy, 2 tablespoons): 180
30. Pizza (pepperoni, regular crust, one slice): 298
31. Pork chop (center rib, boneless, broiled, 3 ounces): 221
32. Potato, medium (baked, including skin): 161
33. Potato chips (plain, salted, 1 ounce): 155
34. Pretzels (hard, plain, salted, 1 ounce): 108
35. Raisins (1.5 ounces): 130
36. Ranch salad dressing (2 tablespoons): 146
37. Red wine (cabernet sauvignon, 5 ounces): 123
38. Rice (white, long-grain, cooked, 1 cup): 205
39. Salsa (4 ounces): 35
40. Shrimp (cooked under moist heat, 3 ounces): 84
41. Spaghetti (cooked, enriched, without added salt, 1 cup): 221
42. Spaghetti sauce (marinara, ready to serve, 4 ounces): 92
43. Tuna (light, canned in water, drained, 3 ounces): 100
44. White wine (sauvignon blanc, 5 ounces): 121
45. Yellow cake with chocolate frosting (one piece): 243[5]

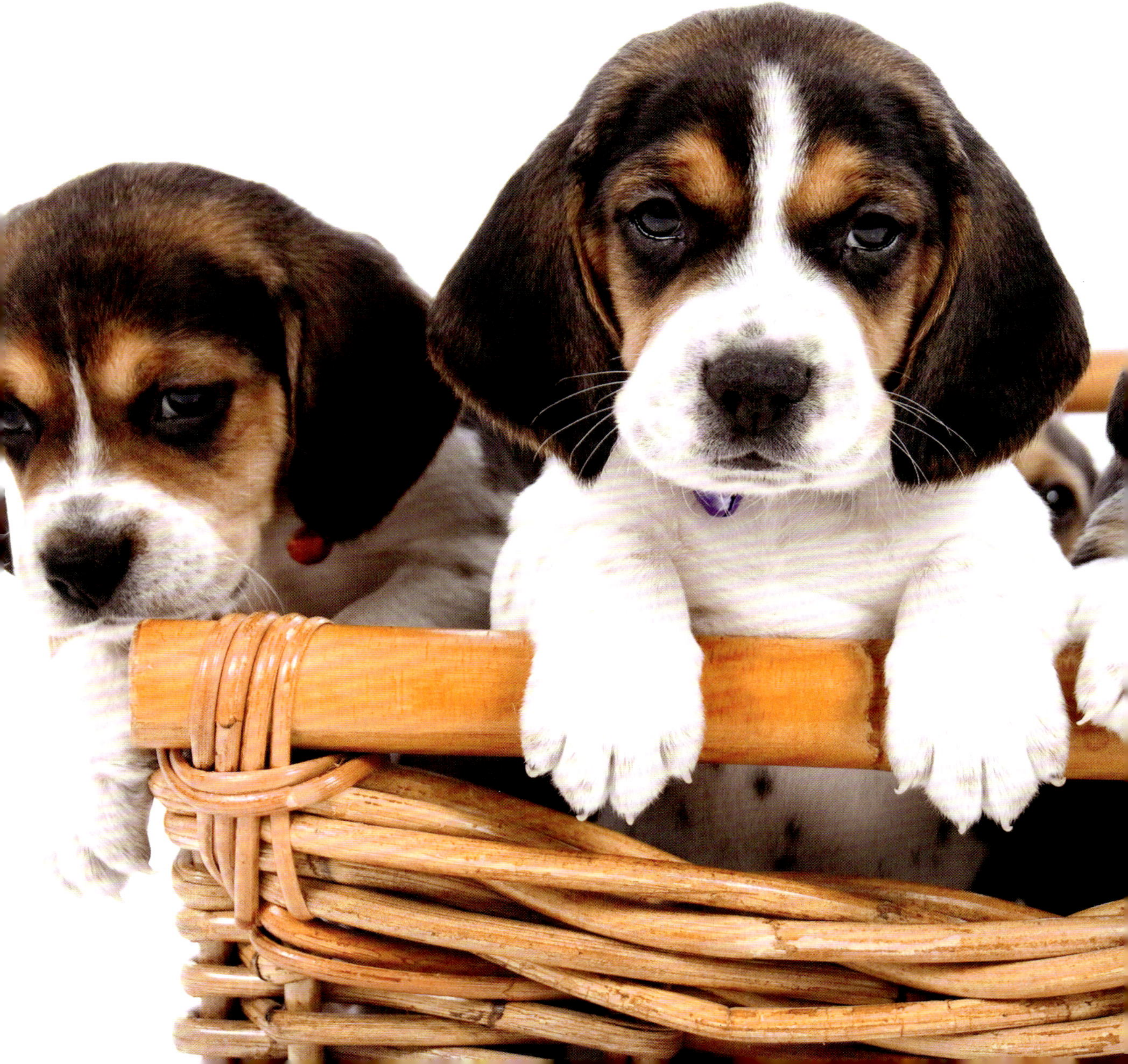

# CHAPTER 3
# GOD LOVES REPRODUCTION

> Then God spoke to Noah, saying, "Go out of the ark, you and your wife, and your sons and your sons' wives with you. Bring out with you every living thing of all flesh that is with you: birds and cattle and every creeping thing that creeps on the earth, so that they may abound on the earth, and be fruitful and multiply on the earth." (Genesis 9:15-17)

Our Creator God is a God of life, and He wants His world full of life. The Creator loves reproduction, and He has made a world teeming with life. The world is designed such that new life reappears constantly to take the place of the dying and the dead. Reproduction happens when something makes more of itself: when, for example, a rabbit produces another rabbit or two. One female rabbit and her female children could potentially produce 185 billion rabbits in seven years. The rabbit wasn't equipped with much to defend itself against predators, but it can reproduce, making sure the world always has plenty of bunnies.

Each pine tree yields hundreds of pine cones over a lifetime, and each cone has a hundred or so seeds under the scales. That's tens of thousands of seeds from a single tree! The Lord has made sure that His creation will continue to reproduce even if people are not planting trees and raising rabbits.

## Reproduction

What if you could reproduce yourself by making a copy of your whole body—a younger and fresher you? What if you could create a clone like a Siamese twin and then easily sever the twin from your body? What

# GOD MADE LIFE

Pine cones

by other animals. The **hydra** (of the jellyfish family) can keep reproducing itself using its own cells. This little creature testifies to the power of God.

> For we know that if our earthly house, this tent [our body], is destroyed, we have a building from God, a house not made with hands, eternal in the heavens. (2 Corinthians 5:1)

if humans could make a plastic 3D manufacturing printer that could reproduce itself? What if cars could reproduce themselves? Then, there would be no shortage of cars in the world. There would be no need for humans to work hard to produce more cars. In short, it would be a powerful, impressive act of pure genius to create something that could reproduce on its own. God's creation is simply amazing!

If cells can constantly reproduce themselves, what would prevent an organism from living forever? When Adam fell into sin in the garden, death came to all mankind. And that death comes in the gradual breakdown of cell reproduction. As people get older, the DNA code is no longer capable of making perfect copies of itself in reproduction.

Yet, we know that God can create a being to live forever. When we get to heaven, He will give us a body that will never die.

Interestingly, the Lord has included some creatures around us that do not age—and could live forever if they were not eaten

Hydra

## Reproduction by Cell Division

God designed the cell to reproduce itself. We have already compared the cell to a city with streets, communication systems, and factories. When the cell reproduces, it makes a cell as an exact copy of itself. This happens in three phases.

**Phase 1: Interphase**. To build a city, engineers and contractors must use blueprints or plans. So the new cell will need its own

## Stages of Mitosis

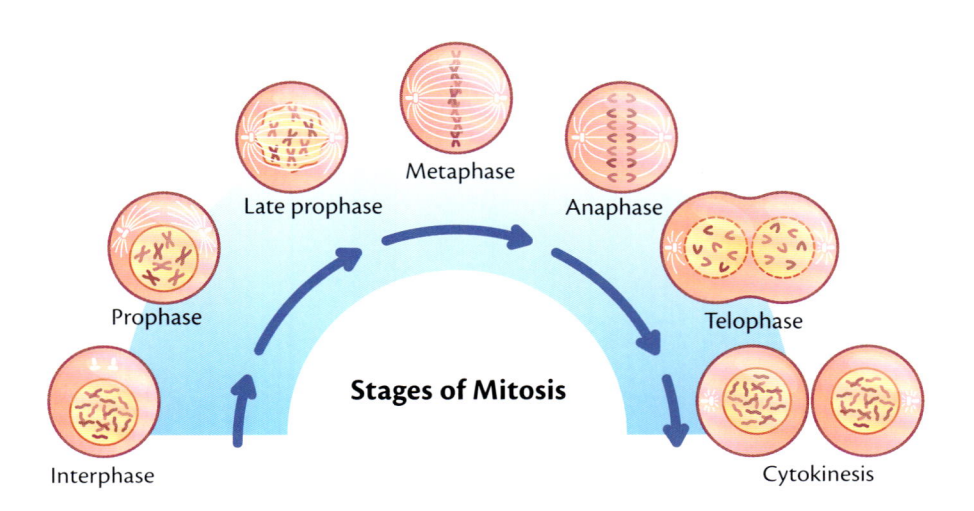

set of blueprints or DNA code. During interphase, the cell copies the DNA genes inside the nucleus. At the end of interphase, there are two identical DNA strands called **sister chromatids**, joined at the middle.

**Phase 2: Mitosis.** During this phase, the two identical chromatids separate and move to opposite sides of the cell. Then two brand-new nuclei membranes form around the chromatids.

**Phase 3: Cytokinesis.** During cytokinesis, the original parent cell brings the cell wall between the two nuclei, pinching off two daughter cells. That is how the cell reproduces and divides itself.

There may be an unequal division between the amount of cytoplasm and quantity of organelles given to each of the two new daughter cells. But that's okay because each cell can make its organelles according to the blueprints in the DNA.

## Growing Kids into Adults

How does a baby grow into a little boy? How does a little girl grow into a full-sized woman? While you still have two ears, two eyes, two hands, and two feet, all of these grow bigger over time. The organism is the same. You are still the same person you were when you were a baby. But, everything has grown bigger through the years. This happens by **mitosis.** A newborn baby has about 26 billion cells, but an adult has 2,000 times that number of cells—somewhere around 50-100 trillion cells.

The cells must work hard to not only grow the body but also to replace the cells that are dying. The adult body is producing 2 million cells every second, or 240 billion cells a day. All of your skin replaces itself in about a month. This is something like building 2 million fully operational cities in one

A person grows bigger over years

day. And that's for just one human body. Remember, every cell has millions of things going on at the same time. Tens of thousands of ribosomes are making new proteins. Millions of enzymes and proteins are doing their jobs inside the cell. Like truck drivers, motor proteins are transporting cargo around the cell. God's work in the cell is infinitely more impressive than anything man has done with all of his factories and roads. On top of all of that, man can't even create a contraption that can reproduce itself. Praise God for His infinite wisdom that is shown in the creation of life!

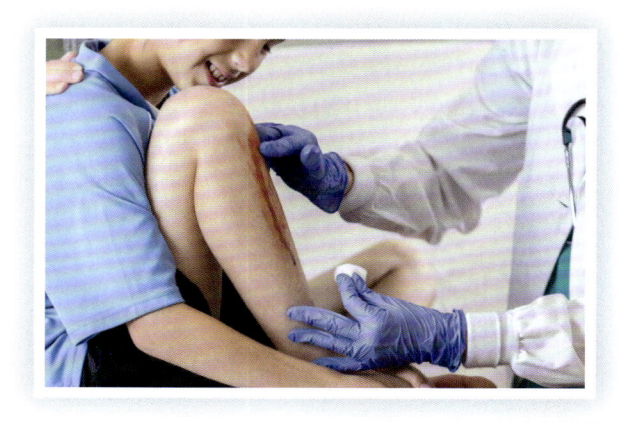

## Cells Repairing Wounds

When you get a gash on your leg, millions of cells die in the process. These cells are going to have to be replaced by new cells. Within a month or so, your skin looks almost like it did before it took the gash. Have you ever seen a car repair itself after getting into an accident? The process for wound repair is extremely complicated. The most brilliant scientists in the world cannot explain how it works exactly.

God has designed a four-step process for wound repair:

- Step 1: Clotting
- Step 2: Scabbing
- Step 3: Rebuilding
- Step 4: Strengthening

CHAPTER 3: GOD LOVES REPRODUCTION

Clotting is the most complicated part of the whole process. It is a highly engineered, delicate process. Every time an animal's skin is punctured, God wants to keep the creature from bleeding to death. Animals out in the wild don't have doctors who can sew up their wounds, so their bodies need to do something to stop the bleeding right away. Of course, the Creator also designed our bodies to stop bleeding by clotting. Some scientists say that this clotting is as difficult as landing an airplane. If the blood congeals in the wrong spot, that might block up the blood circulating in the animal's body. The animal would have a heart attack and die. Scientists have identified over 40 processes that must work properly in blood clotting if the animal will survive.

You can witness the clotting process on your own body when you fall down and skin up your knee. When your body senses that you've broken the skin, an army of blood cells called **platelets** immediately rush to the scene of the accident. These platelets stick to each other like glue, forming a gooey covering. This keeps you from bleeding to death, and it keeps germs out of the cut. The clot is sewn together with thread-like "fibrin" cells, and then it hardens into a scab. Underneath the scab, the Creator has arranged for your body to produce new skin cells to fix the cuts. Those delicate blood vessels that were torn in the accident are fixing themselves too. This is all very wonderful!

The blood-clotting process is another example of irreducible complexity. If one part of the team is missing, the clotting won't work. Using the example of an airplane landing, suppose it would take forty operators to land the craft without crashing. Each operator must do his job, and in the right order. If one operator acts out of turn, the plane will crash. Likewise, if one of the little guys in the blood clotting business doesn't do his job in turn, the patient will bleed to death. Actually, there are at least thirteen different proteins involved in forty blood-clotting operations. God has also provided explicit instructions on how to make up these proteins in the DNA. There are 186,000 different **nucleotides** in the DNA, all of which must be arranged in just the right order for the proper production of a

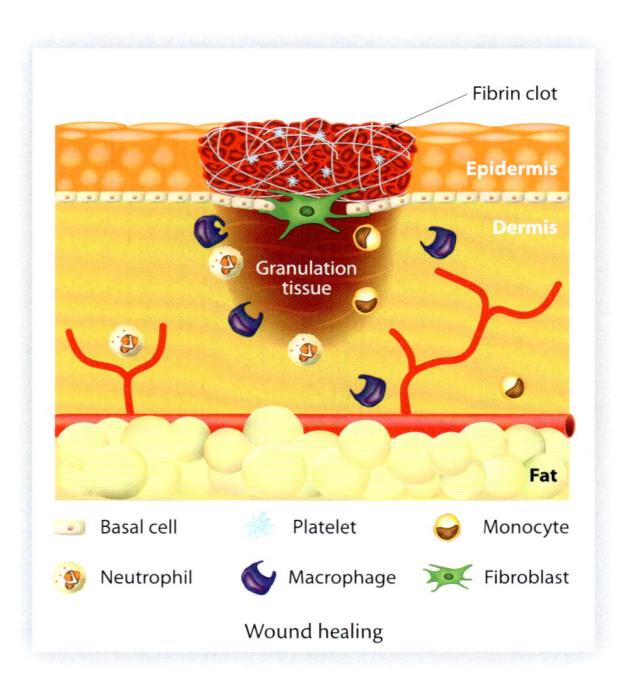

63

# GOD MADE LIFE

single protein. For example, Protein Number Eight (Factor VIII) requires 2,400 amino acids to properly clot a wound. If one nucleotide in the gene is out of order, then the human body won't produce Protein Factor VIII. A child who inherits a bad gene with the wrong set of nucleotide instructions from a parent will suffer from **hemophilia**. And so, if a child with hemophilia gets a cut, it can take a very long time to heal.

Even after the clot forms, our Creator has provided another chemical called FSF to fix the clot if it gets punctured. As the wound heals, He designed a protein called **plasmin** which works like little scissors. It slowly cuts out the scab and prepares for the next stage—rebuilding and strengthening.

Billions of wild animals survive wounding every day. They don't need to see a veterinarian to fix them up. Their bodies are designed to heal themselves, and they go on living as if nothing happened. Some animals like deer are especially robust against wounds—in some cases clotting twice as fast as humans.[1]

Then the Lord answered Job out of the whirlwind, and said:

"Who is this who darkens counsel
By words without knowledge?
Now prepare yourself like a man;
I will question you, and you shall answer Me.
Where were you when I laid the foundations of the earth?
Tell Me, if you have understanding. . . .
Do you know the time when the wild mountain goats bear young?
Or can you mark when the deer gives birth?
Can you number the months that they fulfill?
Or do you know the time when they bear young?
They bow down,
They bring forth their young,
They deliver their offspring.
Their young ones are healthy,
They grow strong with grain;
They depart and do not return to them."
(Job 38:1-4, 39:1-4)

## How Organisms Reproduce

Sometimes it takes two organisms to reproduce. In this case, two organisms are needed to make one or more additional organisms. So where there were once two organisms, now there are three or more organisms. This is called **sexual reproduction**.

Sometimes it takes one organism to reproduce another organism. This is called **asexual reproduction**. So where there was once one organism, there are now two organisms. This occurs by mitosis. But there are several different kinds of asexual reproduction.

Yeast will reproduce in a process called **budding**. After mitosis, the daughter will form a little bump or bud at the yeast's cell wall. Sometimes this will break off, but sometimes it stays attached. Eventually a granddaughter bud will form on the daughter bud, and extended chains of yeast develop.

**CHAPTER 3: GOD LOVES REPRODUCTION**

Some organisms reproduce by **sporing**. After cell division, these organisms (like bread mold) will create a protective coating around the new cells. This helps them to survive and to multiply in other areas. Often the **spores** are blown around by the wind, but the cell remains alive because it is protected inside the spore covering. Be careful with bread molds. If the mold spores blow around the kitchen, they can survive for months.

## Genetics

*Therefore, just as through one man sin entered the world, and death through sin, and thus death spread to all men, because all sinned... (Romans 5:12)*

A son will look a lot like his father. His nose might have the same shape or he might have the same color eyes. As organisms reproduce, they pass along common characteristics through the genetic code. **Genetics** is the study of inherited characteristics.

The most important trait passed on from father to children is Adam's sinful nature. No ordinary child born on this earth has been

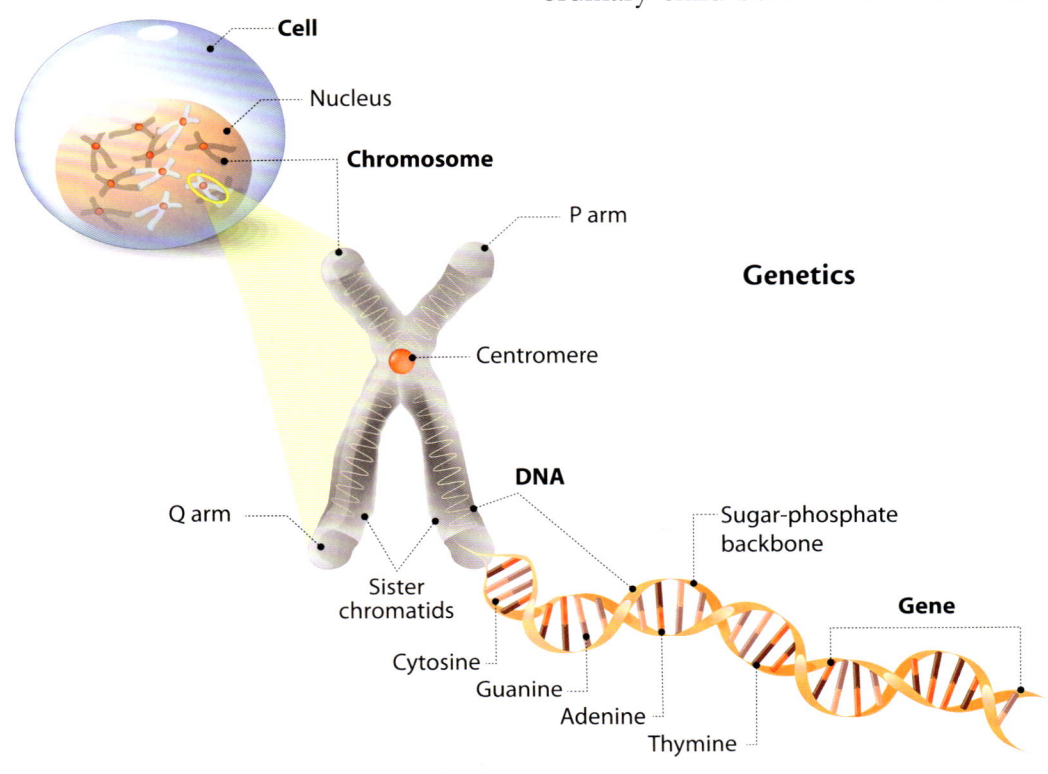

**Genetics**

- Cell
- Nucleus
- Chromosome
- P arm
- Centromere
- Q arm
- Sister chromatids
- Cytosine
- Guanine
- Adenine
- Thymine
- DNA
- Sugar-phosphate backbone
- Gene

# GOD MADE LIFE

free of sin. As soon as a child comes out of the womb, he speaks lies. Babies will disobey their parents. They receive these traits from Adam. Because these are moral traits, they probably come from some other source than DNA. Animals and plants do not sin against God, and so they do not inherit these traits from their parents.

## Other Terms You Need to Know

There are a lot of terms used in genetics, so here is a simple summary.

> **Chromosome**: Before the two sister chromatids separate, they are called the chromosome. It is a long thread-like structure containing DNA, tightly coiled around another protein called the **histone**. Humans have 46 chromosomes, or 23 pairs. A rice plant has 12 pairs, and a dog has 39 pairs. This is a neat way of organizing the DNA in the cell.
>
> **Gene**: A section of a chromosome that contains instructions for building a specific protein. The gene will determine a particular trait for the living organism. For example, blue-eyed genes will make sure that a child has blue eyes. Brown-eyed genes make sure a child will have brown eyes.
>
> **DNA** (Deoxyribonucleic acid): These are long molecules containing instructions for the building of various parts of your body.

The DNA in your cells are super-long strings made up of chemicals called **nucleotides**, arranged in pairs, called **base pairs**. If all the DNA in your body was put end to end, it would go for 110 billion miles, stretching from Earth to the sun and back six hundred times.

Every cell in your body contains DNA in the nucleus. Each cell has about 3.5 billion of these base pairs. That's about 875 billion megabytes of information, which would fill up about 106 volumes of a regular encyclopedia set. Imagine a whole bookshelf-worth of information stuffed into a tiny microscopic cell like that!

The DNA in one cell contains a library of information.

Scientists from Harvard University stored 700 terabytes (or 700 billion copies of a large book) in one gram of DNA. That would be about the same as the memory contained on 700 laptop hard drives made in 2020. If DNA was used for computer memory, all of the information collected

## CHAPTER 3: GOD LOVES REPRODUCTION

and produced in the whole world over one year could fit into four grams of DNA (four cubic centimeters or .25 cubic inch)![^12] If you think about it, God's system of information storage is still 1,000 times more efficient than the best computer storage of our day.

---

Praise the LORD!
Praise the LORD from the heavens;
Praise Him in the heights!
Praise Him, all His angels;
Praise Him, all His hosts!
Praise Him, sun and moon;
Praise Him, all you stars of light!
Praise Him, you heavens of heavens,
And you waters above the heavens!
Let them praise the name of the LORD,
For He commanded and they were created.
He also established them forever and ever;
He made a decree which shall not pass away. (Psalm 148:1-6)

---

## Basic Genetics

Your body has a unique set of DNA instructions different from other children. When you were only a single cell floating around in your mother's womb, there were 46 chromosomes (23 pairs) in that cell. Twenty-three came from dad, and 23 came from mom. Each time the cell reproduced itself, the same 46 chromosomes were replicated. The same DNA instructions were copied for every cell in your body. These instructions would tell what your toes and fingers would look like, what your face would look like, the color of your hair, and the size of your nose.

The English language uses 26 different letters from which millions of words are put together. God's DNA strands are made up of five different nucleotides: adenine (A), thymine (T), guanine (G), cytosine (C), and Uracil (U). These are similar to alphabetical letters meant to communicate important messages about the design of you. The Creator lays out these nucleotides in a meaningful order. The order in which they appear on the DNA strand will determine the kind of protein to be produced. These proteins are the building blocks for cells, and for animal and human life.

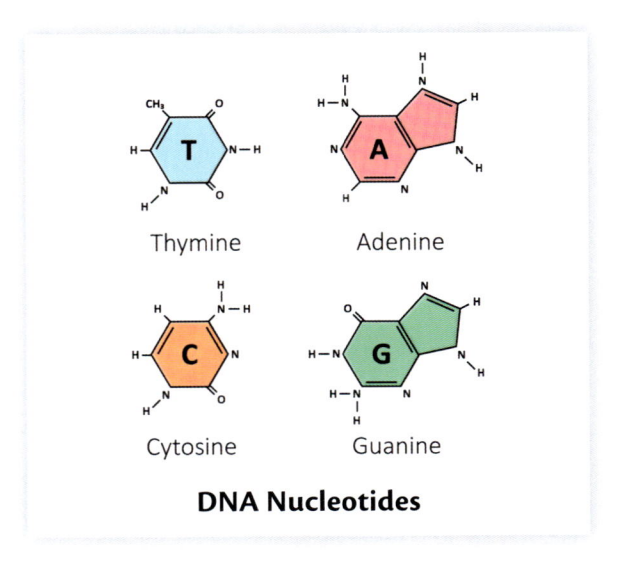

**DNA Nucleotides**

The DNA is expressed in three-letter words called **codons**. Thousands of these

GOD MADE LIFE

"words" are strung together to put together the instructions necessary to build life and keep it going. Here are examples of the codon words and the nucleotides called for by the code:

| Codon | Nucleotides Called For |
|-------|------------------------|
| CCA | Cytosine, Cytosine, Adenine |
| CCG | Cytosine, Cytosine, Guanine |
| CCC | Cytosine, Cytosine, Cytosine |

Each of these three codons describes an amino acid, which will be put together using the nucleotides. Also, tucked in with the codon are hidden instructions. These instructions tell the ribosomes exactly how to make the protein, including when to pause at points in the process. It is critical that the protein put together in the ribosome folds comes together in the right position. The shape of the protein is very important—just as important as using the rights parts to construct it.

The RNA (Ribonucleic acid) is a replicated piece of the long DNA strands. Building a large skyscraper requires thousands

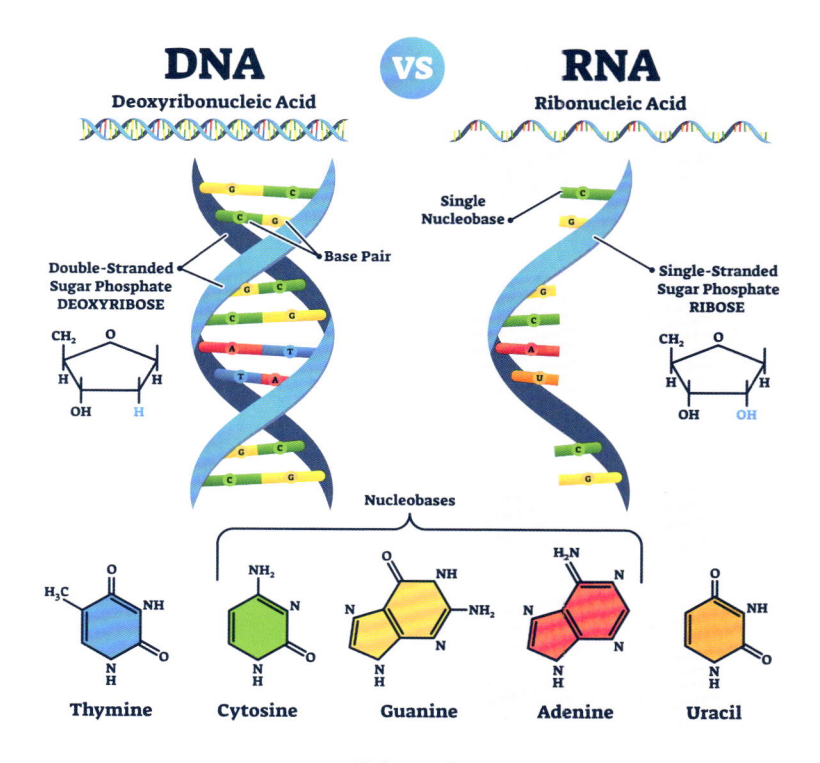

**DNA** vs **RNA**
Deoxyribonucleic Acid | Ribonucleic Acid

Double-Stranded Sugar Phosphate DEOXYRIBOSE — Base Pair

Single Nucleobase

Single-Stranded Sugar Phosphate RIBOSE

Nucleobases

Thymine | Cytosine | Guanine | Adenine | Uracil

68

of pages of blueprints and instructions. Suppose a contractor is working on a small part of the building—the bathroom on the 20th floor. He doesn't need all of the instructions for the small part he is working on. He would need a copy of the instructions that applied to his project. Similarly, the RNA is a copy of a small part of the DNA for the ribosome contractor who will build a particular protein. This protein becomes a building block for a cell, which will provide for some part of your body.

Praise God for His wonderful design of DNA! His creation is too complicated for the most brilliant scientists in the world to figure out! What can we say but that God's creation of this vast inner world of the cell and the DNA is amazingly elegant, extremely complicated, and ultimately intelligent and purposeful. The creation of life really does display God's perfection, showing His great power, wisdom, and goodness. Indeed, as the Lord God said at the end of His creation on the sixth day, "It was very good."

## Evolution Is Impossible

If you have any common sense, the idea that a rock turned into a human being is just plain ridiculous. It is as foolish as the notion that a princess kissed a frog and turned him into a prince. But, evolutionists desperately want it to be so. They want to believe that non-life turned into life by chance processes. They don't want God. They don't want to be morally responsible to Him. They don't want God to tell them what to do.

So they believe that all of the processes in the cell, the DNA, the RNA, the defense mechanisms, the clotting, the little flagellum motors, the reproductive systems, and all of the other complicated systems came about by chance. They call it an accident. They think all of this was unintended. They deny that there could have been an intelligent Designer. These "scientists" cannot explain all the complexities of life. It is complicated for the human mind to comprehend. Yet, they still say it all came about by unintelligent cause or dumb chance. This is the most radical illustration of spiritual blindness and willful stupidity the world has ever seen.

The fool has said in his heart, "There is no God." (Psalm 14:1)

Let's take for example what it would take to bring about the simplest reproducing life form. You would need 124 proteins constructed with 400 amino acids each, every part of which would be critical for life. The whole living cell with all its pieces and functions has to appear at the same time. Creationists tell us that the probability to get something as complicated as life is 1 in 10 with 14,184 zeroes following the number. In other words, a quadrillion has just 12 zeroes following the 1. Add another 14,172 zeroes and you have the odds of a single reproducible life form appearing by chance.

Most scientists do not doubt that factories making jet airplanes were built by intelligent human beings. If a scientist walked into

Aircraft assembly

a factory where jet airplanes were assembled, he would know that there was an intelligent design for the factory and the airplane. If some skeptic laughed at him for believing the "false" stories about the intelligent design of jets and assembly plants, he would brush it off. "Of course," he would say, "This was a product of intelligent design. Man has produced these wonderful designs." Now, the human body is a million times more creative, more intelligent, more complex, and more awe-inspiring than an airplane factory. But still, so many scientists refuse to recognize a very intelligent Designer. And they mock those who believe God made all of this.

As explained in Romans 1:18-21, these "intelligent" scientists have:

a. Suppressed the truth in unrighteousness,
b. Refused to glorify Him as God,
c. Refused to be thankful to God,
d. Their foolish hearts were darkened.

## Basic Heredity

Two scientists studied genetics in the 19th century—Charles Darwin and Gregor Mendel. Darwin did not fear God, while Mendel feared God in his studies. Darwin's theory was very bad, and it has never been proven by hard evidence. Yet, Mendel worked very hard to test his theories using pea plants in his garden. His conclusions have been very helpful to understand God's principles of heredity that work in plants, animals, and human beings.

Mendel studied seven physical traits of the pea plant:

- **Height:** Tall or short
- **Flower Position:** On the top or side of the stem
- **Pod Shape:** Inflated or shrunk
- **Pod Color:** Yellow or green
- **Seed Color:** Yellowish or green

70

## CHAPTER 3: GOD LOVES REPRODUCTION

Gregor Mendel (1822-1884)

- **Seed Shape:** Round or wrinkled
- **Flower Color:** Colored or white

Mendel carefully mated plants with different characteristics like tall plants with short plants. Then he kept track of what the offspring looked like. He did this for many generations over many years. To mate the different plants, he would take the pollen from tall plants and drop it into the pistils of a short plant. The pollen contains the male cell, and the pistil held the egg. Then, he would cover the plants so there wouldn't be any other pollen that might mess up his experiment. Finally, he planted the pea seeds produced by each plant.

So Mendel mated tall plants with short plants as he studied the height **trait**. Right away, he discovered that mating a tall plant with a short plant would not produce medium-sized plants. Sometimes the new plant was tall and sometimes it was short. The plant's height depended on which genes its seed received from the parent plants.

Each parent contributes one possible gene for each trait. Each gene contained two **alleles** (possibilities). One parent might provide two tall alleles, two short alleles, or one tall and one short allele. This depended on which genes the parent plant had inherited from its parents.

If the new plant's seed had received all short alleles from each parent, it would grow into a short plant. If the new plant's seed had received all tall alleles from each parent, it would become a tall plant. But what would happen if the new plant's seed received two short alleles (a short gene) from one parent and two tall alleles (a tall gene) from the other parent? Remember that Mendel didn't get any medium-sized plants. The new plants would be either tall or short. This experiment produced 100% tall plants for Mendel. That's because for the size trait, the tall gene is **dominant**, and the short gene is **recessive**. So in this case, all of Mendel's first-generation pea plants were tall. The short gene inherited from the one parent seemed to disappear. However, when Mendel interbred the second generation of these tall plants, three quarters of the plants were tall and one quarter of the plants

# GOD MADE LIFE

were short. That's because there was still a hidden short allele tucked away in the genes.

Charles Darwin theorized that the combining of the tall and short plants would yield a very different plant—perhaps a medium-sized plant. This didn't happen. Instead, what Mendel found was that the new generations would retain the original genetic information from parents and grandparents—thus producing tall and short plants, either by a dominant or recessive gene.

| Dominant Trait in Humans | Recessive Trait in Humans |
|---|---|
| A blood type | O blood type |
| B blood type | O blood type |
| Baldness (in the male) | Not bald |
| Hazel or green eyes | Blue or gray eyes |
| High blood pressure | Normal blood pressure |
| Large eyes | Small eyes |
| Nearsightedness | Normal vision |
| Second toe longest | First or big toe longest |
| Short stature | Tall stature |
| Tone deafness | Normal tone hearing |

**Mendel's Law**

|  | Flower color | Seed shape | Seed color | Pod color | Pod shape | Plant height | Flower position |
|---|---|---|---|---|---|---|---|
| Dominant | Purple | Round | Yellow | Green | Inflated | Tall | Axial |
| Recessive | White | Wrinkled | Green | Yellow | Constricted | Short | Terminal |

## The Beauty and Order of God's Design

As we investigate the genetic variations in people and peas, we learn that the Creator wanted variety in His world. He didn't want everybody to look alike. But at the same time, God did not want chaos and absolute dissimilarities. He wanted children to look a little bit like their parents. So He created the world with differences and with patterns. He loves variety, but He doesn't want there to be too much variety. If there were no patterns, the world would be chaotic and disorderly. This reflects the nature of God. For, God is unity. He is One God. But, He is also three Persons—the Father, the Son, and the Holy Spirit.

If creation was a pure unity, there would be no diversity. On the other hand, if there were many gods and there was no unity in the universe, the world would only reflect diversity and disorder. Creation is a beautiful combination of unity and diversity. The orderly patterns are produced by genetics.

# CHAPTER 3: GOD LOVES REPRODUCTION

When a father and mother combine their genetic information, patterns of similarities appear in the children and grandchildren.

## Genetic Mutations and Problems

Ungodly evolutionists have a very wrong view of the world. They believe that things can evolve into better or more complex forms. Some evolutionists even think that someday man will evolve into a superman.

However, the Bible teaches something very different. The Bible teaches truths that better reflect what really happens. Consider that most people live 70-90 years today. When God created Adam and Eve, they were made to live forever. There was no death in the world. But after Adam and Eve ate the forbidden fruit, they introduced death into the world. Adam died after 930 years, and Noah lived about 950 years.

However, the flood brought a severe drop-off in lifespans. Shem died at about 600 years of age, and Shem's son Arphaxad died at 438 years of age. Abraham died at 169 years of age, and Moses died at 120 years of age. Since then, humans' lifespans varied between 70 and 80 years. And that hasn't changed to the present day. So you can see that the human body has been degrading instead of evolving after the worldwide flood. According to God's plans, bad genetic mutations were introduced into the human cells after Adam fell into sin. This might have continued after the flood.

## Mutations

A **mutation** is a change in an animal's or plant's DNA. Evolutionists hope that mu-

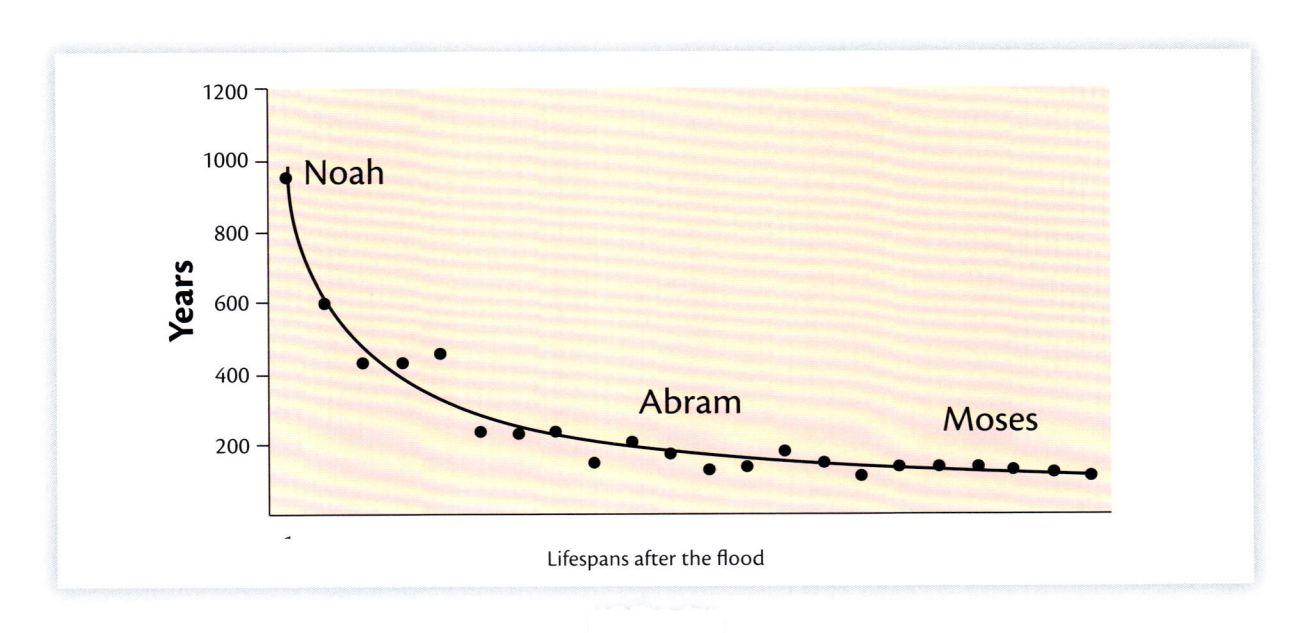

Lifespans after the flood

tations in the genetic code might produce increasing levels of complexity, better reproductive systems, and longer lifespans. Actually, mutations do just the opposite. In humans, mutations produce diseases like cancer, albinism, hemophilia, cystic fibrosis, and **sickle cell anemia**. Nobody considers these things as evolutionary improvements. **Albinism** is a mutation that hurts the body, preventing it from making skin pigment (coloring). Darker skin protects a person's body from heavy sunlight, so albinism is not an improvement. It is a handicap.

By mutations in the genetic code, a beetle on a windy island could lose its wings. This might keep him from getting swept off the island into the ocean waters where he would drown. This might help him to survive life on a windy island. But, this is not an additional improvement that makes the beetle's body more complicated.

A mutation might help an antibiotic-resistant bacteria to survive the antibiotic. But, how does this help the bacteria to survive all of the other threats to its well-being? Or, how might the mutation help the bacteria to adapt to other environmental conditions?

Mutations don't change the basic kind (or species) of the animal. Some dogs are born without hair because of a mutation. But they are still dogs, albeit very ugly dogs. It takes mutations to bring about the tiny chihuahua or the poodle. Although some people like the chihuahua as a pet, the mutations create more problems for the little dog. By selectively breeding wolves over many gen-

Genetic mutations: Albino Wallaby

## CHAPTER 3: GOD LOVES REPRODUCTION

erations, you might eventually get a poodle. This comes about by a lot of mutations. But you would never get a healthy wolf absent of harmful mutations by breeding poodles over many generations.

If you think of the DNA as a typed story, mutations are changes which introduce mistakes to the script. They don't add any improvements to the story. They take away from the original story by adding mistakes to it.

Chihuahua

When the messenger RNA peels off a mutated DNA for making red blood cells, the recipe calls for too much **hemoglobin**. So, the blood cells manufactured according to the bad recipe will be defective. They can't carry enough oxygen to service the body. This is what causes sickle cell **anemia**. Both parents must pass along the genetic mutation for the child to have this malady.

## Genetic Engineering

Scientists have tried to use their knowledge in genetics for improvements. But, this has introduced many other problems through the years. Because most scientists do not fear God in the present day, the result of their work is very questionable.

Some scientists want to clone human beings and animals. Thus far, they have managed to clone sheep, pigs, and cats. Other scientists have new human babies in test tubes, keeping them in freezers for many years. Still other scientists are trying to design perfect genes before they make the baby in a test tube.

More commonly, doctors will perform tests on little babies in the test tube for the genetic diseases of sickle cell anemia, Tay Sachs disease, and cystic fibrosis. If the babies have these conditions, the doctors will throw them away before they have a chance to grow in their mother's wombs. This is murder even in the case of small babies who

Child with Down syndrome

75

# GOD MADE LIFE

are stored away in test tubes. A recent study found 53% of British children with Down syndrome were killed before they were born.[3] God is always watching doctors when they do such things.

Recently, the first genetically modified salmon have been produced for human consumption. There is a huge business created around genetically modified grains, fruits, and vegetables.

When ungodly scientists do not fear God, they will use science to sin against God. Nowadays, there is much evil going on in the area of genetic engineering.

## Eugenics

After Charles Darwin wrote his book *The Descent of Man*, others made a popular new teaching called **eugenics**. Darwin's cousin Francis Galton coined the term and wrote several books on the subject. These were attempts to use Darwin's theory of evolution to improve on human genetics. These men thought that if people with bad genetics don't have children and die off, then the world would be better for it. Some scientists and political leaders thought that a whole tribe of people should die and never reproduce. The Nazis killed many Jews because of these bad "scientific" ideas.

In 1927, the US Supreme Court forcibly sterilized a mentally-challenged woman named Carrie Buck. This means that she could never have children after that. The judges thought that Carrie would pass her mental illness on to her children, and there

Francis Galton (1822-1911)

would be too many mental problems in the country. But since then, new scientific studies have shown that mentally-challenged parents don't pass their problems on to their children very much. The genetic influences "account for only a small amount of risk for mental illness."[4]

Proud scientists thought they could fix problems brought to the world by sin. They thought they might eliminate bad DNA by killing people or not allowing them to have children. These godless scientists also did not believe that God is in control of the genetic code. They did not believe that Jesus Christ could redeem the genetic code by supernat-

## CHAPTER 3: GOD LOVES REPRODUCTION

ural means. They did not believe that man would have to live with genetic problems until the final redemption at the end. You can see how the false ideas and bad worldview of the ungodly resulted in bad decisions.

Actually, there aren't that many life-threatening genetic problems in this country in any given year.

Altogether, these common forms of genetic diseases appear in 10,000 babies a year in the US. That's only about 0.25% of the total babies born. We do not live in a perfect world. It is a sinful world, and we still wait for the final redemption when death will be conquered forever. Everybody will die. Some people die at 78 years of age, and other people die at 45 years of age. But, those who believe in Jesus Christ will live forever with Him. He has already risen from the dead, and He has conquered death forever for Himself and for us. After we have lived and died, we will be resurrected and live forever in our perfected bodies. That applies to everybody who believes in Christ, the Savior of the world.

Instead of working on ways to kill babies with these problems, it is always better to find ways to improve their lives on this earth. Thus, the average lifespan of children with Down syndrome has improved from 25 years to 60 years just since 1983. Much of this improvement comes from taking children out of institutions and keeping them in loving homes with their families.

| Genetic Disease | Births effected per year in America | Lifespan |
| --- | --- | --- |
| Tay Sachs | 12 per year | 5 years |
| Muscular Dystrophy | 275 per year | 30 years |
| Hemophilia | 400 per year | 63 years |
| Cystic Fibrosis | 1,000 per year | 44 years |
| Sickle Cell Anemia | 2,000 per year | 54 years |
| Down Syndrome | 6,000 per year | 60 years |
| Average American | | 78 years |

GOD MADE LIFE

## What the Bible Tells Us

The Bible is concerned with the "seed." To play around with the seed of plants, animals, and humans is a risk. The Scriptures warn about the problem of defilement. Christians must be very careful when experimenting with the genes God has created. Mixing seeds can result in problems for future generations of plants and animals. When breeding animals especially, God does not want us to create animals that cannot reproduce. When a horse breeds with a donkey, for example, they produce a mule. It is a reproductive dead-end. Mules cannot have babies. The world doesn't pay any attention to these rules, but Christians find wisdom in the principles of God's Word. They will respect God's creation and fear the Creator as they set out to do their scientific work.

> "'You shall keep My statutes. You shall not let your livestock breed with another kind. You shall not sow your field with mixed seed.'" (Leviticus 19:19)

> "You shall not sow your vineyard with different kinds of seed, lest the yield of the seed which you have sown and the fruit of your vineyard be defiled." (Deuteronomy 22:9)

The mule is a mix of a donkey and a horse.

# GOD MADE LIFE

## Sing

Having seen the marvelous work of God in this spectacular creation, an appropriate response is always worship and praise. If the student is unfamiliar with the hymn or psalm, some version of it is available on the internet and may be accessed (with supervision) for singing along.

*Exalt the Lord!*

Exalt the Lord, His praise proclaim;
All ye His servants, praise His name,
Who in the Lord's house ever stand
And humbly serve at His command.
The Lord is good, His praise proclaim;
Since it is pleasant, praise His name;
His people for His own He takes
And His peculiar treasure makes.

I know the Lord is high in state,
Above all gods our Lord is great;
The Lord performs what He decrees,
In heaven and earth, in depths and seas.

He makes the vapors to ascend
In clouds from earth's remotest end;
The lightnings flash at His command,
He holds the tempest in His hand.

Exalt the Lord, His praise proclaim;
All ye His servants, praise His name,
Who in the Lord's house ever stand
And humbly serve at His command.
Forever praise and bless His name,
And in the church His praise proclaim;
In Zion is His dwelling place;
Praise ye the Lord, show forth His grace.

## Do

Choose at least one of the following activities and apply the lessons you have learned in this chapter.

1. Make a list of the most complicated machines made by man. Then make another list of the most complicated designs God has made in His creation. Show how God is much wiser than men.

2. Make a list of the questions that evolutionists cannot answer. What are the most serious problems with Charles Darwin's theory of evolution?

3. Find ways to oppose eugenics and cruel applications of bad science. Research ways in which you might be able to help children with genetic problems to live longer. How might we encourage mothers pregnant with Down Syndrome babies not to abort their children? As of 2020, only six states have banned Down Syndrome-based selection abortions (Indiana, Ohio, North Dakota, Louisiana, Kentucky, and Tennessee). Does your state ban this practice? _____

Here are some ways you might advocate for children with special genetic challenges:
- Volunteer for a Special Olympics program
- Volunteer to help a parent with a special-needs child once a week
- Support organizations like *joniandfriends.org* or *lacasadefe.org*
- Organize a birthday party for a child with Down Syndrome or some other special genetic challenge
- Contact a pro-life legislator and encourage him to sponsor legislation to prevent abortions based on birth defects

## Watch

To watch the recommended videos for this chapter, go to **generations.org/GodMadeLife** and scroll down until you find the video links for Chapter 3. Our editors have been careful to avoid films with references to evolution; however, we would still encourage parents or teachers to provide oversight for all internet usage. The producers of these videos may not themselves give God the glory for His amazing creative work, but we encourage the student and parent/teacher to respond with prayer and praise.

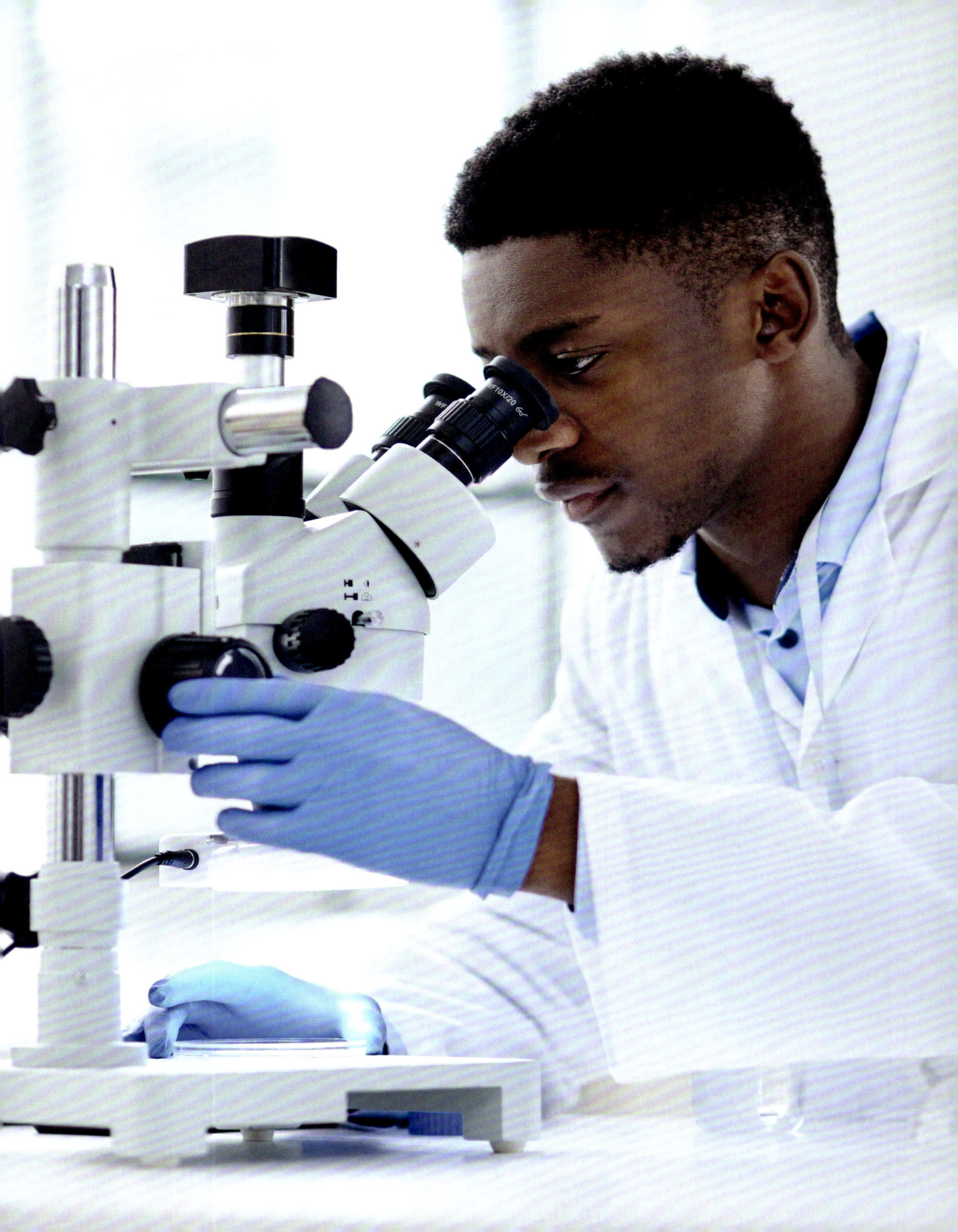

# CHAPTER 4
# GOD MADE MICROSCOPIC ORGANISMS

> Out of the ground the LORD God formed every beast of the field and every bird of the air, and brought them to Adam to see what he would call them. And whatever Adam called each living creature, that was its name. So Adam gave names to all cattle, to the birds of the air, and to every beast of the field. But for Adam there was not found a helper comparable to him. (Genesis 2:19-20)

From instructions given in Psalm 8 and Genesis 1, we know that God has assigned for us to rule over creation. A very important part of this job is to name all of God's distinct creatures. Adam started by naming the big animals: those he could see with the naked eye. Because Adam didn't have a microscope, he couldn't name the little guys—the protozoa, the bacteria, and other single-cell organisms.

This world is teeming with life all around you. And, you can't even see most of it. Scientists have counted over 9,000 kinds of microbes, bacteria, and fungi living in the average home. Thankfully, 99.9% of them are harmless. Some of these microbes come from your body. About 37 million crawl off your body every hour. Most surfaces in the room contain about 1,000 little critters per square inch, while closer to 300,000 critters per square inch are crawling around the kitchen sink. Also, don't forget there are also at least 100 different varieties of mites, spiders, and other arthropods living in most homes. Can you imagine 37,000,000 little animals crawling around the room right now? God made this world full of life. This reflects His glory, wisdom, and power. It is a picture of His endless store of life. God is eternally and ultimately life itself.

# GOD MADE LIFE

Some microorganisms are edible for animals. Some microorganisms clean up God's world of manure and other waste products. Others help us to digest our food. Not all of these organisms are friendly, however. Many harmful types of bacteria and fungi can cause disease, destroy crops, and even damage buildings. When man fell into sin, the creation was cursed, and death came into the world. And God uses these microorganisms to bring this death to man.

Mold colony (fungus)

The study of tiny, microscopic organisms is called **microbiology**. If God wants us to rule over animals that could kill us like lions and bears, He certainly wants us to rule over the micro-creatures that can be just as deadly.

So God created man in His own image; in the image of God He created him; male and female He created them. Then God blessed them, and God said to them, "Be fruitful and multiply; fill the earth and subdue it; have dominion over the fish of the sea, over the birds of the air, and over every living thing that moves on the earth." (Genesis 1:27-28)

## Classifications

Over 300 years ago, a Christian man named Carolus Linnaeus studied God's creation of plants and animals. He came up with the categories of plant and animal life, organized by the characteristics God gave to each of them. Mr. Linnaeus first divided life into groups called kingdoms. To this he added phylum, class, order, family, genus, and species. Each kingdom of God's creatures is broken up into different phyla, each phylum is broken

Carolus Linnaeus (1707-1778)

up into different classes, and so forth. This type of organization is called **classification**.

To better understand the classifications of God's creation, the table below provides several examples. This chapter focuses on the four kingdoms of **Fungi**, **Protista**, **Archaebacteria**, and **Eubacteria**. None of these received the breath of life, and none of these contain the life of the blood (hearts and lungs). Yet God made these creatures to move over the earth. These four groups of creatures are termed **microbes** or **microorganisms**. They are creatures which can only be seen under a microscope.

Usually, biologists limit the number of kingdoms to six (or sometimes five). Since the Scriptures are given by the authority of the Creator Himself, this will change how we look at God's world. This study will add two kingdoms for a total of eight. The human being must be considered separately from the animals, because Genesis 1:26 states that God made man in His image. And, there are two kingdoms of animals—those with the breath of life and those without the breath of life.

Then God said, "Let Us make man in Our image, according to Our likeness; let them have dominion over the fish of the sea, over the birds of the air, and over the cattle, over all the earth and over every creeping thing that creeps on the earth." (Genesis 1:26)

## Examples of Classification for the House Cat, the Wolf, and the Housefly

|  | Cat | Wolf | Fly |
|---|---|---|---|
| Kingdom | Animalia (with breath of life) | Animalia (with breath of life) | Animalia (without breath of life) |
| Phylum | Chordata | Chordata | Arthropoda |
| Class | Mammalia | Mammalia | Insecta |
| Order | Carnivora | Carnivora | Diptera |
| Family | Felidae | Canidae | Muscidae |
| Genus | Felix | Canis | Musca |
| Species | F. Domesticus | C. Lupis | M. Domestica |

# GOD MADE LIFE

**Eight Kingdoms of Life**

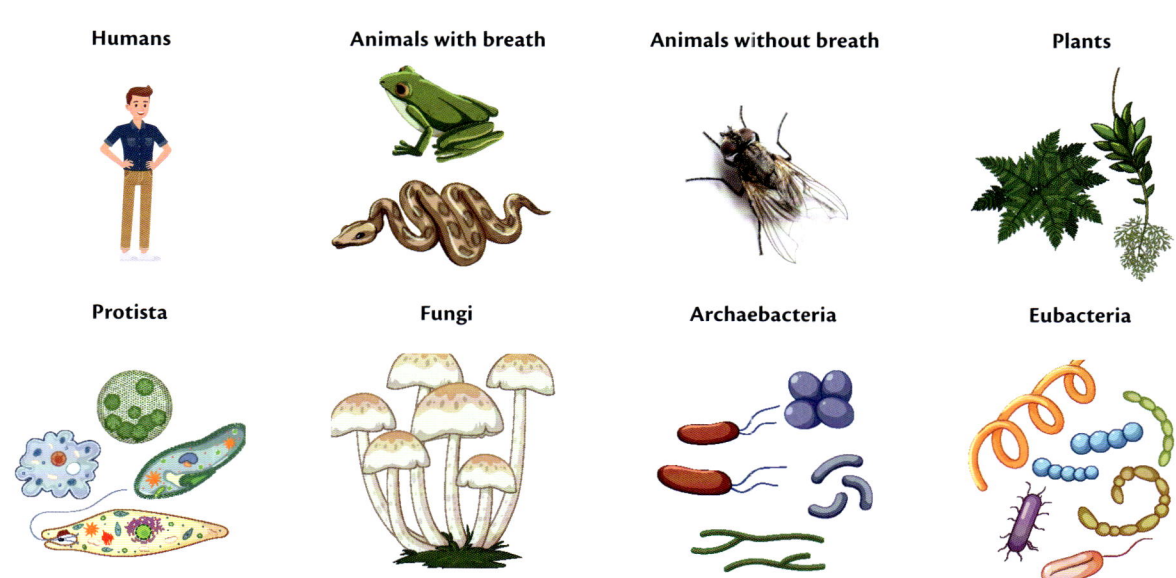

| Humans | Animals with breath | Animals without breath | Plants |
| --- | --- | --- | --- |

| Protista | Fungi | Archaebacteria | Eubacteria |
| --- | --- | --- | --- |

## Eight Kingdoms of God's Creation

- Human
- Animals with the breath of life
- Animals without the breath of life
- Plants
- Protista
- Fungi
- Archaebacteria
- Eubacteria

## The Differences Between Fungi, Protista, Archaebacteria, & Eubacteria

There are also various divisions in function for these microbes. They are created by God to do different things. There are **heterotrophic** microorganisms, meaning they eat other things around them. There are also **autotrophic** microorganisms, meaning they make their own food. Some protozoa like amoebas are solitary, and they work alone. Others work together in colonies. Some reproduce sexually, requiring the DNA of two microbes to mix before a third microbe develops (**sexual reproduction**). Sometimes the microbe splits itself into two identical pieces (**asexual reproduction**). Some organisms are **prokaryotic.** These are made of only one cell that has no nucleus. Others are **eukaryotic**, made up of many cells that have nuclei.

More differences between the microbes may be seen in the following table.

## God's Creation of Microbes

|  | Archaebacterium | Eubacterium | Protista | Fungi |
|---|---|---|---|---|
| Harmful to Humans | No | Sometimes | Sometimes | Sometimes |
| Where They are Found | Extreme Environments | Everywhere | Everywhere | Everywhere |
| Reproduction Method | Fission and budding | Spores | Mitosis | Spores |
| Cell Type | Prokaryotic | Prokaryotic | Eukaryotic | Eukaryotic |

## Diseases Caused by Microbes

The term **germ** is used to describe any kind of microbe that can cause a disease. A virus is not considered a microbe by scientists. But that is only because it cannot reproduce. According to biblical definitions, we would still include it under the category of that which has DNA but does not have the breath of life or life in the blood.

Viruses are not cells either. They do not have a cell membrane, cytoplasm, or any organelles. The virus is just an unfriendly nucleic acid wrapped in a protein coating. It attaches itself to otherwise healthy cells and infects them with its bad acid (DNA or RNA).

The following are examples of diseases caused by viruses, bacteria, Protista, and fungi:

- **Viruses:** Colds, flu, HIV/AIDS, COVID-19, polio, rabies, hepatitis, measles, mumps, rabies, smallpox, warts
- **Bacteria:** Cholera, leprosy, tuberculosis, plague, syphilis, food poisoning, meningitis, pneumonia, anthrax
- **Protista:** Malaria, amoebic dysentery, sleeping sickness
- **Fungi:** Athletes foot, ringworm, thrush, valley fever, mold allergies

Once these germs get into the body, they eat up the nutrients and produce poisons (toxins). These proteins invade the body and cause fevers, coughs, vomiting, and diarrhea. Doctors have to pinpoint the virus or bacteria using a microscope. They will look at samples of urine, blood, or spit to identify the root problem and suggest a remedy.

The most common sickness is the common cold. But the most deadly disease (caused by microbes) that kills the most people worldwide is pneumonia. The most deadly viruses that kill the highest percentage of people infected with them are HIV/AIDS and Ebola.

# GOD MADE LIFE

## Virus Forms

| | | | | | |
|---|---|---|---|---|---|
| Human immunodeficiency virus | Helicobacter pylori | Coronaviridae | Staphylococcus aureus | Pneumococcus | Chlamydia pneumoniae |
| Varicella-zoster virus | Lactobacillus | Vibrio cholerae | Rabies lyssavirus | Clostridium tetani | Treponema pallidum |
| Haemophilus influenzae | Meningococcus | Yersinia pestis | Hepatovirus A | Bacillus anthracis | Mycobacterium tuberculosis |
| Anaplasma | Salmonella | Rotavirus | Measles virus | Zika virus | Adenovirus |
| Bacteriophage | West Nile virus | Influenzavirus | Bifidobacterium | Saccharomyces boulardii | Streptococcus Thermophilus |

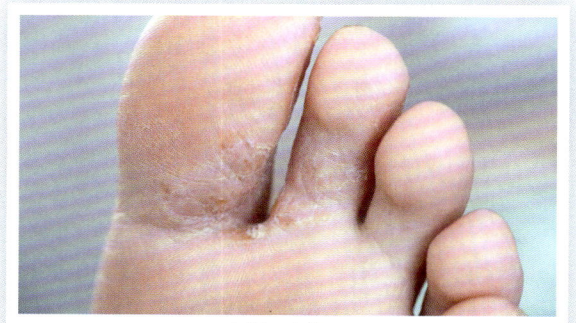

Athlete's foot

Ebola virus

88

## Viruses

Viruses cannot reproduce on their own. Without getting energy from a host cell, these viruses cannot reproduce. Like a little leach or tick, the virus attaches itself to the host cell. Enzymes from the cell are released that strip off the coating around the virus. The nucleic acids from the virus crawl into the cell and begin reproducing themselves in the organelles of the cell. New viruses are then released to attack other cells and repeat the process.

If a cell containing a virus reproduces, the viral DNA is copied and tags along with the new cell. Sometimes the virus can hang out in the host cell without reproducing for a while. This is called a "latent virus." **Chickenpox** is an example of a virus that can become latent. That's why elderly people will contract shingles. It is the virus reappearing many years later.

The most common virus is the **rhinovirus** (it has nothing to do with the rhinoceros). This is the virus that causes the common cold. This virus causes inflammation in the tissue, which is what irritates your throat. The rhinovirus usually gets its start in the nose because it spreads when the temperature hovers around 91–95 °F (33–35 °C).

The most deadly viruses in history are influenza, smallpox, and HIV/AIDS. The 1917 Spanish flu outbreak killed about 50 million people worldwide, and HIV/AIDS has killed about 35 million since 1981. About 600,000 people have died of AIDS

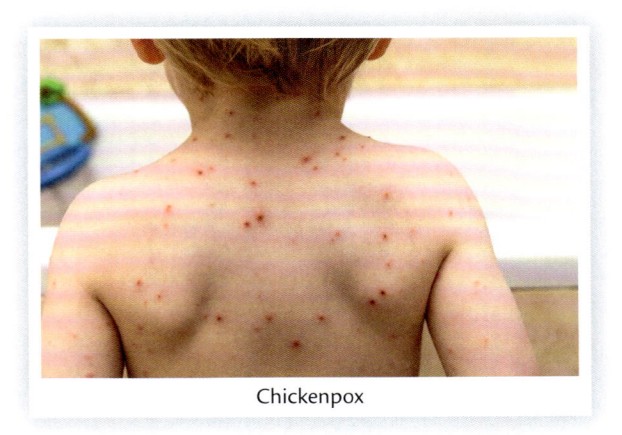

Chickenpox

in the US, making this disease the worst pandemic in a hundred years. This was an unusual disease because 99% of those who contracted it were involved in sinful lifestyle choices, such as the use of illegal drugs. Less than 1% of AIDS patients contracted the disease by non-sinful means.

## Vaccines and the Body's Wonderful Built-In Defenses

Bless the Lord, O my soul;
And all that is within me, bless His holy name!
Bless the Lord, O my soul,
And forget not all His benefits:
Who forgives all your iniquities,
Who heals all your diseases... (Psalm 103:1-3)

Nothing displays the genius of God's creation like the body's system to defend itself from disease. It is an extremely involved network of defenses that the most intelli-

# GOD MADE LIFE

## Diseases Caused by Microbes in the World Each Year

| Disease | How Many People Get It | How Many People Die From It* |
|---|---|---|
| Pneumonia (bacteria or virus) | 1.4 billion | 2.6 million |
| Acute Diarrhea (Rotavirus, etc.) | 4 billion | 1.6 million |
| Tuberculosis (bacteria) | 1.7 billion | 1.7 million |
| Coronavirus (COVID-19 - 2020) | 60 million | 1.4 million |
| Dengue Fever (virus) | 50 million | 1.25 million |
| Malaria (protozoa) | 500 million | 500,000 |
| Influenza (virus) | 1 billion | 650,000 |
| HIV/AIDS (virus) | 1.7 million | 770,000 |
| Cholera (bacteria) | 4 million | 143,000 |
| Measles (virus) | 400,000 | 140,000 |
| Amoebic Dysentery (protozoa) | 500 million | 100,000 |
| Meningitis (bacteria) | 310,000 | 35,000 |
| Ebola (virus) | 30,000 | 11,000 |
| Chicken Pox (virus) | 7 million | 3,000 |
| Common Cold (virus) | 21 billion | Rare |

CHAPTER 4: GOD MADE MICROSCOPIC ORGANISMS

Red and white blood cells

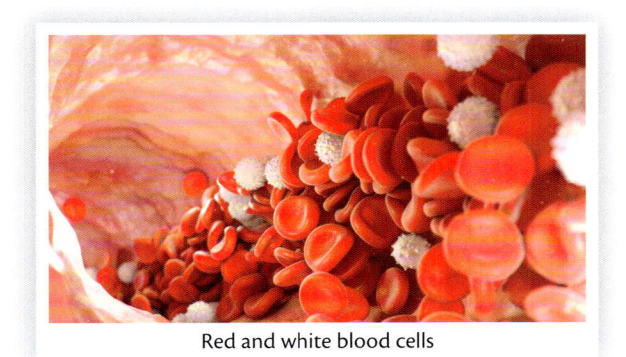

gent scientists in the world have yet to figure out. Here are some of the basic elements you need to know:

1. There are cytotoxic t-cells (**white blood cells**) whose job is to patrol the body and identify the bad viruses when they get into the body and penetrate a cell. The t-cells start firing **cytotoxic granules** at the infected cell like bullets from a gun. One particular cytotoxin called **perforin** actually makes holes in the cell to introduce enzymes. These enzymes are tasked with destroying the cell and the virus within it.

2. God has also provided the body with a defense mechanism called **interferons** (in-ter-FEER-ahns). These little protein soldiers prevent the virus from multiplying. They also signal to neighboring cells that a virus has penetrated the cell.

3. Antibodies will gang up on a virus. They attach themselves to the virus and make it impossible for the virus to attach itself

to another cell. What a genius idea for stopping a disease in the body!

In the spring of 1721, a sailor brought the dreaded disease of smallpox into the city of

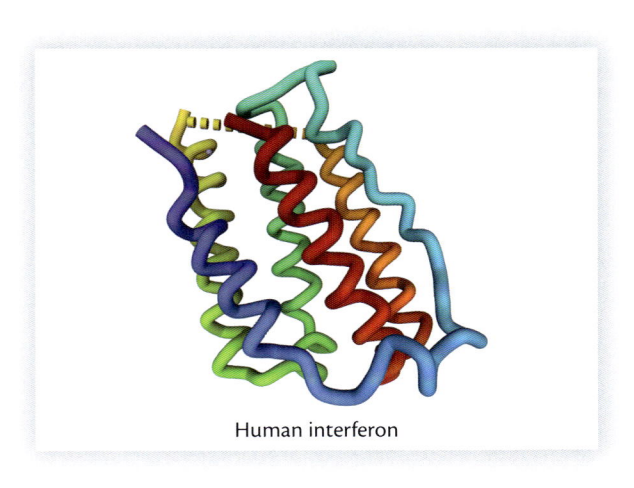

Human interferon

Boston. This resulted in a pandemic infecting about 5,000 people and killing 800. At this time, a puritan pastor named Cotton Mather became the first person to promote the idea of inoculation in America. The process he recommended was crude. He would take the pus from a person infected with smallpox and introduce it into the skin of a healthy person. Although Cotton Mather, his father Increase, and a few other pastors strongly advocated inoculation, the doctors vigorously opposed it. However, the experiment worked. Of the 400 who received inoculation, only six died. Mortality for the rest of the city's population was 1 in 6.

Another Christian would develop the

# GOD MADE LIFE

Cotton Mather (1663-1728)

Edward Jenner (1749-1826)

safer vaccine later in 1796. Edward Jenner (1749-1826) noticed that milkmaids who worked with cows on a dairy farm would not contract smallpox if they had the more mild disease of cowpox. He developed the first vaccine from cowpox and administered it successfully to an 8-year-old boy. It is estimated that about 10% of the English population died from viruses like smallpox before the vaccine came along. Cotton Mather and Edward Jenner's invention of the vaccine has saved about 500 million lives worldwide since then.

Towards the end of his life, Dr. Jenner told a friend: "I am not surprised that men are not grateful to me; but I wonder that they are not grateful to God for the good which He has made me the instrument of conveying to my fellow creatures."[5]

Edward Jenner boiled down his principles of scientific investigation to this: "The Sacred Scriptures form the only pillow on which the soul can find repose and refreshment. . . . The power and mercy of Providence is sublimely and awfully displayed in lightning and tempest. . . how beautifully is power here seasoned with mercy."[6]

A vaccine doesn't really cure a disease. It prepares the body to fight the disease by introducing a weaker virus into the body. The weaker virus is a little bit similar to the really bad virus (like smallpox or influenza).

Antibodies attacking virus

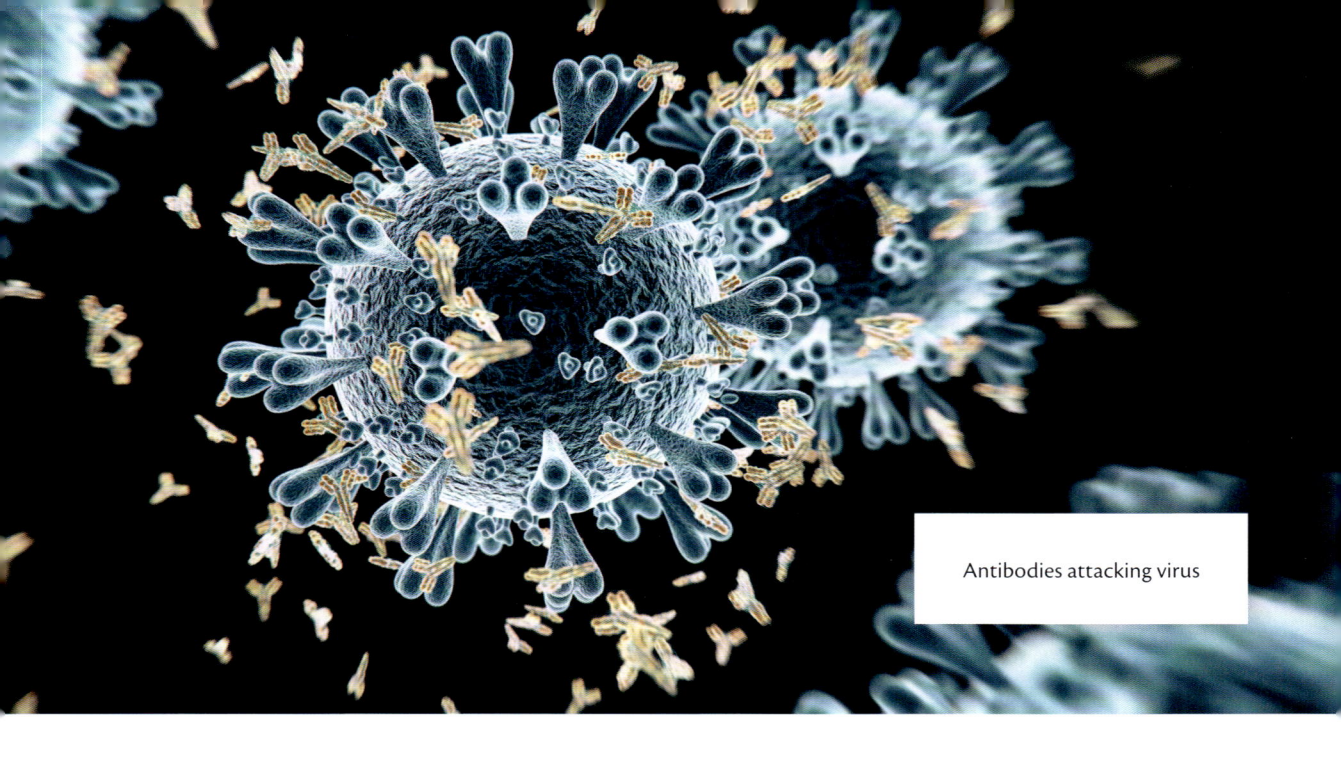

Then, as the body encounters the weaker virus, it creates for itself **antibodies** which are capable of beating up on the bad virus when it comes.

When sin brought death into the world, the human body would be threatened with very dangerous microbes. In His mercy, God still provided our bodies with defense mechanisms to fight off these enemies. Every day, trillions of cells are on the alert and fighting off disease in your body and the bodies of all the other creatures around the world. The Creator has equipped them to do battle.

## How New Viruses Spread to Man

Since the 20th century, the world has become very aware of the introduction of new viruses from the animal kingdom. Initially, the viruses live off their animal hosts. As the animal comes into contact with humans, the virus finds ways to adapt so it can attach to humans. A number of these viruses have been introduced to humans through China, where the people shop for freshly killed meat and consume exotic animals.

You can see how important it is to stay away from sick animals. The Scriptures also warn about consuming certain animals, particularly carnivores, rodents, cats, monkeys, and bats. These unconventional diets can be very risky and bring much suffering to mankind. Farmers also should be careful to test their sick animals and maintain careful cleanliness standards when handling them. Although the Old Testament food laws do not apply to our situation today as they did then, there may be some wisdom to glean from Deuteronomy 14 and Leviticus 11. This is what God told Israel:

# GOD MADE LIFE

| Outbreak | Year | Animal Carrier | How it Spreads to Man |
|---|---|---|---|
| Marburg | 1967 | Monkeys or fruit bats | Exposure to sick animal |
| Lassa fever | 1969 | Rodents | Contact with animal urine |
| Ebola | 1976 | Bats or monkeys | Consumption of bushmeat |
| H1N1 | 1930 | Swine (pigs) | Exposure to sick pig |
| H3N2 | 1998 | Swine (pigs) | Exposure to sick pig |
| SARS - CoV | 2002 | Civets | Raising or eating civet meat |
| West Nile | 2002 | Mosquitoes or birds (crows) | Mosquito bite |
| Japanese Encephalitis | 2003 | Mosquitoes, birds, or pigs | Mosquito bite |
| H5N1/Avian Flu | 2004 | Chickens or pigs | Exposure to sick animal |
| COVID 19 | 2020 | Bats or pangolins | Consumption of pangolins or bats |

Bat

Pigs

"You shall not eat any detestable thing. These are the animals which you may eat: the ox, the sheep, the goat, the deer, the gazelle, the roe deer, the wild goat, the mountain goat, the antelope, and the mountain sheep. And you may eat every animal with cloven hooves, having the hoof split into two parts, and that chews the cud, among the animals." (Deuteronomy 14:4-6)

Bacteria

## Kingdom Bacteria

Bacteria are single-celled organisms. **Bacterium** is the word used for just one cell of bacteria. Don't forget that the bacterium is a prokaryotic cell, which means it doesn't have a nucleus. Its DNA swims around inside the cell membrane. Bacteria can reproduce like crazy, and that's why bacterial germs are so dangerous to the human body. The DNA replicates itself, and the cell reproduces by **binary fission**. This is kind of like mitosis in a cell. The cell collapses in on itself, and half the genetic material moves to one side and the other half moves to the other side. When the cell pinches off completely, a second cell is created with the very same genetic material. It is a clone of the first cell. You could watch this using a microscope set at 1000X magnification. A bacterium can reproduce in just 20 minutes. At this rate, you would see 500 new cells develop in 3 hours, 32,000 in 5 hours, and 1 billion in 10 hours. In a week, there would be so much bacteria in the world it could fill up the earth.

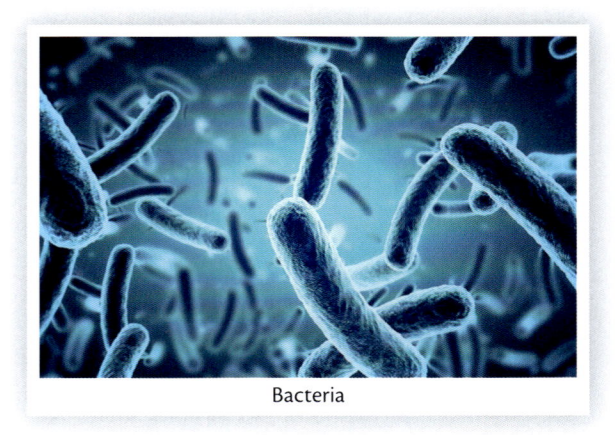

However, there are some limitations to bacterial growth. First, bacteria need food to live, and if they don't have access to food or oxygen, they will die out. If a colony of bacteria is sitting on a food source, the cells in the middle or towards the top of the heap don't have access to the food. It gets too crowded in the colony, and so there comes a point at which the cells die faster than they reproduce. Consequently, the largest bacterial colonies are no bigger than a pea.

Also, bacteria are pretty picky when it comes to temperature and pH range. There needs to be a high degree of moisture in the environment for a bacterium to grow. Most bacteria like tropical temperatures: 80—100 °F (27—38 °C). Also, many forms of bacteria grow best in darkness or limited light. **Aerobic bacteria** need oxygen to survive to convert food into energy. However, **anaerobic bacteria** don't like oxygen.

**Staphylococcus** and **streptococcus** are

95

# GOD MADE LIFE

some of the most dangerous bacteria for humans. Although the words are long and complicated, these describe different types of bacteria.

"Staph" refers to the appearance of the bacteria when you look at it under the microscope. These are bacteria arranged in clusters. "Strep" describes bacteria that is arranged in rows. "Coccus" refers to its spherical shape.

Sometimes, streptococcus bacterial infections can become extremely serious. The bacteria can begin to eat away at muscles and organs. This is called flesh-eating bacteria. Patients can lose arms and legs from this infection, and the death rate is as high as 30%. Immediate treatment is needed for these serious infections. The bacteria manufacture toxins that destroy the body tissue.

The most deadly disease in history was the **bubonic plague**, caused by bacteria called **Yersinia pestis**. As is often the case, this disease first spread to animals like rats and squirrels. Rats were carried on boats from China, which then ported in Egypt and Italy. That's when the plague began in 1346. Fleas would carry the bacteria from the dead rats to humans. Once bitten by these fleas, people would get sick with fevers within three to five days. About 80% of the people who contracted the bubonic plague would die within five days. Over 50 million people died of this terrible disease.

## Antibiotics

Is anyone among you sick? Let him call for the elders of the church, and let them pray over him, anointing him with oil in the name of the Lord. And the prayer of faith will save the sick, and the Lord will raise him up. And if he has committed sins, he will be forgiven. Confess your trespasses to one another, and pray for one another, that you may be healed. The effective, fervent prayer of a righteous man avails much. (James 5:14-16)

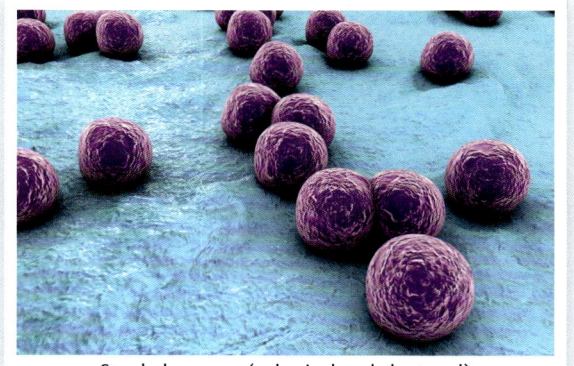

Staphylococcus (spherical and clustered)

Streptococcus (spherical and arranged in rows)

## CHAPTER 4: GOD MADE MICROSCOPIC ORGANISMS

God heals our bodies from diseases in different ways. Sometimes He will do it miraculously without the help of doctors. At other times He will use medicines and the body's immune system to heal. But let us remember that the most important treatment for all illnesses is prayer. This is the encouragement of James 5:14-16.

To fight off diseases caused by bacteria, doctors will use **antibiotics**. Remember from Chapter 1 that the Christian researcher Alexander Fleming came upon this helpful solution by accident—or rather, by God's gracious providence. Antibiotics like **penicillin** keep the bacteria from multiplying by preventing the cell walls from forming during binary fission. Some antibiotics will attach themselves to the proteins and prevent the bacteria from reproducing.

Looking back a hundred years, we can find out the causes of death before Fleming discovered penicillin. We learn that about 20% of annual deaths came from diseases that could have been fixed by antibiotics. In the U.S. alone, these treatments have saved about 45 million lives over 75 years. Applying the same rate for the rest of the world, antibiotics have saved over a billion lives. Thank God for Alexander Fleming's discovery! It has been a lifesaver for many people.

## Superbug

So God blessed Noah and his sons, and said to them: "Be fruitful and multiply, and fill the earth. And the fear of you and the dread of you shall be on every beast of the earth, on every bird of the air, on all that move on the earth, and on all the fish of the sea. They are given into your hand." (Genesis 9:1-2)

After the flood, God warned Noah and his sons of an enmity between animal and man. Many wild beasts are almost impossible to tame, and, as already mentioned, they can be dangerous to humans. While only several hundred people die of shark attacks or lion attacks each year, millions die of bacterial infections.

There is a risk of taking too many antibiotics. Families that overuse the medicine for every little health problem forget this fact. Bad bacteria can adapt to resist antibiotics. God has designed His creation with the capacity for self-preservation. Animals can fight off predators. And even bacteria can adapt so they won't be killed

Antibiotics

97

# GOD MADE LIFE

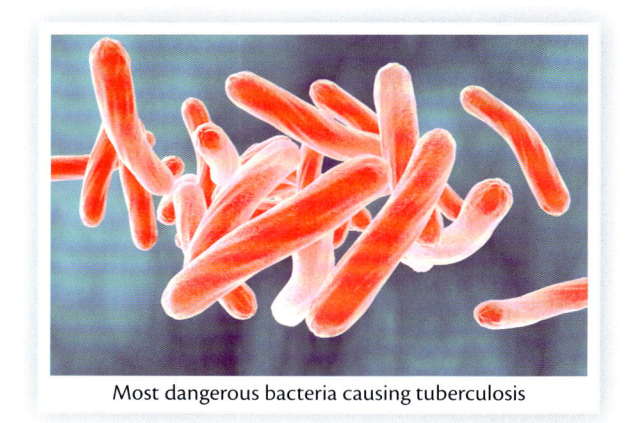

Most dangerous bacteria causing tuberculosis

by the antibiotics. Over time, bacteria will change their outside membranes in order to restrict access to the antibiotic. They can also develop little pumps to remove the invasive antibiotic from the cell. And, they can render the part of the cell being attacked by the antibiotic useless to the cell itself.

These antibiotic-resistant bacteria are sometimes called **superbugs**. One of the most common types is the Staphylococcus aureus. Out of 120,000 Americans who were infected in 2017 by this nasty bug, 20% died. Other very bad superbugs include Pseudomonas aeruginosa and Mycobacterium tuberculosis. The latter is the most deadly bacteria right now, killing about 1.7 million people a year. Over 120 countries have seen instances of this form of tuberculosis (TB), which is most commonly found in Russia, Ukraine, China, Egypt, Peru, and Ecuador.

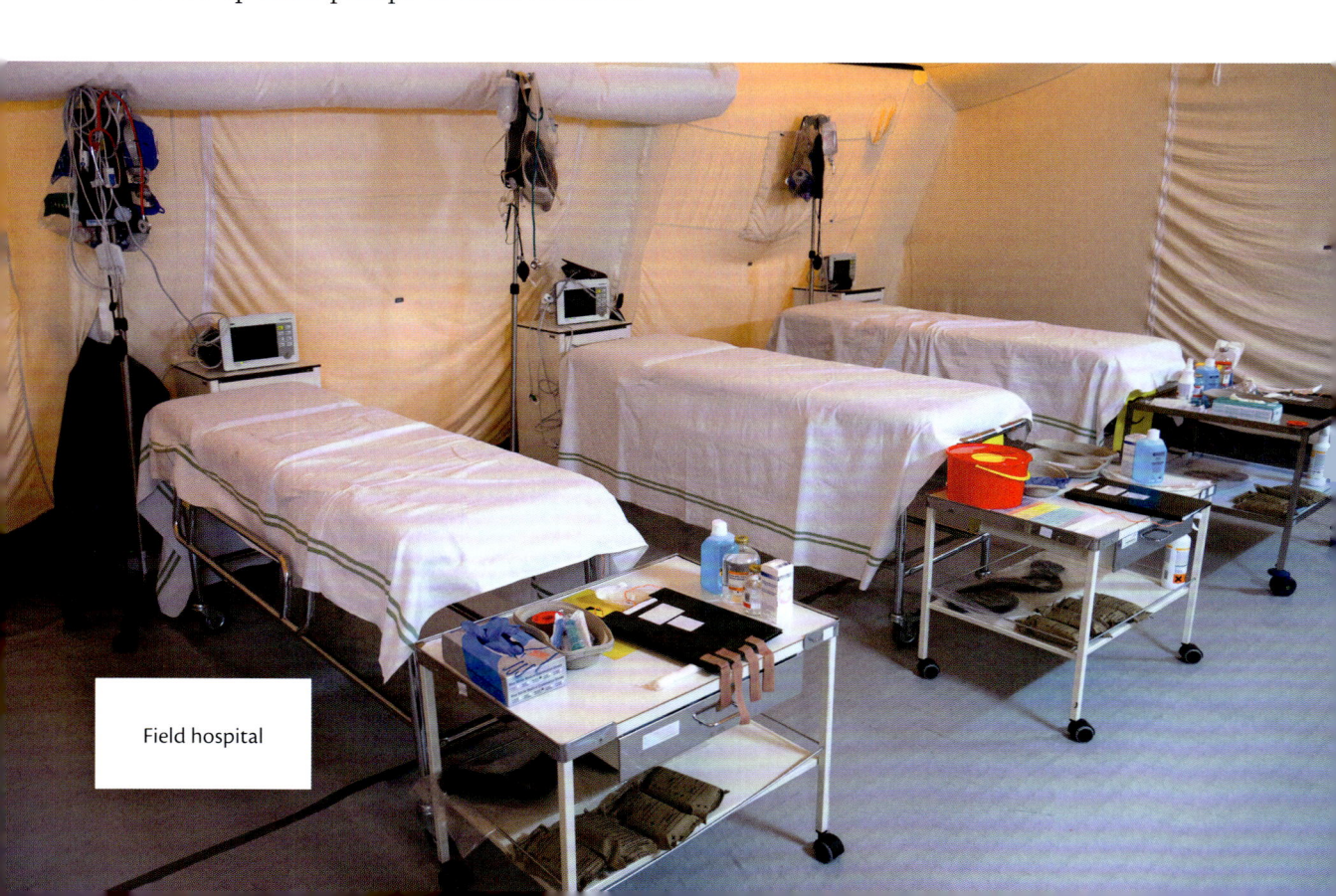

Field hospital

## How Diseases Spread

*Now the leper on whom the sore is, his clothes shall be torn and his head bare; and he shall cover his mustache, and cry, "Unclean! Unclean!" He shall be unclean. All the days he has the sore he shall be unclean. He is unclean, and he shall dwell alone; his dwelling shall be outside the camp. (Leviticus 13:45-46)*

From God's law recorded in the Old Testament, we learn that God is concerned about the spread of dangerous disease. Of course, the most dangerous disease is the spiritual problem of sin. Later in the New Testament, Paul warns the Corinthian church to watch out for the spread of sin in the congregation (1 Corinthians 5).

To keep people from spreading contagious diseases like leprosy God required the lepers to stay outside of the camp. This is called **quarantine**.

Some bacterial and viral diseases spread more quickly than others. Scientists who study diseases use a metric called the **Reproduction Number** to measure the ease with which a disease spreads. This stands for the average number of people who are infected by one person with the disease. From the table below, you can see that measles, mumps, and chickenpox are the most contagious diseases.

| Disease | Reproduction Number | How the Disease Spreads |
| --- | --- | --- |
| Measles | 12-18 | Through the air |
| Chickenpox | 10-12 | Through the air |
| Mumps | 10-12 | By coughing and sneezing |
| Polio | 5-7 | By not washing hands after going to the bathroom |
| Rubella | 5-7 | By coughing and sneezing |
| HIV/AIDS | 2-5 | By kissing or sharing bodily fluids |
| COVID-19 | 2-6 | By coughing and sneezing |
| Common cold | 2-3 | By coughing and sneezing |
| Influenza | 1.4-2.8 | By coughing and sneezing |
| Ebola | 1.5-1.9 | By sharing bodily fluids |

Keep in mind these are the most common ways you might spread diseases like the flu and colds:

# GOD MADE LIFE

1. The most common way of spreading disease is by germs collecting on your hands. When you touch your face, or sneeze into your hands, or rub your eyes, or bite your fingernails, you are transferring germs to your hands. You transfer this to others when you touch them or you open doors or touch a light switch. Avoiding touching your mouth with your hands and washing your hands before you touch others to help a great deal in keeping germs to yourself.

2. Sometimes people pass germs off to others by forgetting to wash their hands before they prepare food. It is especially important to wash your hands after using the toilet. Also, it is critical that you should wash your hands after changing a baby's diaper.

3. Animals can carry germs as well. After petting an animal, you should wash your hands.

4. Some scientists have learned that a sneeze exits the mouth at up to 100 mph (160 kmph), and travels for up to 27 feet (8.2 m). As many as 100,000 germs explode into the air, and they can float around in the air for days. If you are going to sneeze or cough, it would be good to do this apart from everybody else. Sneeze into your elbow, not your hand. But then, you shouldn't hug others because all of those germs have collected into your elbow.

5. Wash your hands when you return from outdoors, or after attending church or going shopping. When washing your

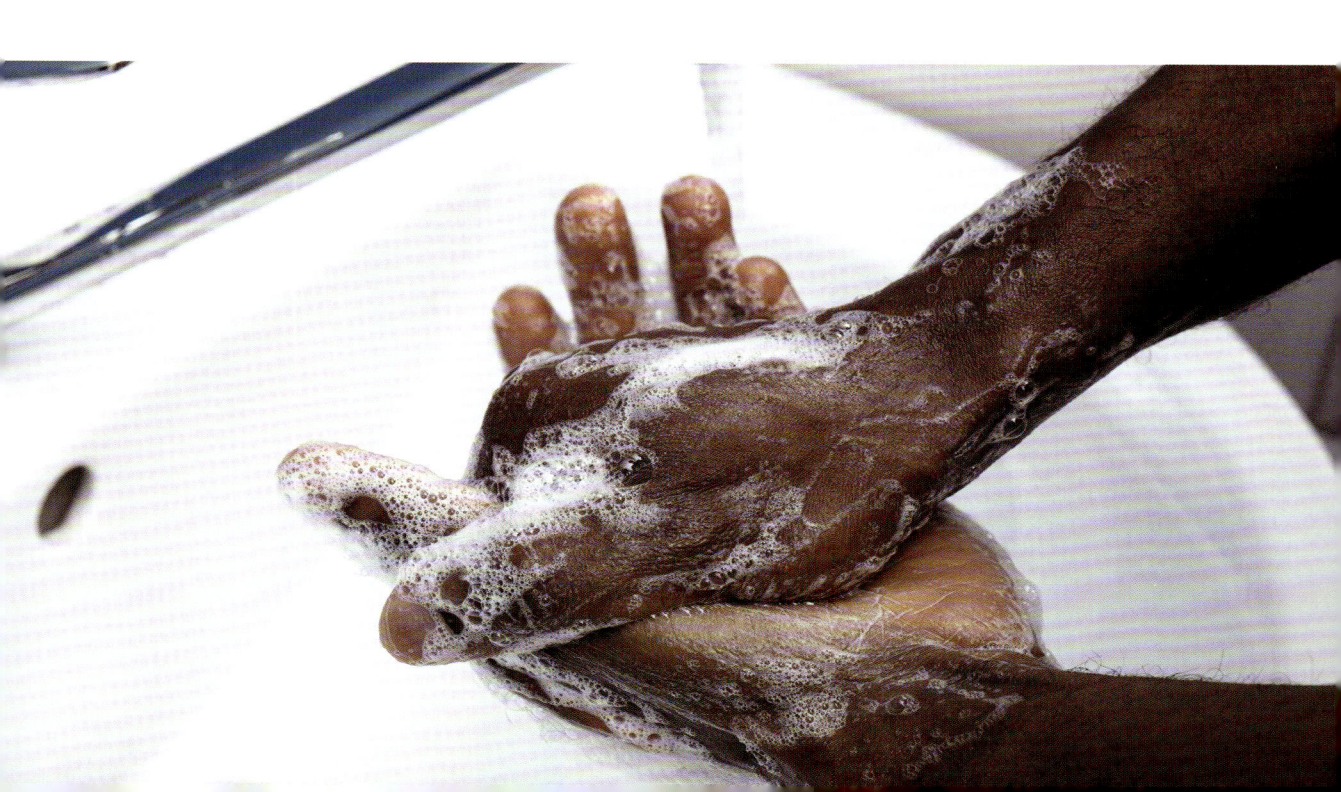

CHAPTER 4: GOD MADE MICROSCOPIC ORGANISMS

Sneezes can go 27 feet.

hands, use both soap and water. Take at least 15 seconds and try to get between your fingers and over your fingertips around your nails. Fifteen seconds is about as long as it takes to sing the "Happy Birthday" song.

6. Throw away your tissues after blowing your nose and wash your hands before touching anything.

By being careful about how we contain our germs, we can love one another as Jesus taught us.

Then one of them, a lawyer, asked Him a question, testing Him, and saying, "Teacher, which is the great commandment in the law?"

Jesus said to him, "'You shall love the LORD your God with all your heart, with all your soul, and with all your mind.' This is the first and great commandment.

And the second is like it: 'You shall love your neighbor as yourself.' On these two commandments hang all the Law and the Prophets." (Matthew 22:35-40)

"And just as you want men to do to you, you also do to them likewise." (Luke 6:31)

## The Dirtiest Places in the House

A 2016 study looked at surfaces in the average home and examined each for microbes.[7] The dirtiest places in the average home were:

1. The dish sponge in the kitchen sink
2. The toothbrush holder in the bathroom
3. Pet water or food bowls
4. The kitchen sink
5. The coffee water reservoir

Incredibly, more than 75% of kitchen dish rags contained traces of Salmonella, E. coli,

Dish rags/sponges collect the most microbes.

101

and fecal matter compared to only 9% found on bathroom faucets.

Less reliance on kitchen dish rags to clean surfaces would keep things a lot cleaner. Consider the following practical guidelines for keeping germs out of the kitchen and bathrooms:

1. Use throw-away disinfectant wipes on kitchen countertops.
2. Soak your sponges in a quart of warm water with a half teaspoon of bleach mixed in.
3. Change dish towels regularly.
4. Close the toilet before flushing. A flushed toilet can splash the gross stuff as far as six feet. And the spray can hang out in the air for up to six hours. If bathroom sinks and toothbrushes are located anywhere near the toilet, you could end up with a lot of bad microbes crawling around where you don't want them.

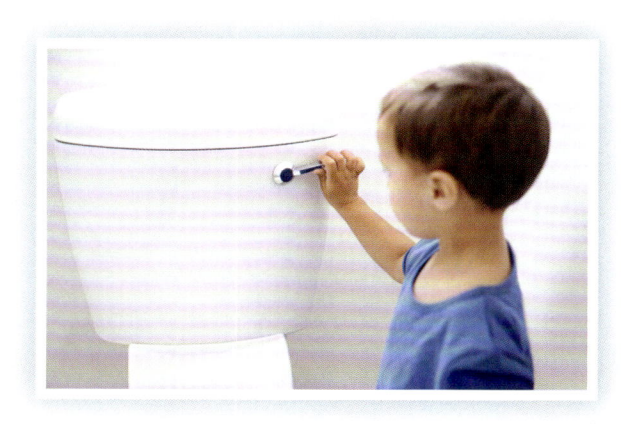

## Jesus Commands the Microbes

Now when Jesus had entered Capernaum, a centurion came to Him, pleading with Him, saying, "Lord, my servant is lying at home paralyzed, dreadfully tormented." And Jesus said to him, "I will come and heal him."
The centurion answered and said, "Lord, I am not worthy that You should come under my roof. But only speak a word, and my servant will be healed. For I also am a man under authority, having soldiers under me. And I say to this one, 'Go,' and he goes; and to another, 'Come,' and he comes; and to my servant, 'Do this,' and he does it."
When Jesus heard it, He marveled, and said to those who followed, "Assuredly, I say to you, I have not found such great faith, not even in Israel!". . . Then Jesus said to the centurion, "Go your way; and as you have believed, so let it be done for you." And his servant was healed that same hour. (Matthew 8:5-10,13)

Here is a wonderful example of the power of Christ. Let us never forget this was the Creator walking on earth. He created all of the microbes in the world, and He exercises all power over all of them. The faith of the Roman centurion is seen in the man's confidence in Jesus's authority over the disease. He says to the Lord, "All you need to do is command the microbes to leave, and they will leave the body of my servant." But the most amazing part of this healing is that Jesus did not need to be present where the sick man was lying. He merely willed the microbes to leave the man's body, and they left. And we read the servant "was healed that same hour."

Kefir with probiotics

## Positive Uses for Bacteria

While there are plenty of bad bacteria in the world, the Lord has provided some good bacteria for man's benefit. Good bacteria populate the nose, mouth, and skin, serving as a guard against bad forms of bacteria. There are also helpful bacteria in the gut to help with digestion. **Probiotics** like yogurt or buttermilk can be helpful to replenish the good bacteria in the gut.

Yogurt is made from at least two different bacteria, and one gram of yogurt contains ten million individual bacterial microbes. Brie cheese is flavored using bacteria, and pickles are produced using several different kinds of bacteria.

## Food Poisoning Caused by Bacteria

Usually, food poisoning is caused by unfriendly bacteria growing in food that was left out of the refrigerator for too long. The most common forms are **salmonella**, **listeria**, and **E. coli**. The bacteria called **Bacillus cereus** is the most common culprit for food poisoning—especially affecting fried rice, meats, milk, vegetables, or fish.

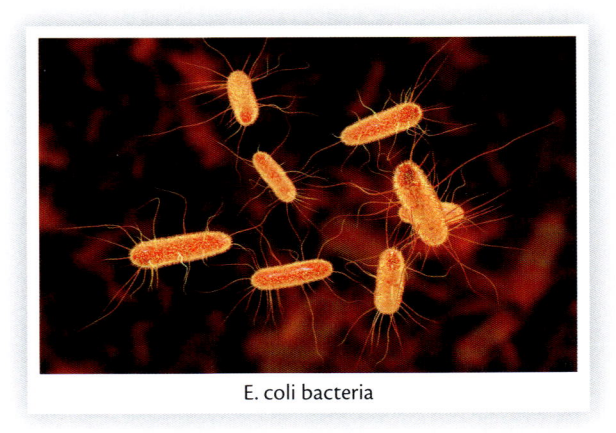

E. coli bacteria

Also, salmonella infects about a million people in this county every year—more commonly in the summertime. This poisoning shows up in raw eggs, imported fruits and vegetables, and raw meats, poultry, and seafood.

The worst food poisoning outbreak in history occurred when a farm in Colorado shipped cantaloupes tainted with listeria. A total of 33 people died from the poisoning.

# GOD MADE LIFE

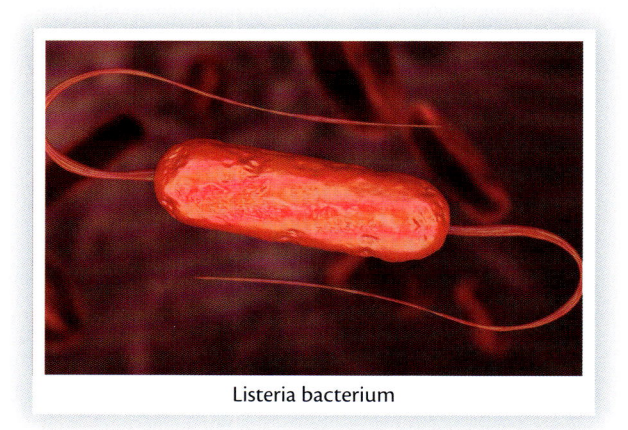

Listeria bacterium

Later, it was found out that the farm had failed to clean equipment that was previously used to process potatoes caked in dirt. Listeria likes to hang out in cool soil at low altitudes. Flooding can bring this toxin to the surface.[8] This is one more reason why you don't want to eat dirt. The listeria would not have been a problem for the potatoes. That's because people cook potatoes at high temperatures, and that kills the bacteria. However, the listeria was smeared over the outside of the cantaloupes. And when people would slice the cantaloupe, the knife would smear the listeria over the face of the cantaloupe. These germs would kill 33 people.

## Preventing Food Poisoning

*"He who is to be cleansed shall wash his clothes, shave off all his hair, and wash himself in water, that he may be clean. After that he shall come into the camp, and shall stay outside his tent seven days. But on the seventh day he shall shave all the hair off his head and his beard and his eyebrows—all his hair he shall shave off. He shall wash his clothes and wash his body in water, and he shall be clean." (Leviticus 14:8-9)*

Food poisoning is a miserable experience. Most of the time, the symptoms will appear anywhere between 30 minutes to 6 hours after ingesting the contaminated food. The victim will experience nausea, vomiting, and diarrhea for 24-36 hours. Sometimes the symptoms don't appear for days or weeks after eating the contaminated food. Salmonella sets in after one to three days, and the effects can last for four to seven days. The best way to prevent food poisoning is to keep cold foods cold (below 40 °F or 4.4 °C) and keep hot foods hot (above 160 °F or 71 °C). To avoid food

Cantaloupe melon

CHAPTER 4: GOD MADE MICROSCOPIC ORGANISMS

poisoning, take into account these very important basic rules:

- Refrigerate leftovers within two hours.
- Toss refrigerated leftovers after three to four days, especially if you have taken it out of the refrigerator several times already.
- Check the internal temperature of meats before you take them off the stove or grill. Minimum internal temperatures should be:
  - Whole or ground turkey or chicken: 165 °F (74 °C)
  - Ground beef, pork, hamburger, or egg dishes: 160 °F (71 °C)
  - Whole cuts of beef, pork, veal, and lamb (roasts, steaks, chops): 145 °F (63 °C). Allow the meat to "rest" for three minutes before cutting or eating.
  - Hot dogs or sausages: 165 °F (74 °C)
  - Fish: 145 °F (63°C)
- Don't put cooked meat back on the same plate that you used for raw meat. Don't rinse your meat.
- Meat thawed out in the open should be cooked immediately. Do not refreeze it.
- Wash all your fruits and vegetables, unless the package indicates they've already been washed.

# GOD MADE LIFE

- Keep all meat juices away from fruits and vegetables.
- Don't cool your leftovers before putting them into the refrigerator. If you divide up the leftovers into smaller containers, they will cool quicker in the fridge.
- Uncooked foods should be placed on the bottom shelf in the fridge.
- Wash your hands with warm, soapy water for 20 seconds before and after handling raw meat, poultry, uncooked eggs, seafood, and their juices.
- Wash cutting boards, knives, dishes, and countertops with hot, soapy water after handling each batch of raw foods.
- Sanitize surfaces in the kitchen where there is contact with raw food using one tablespoon of unscented, liquid chlorine bleach mixed in with one gallon of water.

Canned foods

## Home-Canned Food

Those who can their food need to be aware of the risks involved. If food is not canned correctly, the harmful bacteria living on the food will produce dangerous toxins that will poison the food. While heating the bacteria-laden food might destroy the bacteria, it probably won't destroy the toxins. The food is still dangerous.

One of the more common forms of food poisoning comes with home-canned food which wasn't canned properly. Toxins made by the Clostridium botulinum bacteria in food can leave a person paralyzed, and in some cases kill.

**Botulinum** is present in dirt, and the bacteria is also found growing on vegetables. Keep in mind this is a stubborn bacteria and you can't kill it by just boiling the vegetables. Low-acid foods should be sterilized at 240-250 °F (well above boiling temperature). This can only be done by using a pressurized canner (operating between 10-15 psi, depending on your elevation). If you want to destroy all the bacteria in the food, you should keep the food at these temperatures for 20 minutes minimum. Low-acid foods would include seafood, red meats, poultry, milk, and vegetables (except tomatoes).

When opening a jar of home-canned foods, make sure that the container is not leaking, bulging, or swollen. Spurting foam is a very bad sign. If the food is discolored or smells bad, throw it away. Don't ever try to taste canned food that looks bad.

## Treating Food Poisoning

Most of the time, bed rest for a day or two will give the body time to flush out the toxins from food poisoning. However, if diarrhea and vomiting continue for more than three days or there is extreme abdominal cramping and pain, the victim should contact a doctor. Dehydration is the most serious complication related to food poisoning. If a child or adult is not replacing lost liquids with additional fluids and essential salts and minerals, dehydration may result. Where there is a risk of dehydration, hospitals are well equipped to provide **intravenous fluids**, also known as IVs. For those patients who can't hold liquids or solids down by swallowing, the IV will feed the fluids and minerals directly into their veins.

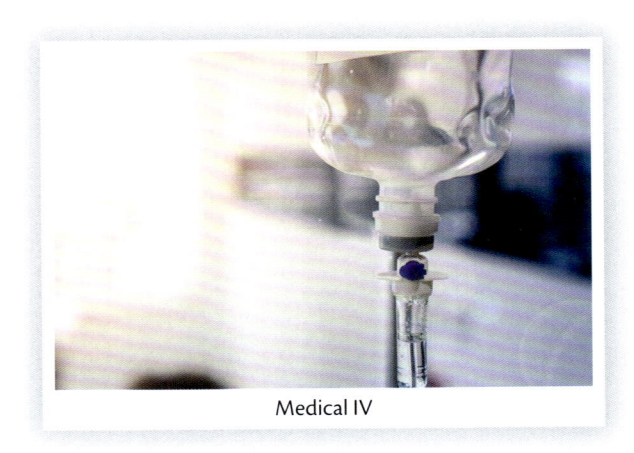

Medical IV

2 Kings 4 tells the story of how the Lord's prophets ate poisonous food during a famine. Miraculously, God had changed the chemi-

# GOD MADE LIFE

cal nature of the stew so it didn't harm any of those who ate it. Certainly if God created all these plants and microbes, He can recreate them to turn them into helpful foods.

And Elisha returned to Gilgal, and there was a famine in the land. Now the sons of the prophets were sitting before him; and he said to his servant, "Put on the large pot, and boil stew for the sons of the prophets." So one went out into the field to gather herbs, and found a wild vine, and gathered from it a lapful of wild gourds, and came and sliced them into the pot of stew, though they did not know what they were. Then they served it to the men to eat. Now it happened, as they were eating the stew, that they cried out and said, "Man of God, there is death in the pot!" And they could not eat it. So he said, "Then bring some flour." And he put it into the pot, and said, "Serve it to the people, that they may eat." And there was nothing harmful in the pot. (2 Kings 4:38-41)

## Kingdom Protista

Unlike bacteria, **Protists** are made up of eukaryotic cells. That means these cells are equipped with a nucleus and various organelles. The most common forms of Protista include algae and protozoa. When you visit ponds and lakes, keep in mind that algae are not plants. These are colonies of Protista.

**Algae** are the most plantlike of this kingdom, but there are different kinds of algae. Normally, these creatures can move themselves around. They can appear as single-celled organisms, multi-celled organisms, or in colonies of single cells. Some

Algae

algae colonies can grow to be 330 feet (100 meters) long. They usually produce their food by photosynthesis. Some algae will also consume food in their environments. **Spirogyra algae** will reproduce by mitosis. God has designed the cell to reproduce itself within the cell. But, the spirogyra cell doesn't divide by itself. It will divide when something disturbs it, as when a frog leaps into the middle of an algae string.

The most well-known **protozoa** include the amoeba, the paramecium, and the euglena. These little critters get around and feed in different ways. For example, the amoeba has pseudopods. These are little legs he creates for himself whenever he needs them. He does this by pressurizing his outside membrane in certain spots. The membrane bulges out, forming something of a makeshift leg. Then, he pushes the rest of his body into the newly-formed "leg," which causes the cell to move forward. He can also use his legs to encapsulate food that is digested into the cell.

The **euglena** gets around using the little motorized flagellum, and the paramecium uses cilia. In fact, parameciums (or paramecia) will use thousands of cilia (plural) to get around. This is another amazing contraption the Creator designed for a tiny microscopic creature. He made nine sets of tubes to run along the outer part of the cilium's tail, with another larger set running through the middle. These are connected by dynein arms. The dynein arms are little muscles that flex, stretching the tubes. While one side of the

Euglena

cilium's tail stretches, the other contracts, causing the tail to move in a serpentine fashion (like a snake). God made this tiny one-celled creature with the microtubule paddles, the dynein motors, and next arms connecting the microtubules to its neighboring tubes. And, He programmed the little tail to do its job—waving the tail about 100 times per second—using the genetic material in the cell. How could something this tiny and complex appear by accident with all the parts working at the same time? Of course, such complicated designs could only have been made by a Creator!

In view of these amazing works of God, the smartest scientist feels himself to be very stupid, like old Agur:

The words of Agur the son of Jakeh, his utterance. This man declared to Ithiel—to Ithiel and Ucal:
Surely I am more stupid than any man,
And do not have the understanding of a man.

# GOD MADE LIFE

> I neither learned wisdom
> Nor have knowledge of the Holy One. . . .
> There are three things which are too wonderful for me,
> Yes, four which I do not understand:
> The way of an eagle in the air,
> The way of a serpent on a rock,
> The way of a ship in the midst of the sea,
> And the way of a man with a virgin.
> (Proverbs 30:1-3, 18-19)

One of the most deadly protozoa is the **Entamoeba histolytica**, a little creature that likes to live in human stomachs. These amoebas swim around in contaminated water, particularly in Africa and Asia. When children drink the water, the creatures end up in their stomachs. If the bug spreads into the liver or the brain, the children die. About 100,000 people die each year from this parasitic disease.

The **plasmodium** is the most common and deadly Protista, spread by mosquitoes in Africa and elsewhere around the world. Malaria kills around 500,000 people a year worldwide.

Another dangerous Protista is carried by the **tsetse fly**. When this fly bites somebody, the little **parasite** Trypanosoma is passed into the bloodstream. This causes a disease called **African sleeping sickness**. If untreated, the disease can lead to death.

## Helpful Protozoa

Hurtful protozoa are usually called **parasites** because they live off the host at the expense of the health of the host. However, God has also created helpful protozoa. His creation offers **symbiosis** between creatures, which means they help one another. For example, the termite consumes wood in the forest but can't digest cellulose. So our Creator provided that termite with a little protozoa called the Trichonympha which helps the termite digest that cellulose.

## Saving Human Life around the World

About 27% of African deaths each year are due to malaria, HIV/AIDS, diarrheal diseases, tuberculosis, and other parasitic diseases. That equates to 2,400,000 deaths per year. Many of the deaths caused by Protista around the world could be avoided. Millions of lives would be saved if poor areas had access to clean water. People die because of drinking dirty water every 10 seconds. Two billion people don't have sanitary toilets, and about 800 million people don't have close access to clean water.

Progress has been made with saving

Tsetse fly

African lives over the last 30 years. Since 1990, deaths by diarrheal diseases have dropped from 2.5 million per year to 1.6 million per year. The best improvement is seen with children under 5 years of age. The number of children dying from these diseases dropped from 1.6 million per year to 600,000 per year. This is mainly due to clean water that is more readily available and sanitary sewers.

## How to Avoid Parasitic Diseases

The following pointers would be helpful towards preventing deadly diseases caused by parasites:

1. Avoid immoral and sinful behavior. Much of the deaths associated with HIV/AIDS is due to sin. The only way to cure this problem is by preaching the Gospel. When men and women are cleansed in the heart, repent of their sin, and believe in Jesus, they stop these harmful life patterns.
2. Provide a source of clean water using wells.
3. Move outhouses away from where people live. Use flushing toilets where possible. Keep all sewage holes well covered.
4. Make soap widely available. Promote hand washing before preparing food, before eating, and after using the restroom.

111

# GOD MADE LIFE

## Kingdom Fungi

Perhaps your mother has exclaimed once or twice in the past, "There's a fungus among us!" While fungi and molds are not as dangerous as some of the other microscopic microbes, they can be an irritation to the human body.

Fungi usually share two major common traits. Their cells have long, thread-like filaments called **hyphae**. And they reproduce using **spores**, which help the new fungi to survive under unfavorable environmental conditions. Fungi cannot manufacture their own food, so they need to consume food outside of themselves. They don't have arms or legs to gather food for themselves, so the Creator provided a unique way for fungi to digest their food. Fungi will leech out digestive enzymes. The enzymes digest the food, and the whole mess gets absorbed back into the hyphae.

Also, fungi form large colonies connected by stems or **stolons**, often under the ground. Mushrooms are a form of fungi. What is the largest living thing in the world? Although the Redwood and Sequoia trees tower upwards of 380 feet tall (116 m), these don't come close to the largest fungi growth on earth. In Oregon and Washington State, huge underground networks of mushrooms have been discovered, covering about 2,200 acres (890 hectares). That's about two miles by two miles (3.2 km by 3.2 km) of continuous growth of honey mushroom. It is truly a humongous fungus!

Yeast is another type of fungus. Once again, the growth of this fungus occurs secretly within bread and other baked goods. Jesus used this to describe the growth of the kingdom of God in the hearts of His people around the world.

> Another parable He spoke to them: "The kingdom of heaven is like leaven, which a woman took and hid in three measures of meal till it was all leavened." (Matthew 13:33)

## The Blessings of the Fungus

God is particularly concerned with cleaning up the world of animal waste and dead bodies. Otherwise, diseases would spread. So fungi and bacteria are really good at decomposing this waste material. Whatever nutrients were in the dead bodies are broken down and used by plants. Nothing is wasted in

Wild mushrooms

God's amazing creation.

Fungi can also be helpful to the roots of plants. While they grow on the roots, they receive energy-rich sugar foods from the plants. In turn, these **mycorrhizae** (MYE-kuh-RYE-zee) **fungi** collect minerals from the soil which help the plants. This is another good example of a symbiotic relationship.

Probably the most helpful use for fungi was found when Alexander Fleming discovered a special mold called **Penicillin**. It was the first antibiotic discovered. The fungus **cyclosporine** is used to help organ-transplant patients. The drug suppresses the immune system so the body won't reject the new organ. It is also used for rheumatoid arthritis, psoriasis, and Crohn's disease.

## Problems Caused by Fungus

Certain skin diseases like **ringworm** are caused by a fungus, as are athlete's foot and thrush. Also, fungi can attack crops and bring about a devastating famine for farmers. **Black stem rust** attacks wheat—the last major problem occurred in 1958-1959, where farmers lost 80 million bushels. In addition, about 10% of the population has allergies to black mold, which grows almost everywhere.

## Conclusion

These three kingdoms of living things made by God are vastly larger in variety and population than the other kingdoms. There are 10,000 species of bacteria, about 250,000 species of Protista, and about 100,000 species of fungi. There are more microorganisms in a teaspoon of soil than there are people living in Africa. One gram of dental plaque contains more individual bacteria than the total number of people who have ever lived. There are more individual viruses on Earth than stars in the whole universe (10 nonillions, or 10 to the 31st power). Scientists estimate there are still millions of different types of viruses hosted in the animal kingdom that have not yet been identified. At this point, the virus is not considered a living thing, although it does possess DNA. Perhaps someday, scientists will see fit to assign the virus to its own kingdom.

We are amazed at the sheer quantity of organisms in this world! We are amazed that despite the many varieties of viruses and bacteria in the world, humans are still living to about 80 years of age. God has equipped the body with amazing defensive and healing systems despite these dangers.

Black fungi

El Yunque National Forest, Puerto Rico

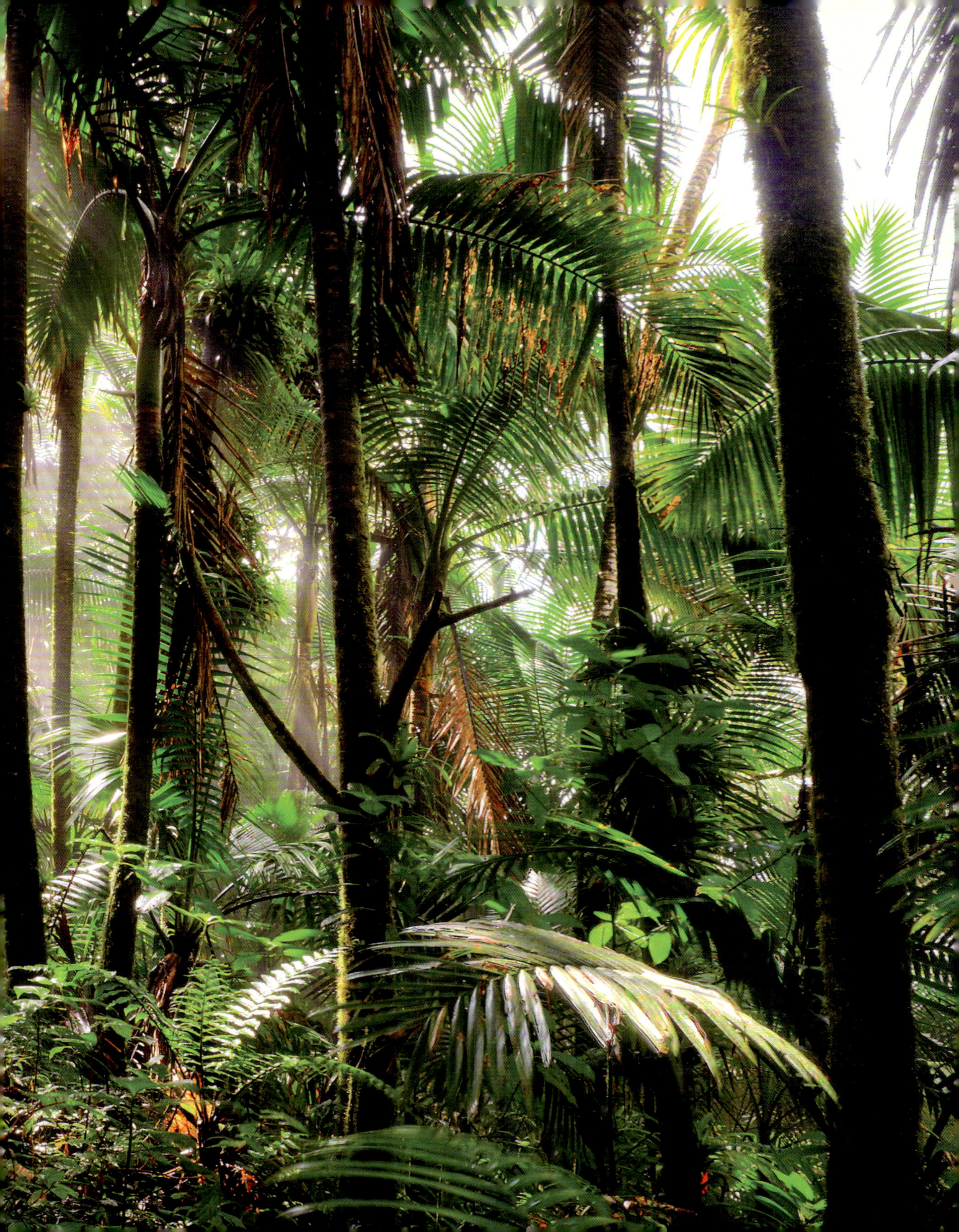

## GOD MADE LIFE

## Sing

Having seen the marvelous work of God throughout this spectacular creation, an appropriate response is always worship and praise. If the student is unfamiliar with the hymn or psalm, some version of it is available on the internet, and may be accessed (with supervision) for singing along.

*I Sing the Mighty Power of God*
I sing the mighty pow'r of God that made the mountains rise,
That spread the flowing seas abroad, and built the lofty skies.
I sing the wisdom that ordained the sun to rule the day;
The moon shines full at His command, and all the stars obey.

I sing the goodness of the Lord, who filled the earth with food,
Who formed the creatures with His Word, and then pronounced them good.
Lord, how Thy wonders are displayed where'er I turn my eye,
If I survey the ground I tread, or gaze upon the sky.

There's not a plant or flow'r below but makes Thy glories known,
And clouds arise and tempests blow by order from Thy throne;
While all that borrows life from Thee is ever in Thy care;
And everywhere that we can be, Thou, God, art present there.

## Do

Choose at least one of the following activities and apply the lessons you have learned in this chapter.

1. Do a quick check on the cleanliness standards in your kitchen.

    a. Are there different cutting boards available for cutting meats and for cutting vegetables?

    b. How often are sponges and rags used for cleaning surfaces changed out?

    c. How long are leftovers kept in the fridge?

    d. Is there a thermometer available to check the inner temperatures of cooked meats?

CHAPTER 4: GOD MADE MICROSCOPIC ORGANISMS

    e. Review the best routine for washing vegetables. Do your cooking practices comply with this?

2. Consider ways in which you could minimize the spreading of colds and flus in your house.

    a. What are the things people touch most of the time, and how would they pass germs around?

    In the bathroom:

    _____

    In the hallways and rooms:

    _____

    In the kitchen:

    _____

    b. Using the studies referred to in this chapter, where are the most common locations for germs in your home?

    _____

    _____

    _____

## Watch

To watch the recommended videos for this chapter, go to **generations.org/GodMadeLife** and scroll down until you find the video links for Chapter 4. Our editors have been careful to avoid films with references to evolution; however, we would still encourage parents or teachers to provide oversight for all internet usage. The producers of these videos may not themselves give God the glory for His amazing creative work, but we encourage the student and parent/teacher to respond with prayer and praise.

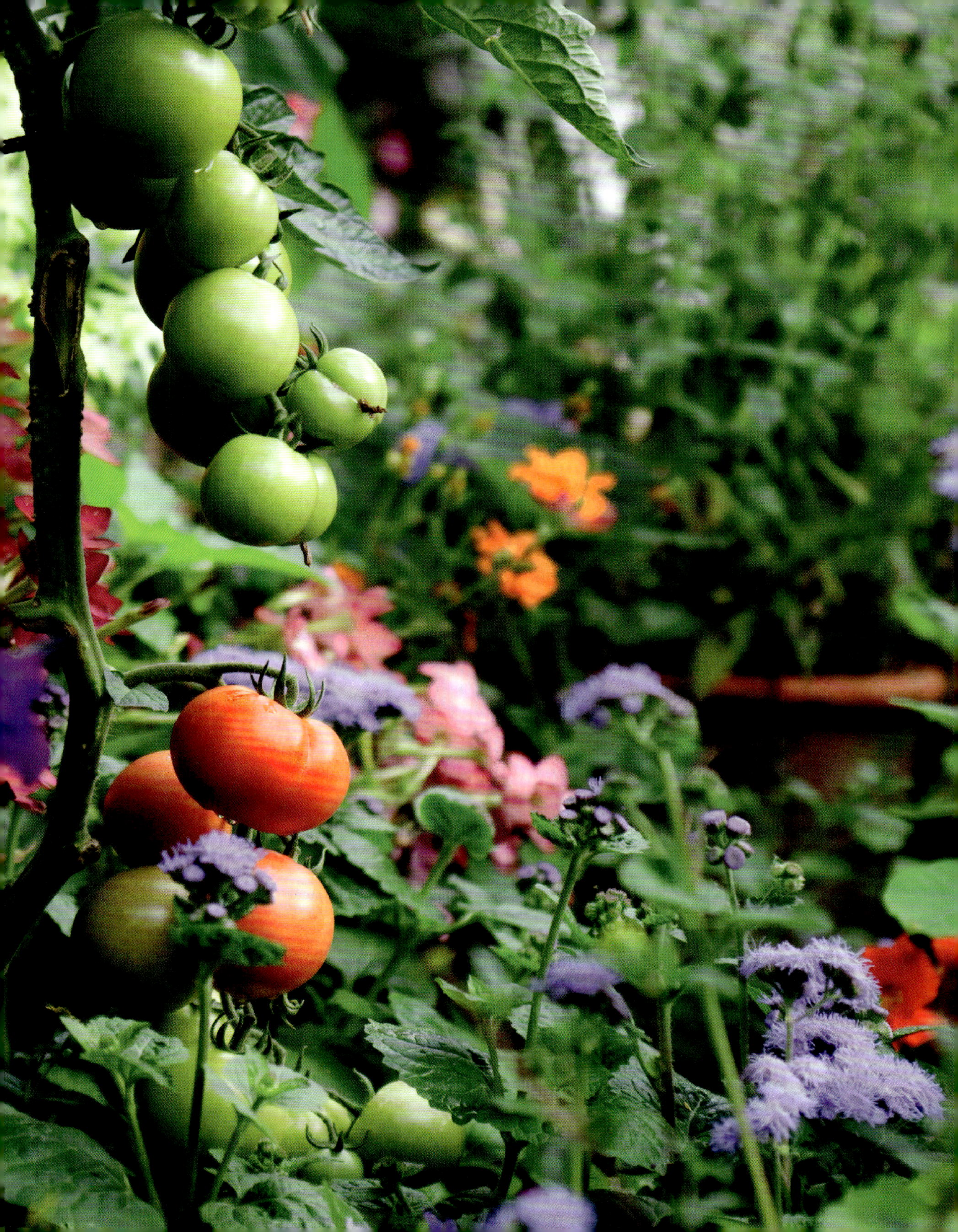

# CHAPTER 5
# GOD MADE PLANTS

Then God said, "Let the earth bring forth grass, the herb that yields seed, and the fruit tree that yields fruit according to its kind, whose seed is in itself, on the earth"; and it was so. And the earth brought forth grass, the herb that yields seed according to its kind, and the tree that yields fruit, whose seed is in itself according to its kind. And God saw that it was good. So the evening and the morning were the third day. (Genesis 1:11-13)

What if God created the world such that food would just grow on trees? What if trees needed only water and sunshine to grow and produce fruit? What if an apple tree could yield 300 pounds (140 kg) of apples every season for 40 years, thus producing 12,000 pounds (5,450 kg) of apples over its lifetime? It turns out that God really did create our world this way! It all came about by His goodness. He contributed His water, His carbon dioxide, and His sunshine. The only thing left to say is that God is amazing. He is good. He is ultimately wise and very powerful to create the apple tree and then produce all that food from a single tree.

The Creator, our Lord Jesus Christ produced about 12,000 pounds (5,450 kg) of food in a few minutes when he fed the 5,000. This was a true demonstration of His limitless power. Whether the Creator produces the food in a few minutes or over forty years by one tree, His work is always wonderful and praiseworthy. We are even more overwhelmed to think of the billions of tons of food produced out of dirt and water each year all over the world.

Man has to plant the seeds, water the

# GOD MADE LIFE

plants, and pick the fruit. But, where does all of this fruit come from? Who does 99% of the work to create 12,000 pounds of fruit out of one tiny apple seed?

To think that a seed planted in dirt could yield 300 pounds (140 kg) of fruit every year is cause for wonder. Where does all the wood come from? Where do the leaves come from? Where does the fruit come from? How does water, minerals, and sunshine turn into all of this greenery and red apples?

One Tree can yield 300 pounds of apples in one season

## God Feeds the Whole World

As we read in Genesis 1, God created plants for food. All these herbs and fruits were meant to feed the whole world filled with billions of animals and human beings. Chickens will eat the grass growing in the backyard. Like little machines, the chickens turn the green grass into little white eggs, and your family cooks up the eggs for breakfast. Yet, all the nutritional sources for the cows and chickens grazing in the fields come from plants, which are all grown by God.

Half of habitable land in the world is cultivated by man. Thirty-seven percent of lands are forested, and eleven percent of lands are made up of shrubs and wild grasses. Farmers produce nine billion tons of food each year out of the earth. That amounts to about a ton of food for each person. On top of this, the Lord produces at least that much grass and fruit for animal consumption each year. For humans, He made 20,000 different kinds of edible fruits and vegetables! Yet, almost 90% of our food comes from only 20 different species of plants. The following table lists the most common foods produced by farmers around the world in a given year.

## How Animals and Plants are the Same. . . and Different

Animals and plants both have cells (and DNA). They reproduce after their own kind, as described in Genesis 1. They both need oxygen to live, which comes from the earth's atmosphere.

## Major Worldwide Crop Yields Annually

| Crop | Global total production value in billions of dollars | Global production in metric tons | Country with highest production value in billions of dollars |
|---|---|---|---|
| Sugarcane | $87.3 | 1,874,611,396 | $40.9 (Brazil) |
| Maize (corn) | $191 | 1,126,990,585 | $61.2 (Mainland China) |
| Rice, paddy | $332 | 751,885,117 | $117 (Mainland China) |
| Wheat | $168 | 748,392,150 | $50.7 (Mainland China) |
| Cow's milk, whole, fresh | $238 | 665,596,536 | $34.7 (United States) |
| Potatoes | $92.7 | 356,952,488 | $32.2 (Mainland China) |
| Soybeans | $107 | 335,613,801 | $40.7 (United States) |
| Vegetables, not elsewhere specified | $89.1 | 292,920,885 | $53.6 (Mainland China) |
| Cassava (yuca) | $34.3 | 288,497,460 | $4.11 (Indonesia) |
| Oil palm fruit | $35.7 | 255,567,218 | $17.7 (Indonesia) |
| Tomatoes | $87.9 | 178,158,747 | $28.9 (Mainland China) |
| Barley | $22.9 | 145,906,773 | $2.08 (Russia) |
| Pig, meat | $280 | 118,956,327 | $167 (Mainland China) |
| Buffalo milk, whole fresh | $56.4 | 115,204,379 | $43.1 (India) |
| Bananas | $38.5 | 112,627,980 | $8.13 (India) |

# GOD MADE LIFE

## Major Worldwide Crop Yields Annually (continued)

| | | | |
|---|---|---|---|
| Chicken, meat | $192 | 106,638,508 | $27.4 (United States) |
| Watermelons | $33.9 | 102,414,316 | $26.1 (Mainland China) |
| Onions, dry | $42.1 | 94,838,690 | $23.9 (Mainland China) |
| Sweet potatoes | $26.1 | 89,985,845 | $17.8 (Mainland China) |
| Apples | $45.9 | 84,743,988 | $20.8 (Mainland China) |
| Cucumbers and Gherkins | $40.2 | 79,844,838 | $29.6 (Mainland China) |
| Eggs, hen, in shell | $93.6 | 74,180,272 | $26.1 (Mainland China) |
| Grapes | $67.8 | 74,089,693 | $14.4 (France) |
| Oranges | $22.6 | 72,771,146 | $5.62 (India) |
| Cabbages and other brassicas | $19.1 | 69,790,058 | $9.3 (Mainland China) |
| Rapeseed | $31.7 | 68,113,132 | $10.4 (Mainland China) |
| Seed cotton | $56.7 | 67,622,193 | $21.4 (Mainland China) |
| Cattle, meat | $269 | 64,568,004 | $52.8 (United States) |

The wonderful thing about plants is that you don't have to feed them. They don't have to hunt for food, because they make their own food. Their cells are all equipped with **chlorophyll**, which keeps their leaves green. They need the sun to survive, and they will not thrive without carbon dioxide. Animals are not dependent on carbon dioxide like this. Also, plants are unable to move around. They are stuck where they are planted. They can't decide to relocate themselves. Furthermore, animals have special instincts to hunt

# CHAPTER 5: GOD MADE PLANTS

for food, communicate with each other, bury their dead friends, mark territory, defend their young, build nests, attract mates, and reproduce. In His wise plans for creation, God did not give plants these instincts.

## The Wonderful Variety of Plants

Consider the lilies of the field, how they grow: they neither toil nor spin; and yet I say to you that even Solomon in all his glory was not arrayed like one of these. Now if God so clothes the grass of the field, which today is, and tomorrow is thrown into the oven, will He not much more clothe you, O you of little faith? (Matthew 6:27-28)

The world of plants represents a huge variety of edible and non-edible organisms. Some of God's trees and plants are intended to reflect beauty, majesty, glory, and life. The variety is stupendous, with some 400,000 different flowering plants as classified by botanists. Our Creator chose not to color all His flowers gray. Every possible color, shape, size, and texture is used for all of these different flowers growing in God's fields.

Moreover, plants are also used for clothing and shelter. Incredibly, the earth produces about one billion tons of wood each year for our homes, furniture, and paper. And 30% of the clothing in the world is made of cotton, wool, and natural fibers. So, in summary, God's plants provide about half

Amazon rainforest

## GOD MADE LIFE

the shelter and clothing, and 100% of the nutrition needed for our diet (directly or indirectly). And, the greatest thing about these plants is that they are reproducible and sustainable. More trees and plants will grow next year and produce more wood, cotton, and tomatoes. God made the world with a wonderful capacity to produce and sustain life for thousands of years.

The lesson from Jesus in Matthew 6 is this. Why should we worry? If God feeds the birds of the air every day, and clothes the lilies of the field, wouldn't He feed His children too? He has turned dirt, minerals, water, and carbon dioxide into fruitful plants for thousands of years. Why should we worry about tomorrow?

Variety adds to the beauty of God's world and our enjoyment of it. Varying types of crops also provide for a division of labor and international trade. Farmers in the Mediterranean areas can grow lots of olives, grapes, and citrus fruits — because the climate is perfect for these crops. These farms will sell products to people in Germany where mainly wheat, corn, and sugar beets are grown. As these countries trade with each other, the peoples of both areas enjoy all the

The lavender fields of Provence, France bear flowers used for perfumes, lotions, and soaps

Iguazu Falls, Argentina demonstrates the beautiful glory of God with waters running through forests and converging in the largest series of waterfalls in the world.

variety the earth yields.

Forests, fields, gardens, and plants are also a demonstration of God's majesty and beauty. The Amazon rainforest is the largest forest in the world, roughly 3,400 miles (5,500 km) long and 600 miles (1,000 miles km) wide. An estimated 300 trillion trees grow in this gigantic forest.

Enormous trees grow in California and Oregon state forests. The General Sherman Sequoia tree is 275 feet tall (83 m) and 36 feet (11 m) wide at the base. That's as wide as a three-lane highway — enough room to pass three cars through a single tree! Recently discovered, the tallest tree in the world is the Redwood tree known as the Hyperion. It is 380 feet (116 m) tall. This single tree weighs an estimated 4,000 tons, and it could provide enough wood to build 40 homes! These trees, 100 times taller than you, are demonstrations of God's matchless power and towering majesty over the world.

Moreover, the awe-inspiring holiness and beauty of God are seen by the verdant fields of flowers and colorful trees throughout the year.

Japanese cherry blossoms are remarkable for their soft beauty, reflecting the elegant majesty of God's creation.

The Keukenhof Gardens in Lisse, Netherlands feature gorgeous tulips, daffodils, and hyacinths, displaying God's eye for beauty in His glorious works.

Forty shades of green beautifully cultivated in this garden in Enniskerry, Ireland

Honor and majesty are before Him;
Strength and beauty are in His sanctuary.
Give to the LORD, O families of the peoples,
Give to the LORD glory and strength.
Give to the LORD the glory due His name;
Bring an offering, and come into His courts.
Oh, worship the LORD in the beauty of holiness!
Tremble before Him, all the earth.
(Psalm 96:6-9)

# CHAPTER 5: GOD MADE PLANTS

## First Job of the Botanist — Classification

The **botanist** is a scientist who studies plants. And one of the primary jobs of the botanist is to classify the different kinds of plants God has created. Roses are not the same thing as redwood trees. A tulip is not the same thing as a lily. We must distinguish between kinds because only the same kind reproduces. As we read in Genesis 1, each plant reproduces "after its kind." Occasionally, different kinds of animals can interbreed,

## The 12 Phyla of the Plant Kingdom

| Type of Plant | Phylum | Common name |
|---|---|---|
| **Nonvascular** | Bryophyta | Mosses |
| | Hepatophyta | Liverworts |
| | Anthocerophyta | Hornworts |
| **Vascular, seedless** | Psilotophyta | Whisk ferns |
| | Lycophyta | Club mosses |
| | Sphenophyta | Horsetails |
| **Vascular, seed, Gymnosperms** | Cycadophyta | Cycads |
| | Ginkgophyta | Ginkgoes |
| | Coniferophyta | Conifers |
| | Gnetophyta | Gnetophytes |
| **Vascular, seed, Angiosperms** | Anthophyta | Flowering plants |
| | class Monocotyledones | Monocots |
| | class Dicotyledones | dicots |

which means they can produce children. But this often produces a reproduction dead end. That means the offspring of these **interbred** species cannot reproduce themselves.

**Classification** is defined as the arrangement of animals or plants in groups according to the similar characteristics we observe. The following is an example of the classification of the California Wild Rose.

## Classification of a California Wild Rose

| Level | Classification | Characteristic |
|---|---|---|
| Phylum | Anthophyta | Vascular tissue |
| Class | Angiosperm | Flower |
| Order | Rosales | Flower grows under ovary |
| Family | Rosacrae | Medium sized flower |
| Genus | Rosa | Rose |
| Species | Californica | California Wild Rose |

Remember that scientists use seven levels of classifications for plants and animals — kingdom, phylum, class, order, family, genus, and species. There are three divisions of the plant kingdom in which you will find 12 phyla. There are non-vascular plants, seedless vascular plants, and seeded vascular plants.

**Non-vascular plants** (mosses, liverworts, and some algae) grow close to moist ground or in water. Unlike vascular plants, they don't need a system of tubes to move water and nutrients through their stems and leaves. Non-vascular plants reproduce by spores, much like fungi as mentioned in the last chapter.

**Vascular plants** use tubes to move nutrients through the plant. These tubes are comparable to the veins and arteries that carry blood and its nutrients through the human body. Vascular plants have two divisions:

1. **Seedless vascular plants** (ferns, horsetails, club mosses, and whisk ferns) reproduce by spores instead of seeds.

2. **Seeded vascular plants** include most of the trees, shrubs, and plants we see. Seeded vascular plants have two divisions:

   a. **Gymnosperms** (conifer trees, cycads, and ginkgo) usually have flat seeds formed in open cones. They don't have flowers or fruits.

   b. **Angiosperms** (flowering plants) have seeds formed in protective fruits. The fruit is formed from the flower when it's finished blooming.

## Non-vascular Plants

Non-vascular plants don't use a system of tubes to distribute nutrients. **Mosses** are non-vascular plants which grow in very moist areas. They are highly resilient. Botanists have dried them out and stored them

CHAPTER 5: GOD MADE PLANTS

away for years. With the application of just a few drops of water, the moss came alive again. They do not have leaves, stems, and roots like other plants. They are provided with root-like **rhizoids**, which keep water and minerals flowing into the plant.

Green moss

a new moss plant. By this means, the moss spreads itself like a beautiful carpet of green across the ground.

During World War I, several British doctors found that two kinds of moss were very good for staunching blood and helping wounds heal: *S. papillosum* and *S. palustre*. They were twice as absorbent as cotton and kept the acidity level high (the **ph level** low). High acidity slows down the growth of bad bacteria in the wound.

## Ferns: Seedless Vascular Plants

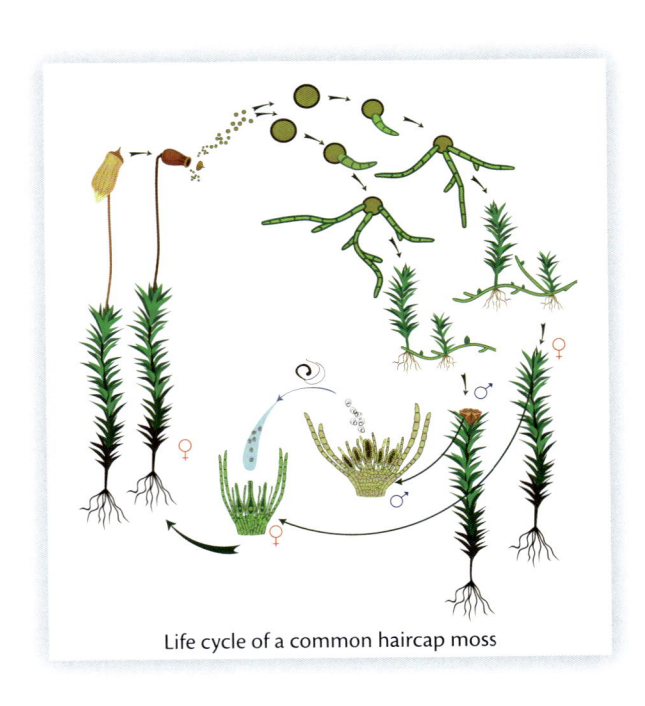

## How Mosses Reproduce

The Creator designed a very interesting means of reproduction for mosses. Remember that mosses grow from spores. When the spore first sprouts, it produces the **protonema**, which develops into leafy shoots. Either male reproductive cells or eggs form on the leafy shoots. When water collects on the shoots, the male cells swim around until they find an egg on a female leafy shoot. Next, stalks grow out of the leafy shoots and extend capsules into the air. The Lord made these capsules to house the reproducing spores. When the timing is right, the capsule opens, and thousands of spores are carried away by the wind. Each spore can grow into

Life cycle of a common haircap moss

Like mosses, **ferns** require a wet environment to reproduce. They do not reproduce by seeds. Instead, little spores form under the leaves. When the spores are released into the air, they land in water and grow into small leaf-like particles. Each particle makes both an

## GOD MADE LIFE

egg and a male reproductive cell. These eggs and cells swim around in the water until they meet up with each other to produce new fern plants. As you can imagine, these ferns are quite productive. A single plant can produce a billion spores in a year.

Young ferns are called **fiddleheads**, which over time unravel into full grown **fronds** (or leaves).

The fern uses an underground root-like stem called a **rhizome** to pipe water up into the fronds. While ferns usually grow a foot or two tall, there are some fern plants in rain forests growing to 60 feet (18 m) high!

We find fossilized ferns all over the earth — an effect of Noah's flood. These ferns must have been covered by mud very quickly and fossilized in a violent, worldwide flood. The fossils show that they were healthy plants with vibrant fronds at the moment they were captured in the mud.

## Seeded Vascular Plants

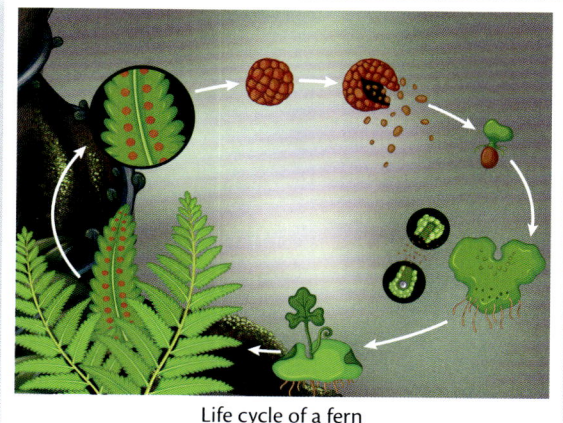

Life cycle of a fern

> But Jesus answered them, saying, "The hour has come that the Son of Man should be glorified. Most assuredly, I say to you, unless a grain of wheat falls into the ground and dies, it remains alone; but if it dies, it produces much grain. (John 12:23-24)

Almost every plant and tree in the world is included in this group of seeded vascular plants.

There are four organs in these plants — roots, stems, leaves, and reproductive organs. The reproductive parts of the plants are usually the seeds, the fruit, and the flower.

**Anthophyta** is the largest phylum with somewhere around 350,000 different varieties, most of which are flowering plants. These are divided up according to seed type — **monocots** and **dicots**. As you can tell from the name, a monocot is a seed that has only one part, and a dicot contains two parts. Monocots are usually flat while dicots are more rounded. When a monocot germinates, it produces a single leaf. When a dicot germinates, it pops out with at least two leaves. Monocot leaves are usually narrower, and dicot leaves are round or elliptical. Monocot stems are also softer and more fleshy, while dicot stems are stiffer. For example, corn and grass seeds are monocots, and bean seeds are dicots.

Seeds are extraordinary little inventions of God. They are more robust than spores— which means they can handle tough con-

CHAPTER 5: GOD MADE PLANTS

|  | Seed | Root | Vascular | Leaf | Flower |
|---|---|---|---|---|---|
| **MONOCOTS** | One cotyledon | Fibrous roots | Scattered | Parallel veins | Multiples of 3 |
| **DICOTS** | Two cotyledon | Tap roots | Ringed | Branched veins | 4 or 5 |

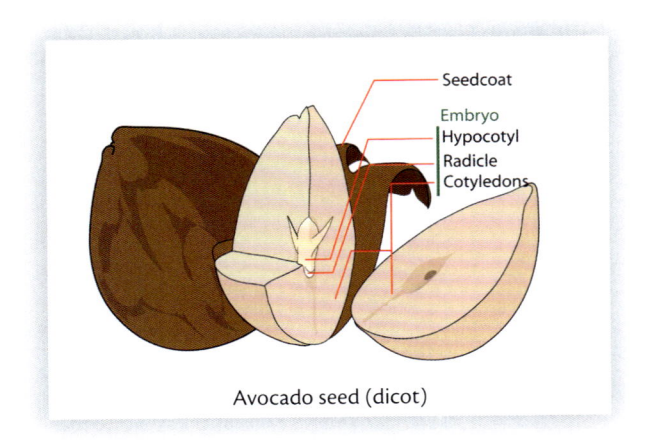

Avocado seed (dicot)
- Seedcoat
- Embryo
- Hypocotyl
- Radicle
- Cotyledons

ditions. Some seeds can stay **dormant** for thousands of years. One date palm seed that was produced around the time Jesus was born was planted in AD 2005, and it sprouted! While the seed is in dormancy, the systems inside the seed shut down so the DNA is protected. If water were to get into the seed, the DNA "seed of life" would be destroyed. Within the seed, there are proteins whose job it is to keep the seed from sprouting. Scientists still can't understand how the whole process works. If the seed has been dormant for 1,000 years, how does it know when it's time to wake up? How does the seed know when it is buried in soil and the conditions are right for sprouting? The Creator made the seed such that it can sense temperature, light, moisture, and nutrients in the soil! When everything is ready to go the seed germinates. **Germination** is the scientific word for the sprouting of a seed.

Seeds are too complex for scientists to figure out. God's creation is very intricate and works at a microscopic level. His intelli-

# GOD MADE LIFE

gence is infinitely greater than man. Foolish scientists think that evolution came up with these complicated workings in seeds. But something this complicated could only have been designed by a Creator who is a million times smarter than foolish scientists.

The DNA "blueprint" that will be used for reproduction hang out in a cell called the gamete. It takes two different gametes to make a new seed — a male gamete and a female gamete. When the male and female gametes meet, the new cell is called a diploid cell or a zygote. You can think of it this way:

One male gamete +
One female gamete = One zygote

## Flowering Plants

I am the rose of Sharon,
And the lily of the valleys. . .
Like a lily among thorns. . .
(Song of Solomon 2:1-2)

There are about 300,000 different kinds of flowering plants in the world. About 95% of all the plants in the world are flowering plants. And most of them are quite beautiful. God made a huge variety of colors, shapes, and sizes in these many flowers. However, the corpse flower (Rafflesia) stinks like a corpse, producing huge, ugly flowers some 40 inches (100 cm) in diameter!

But a flower is known for more than its beauty. God made the flower with a very important job to do. When a plant flowers, the reproduction process has started. This is how God makes seeds which will produce more baby plants. Usually, there are four parts to a flower. Starting from the outside and working towards the inside of the flower, here they are:

1. The **sepals**, which are usually green leaf-like things near the bottom of the flower. The sepals protect the flower when it is in the bud.

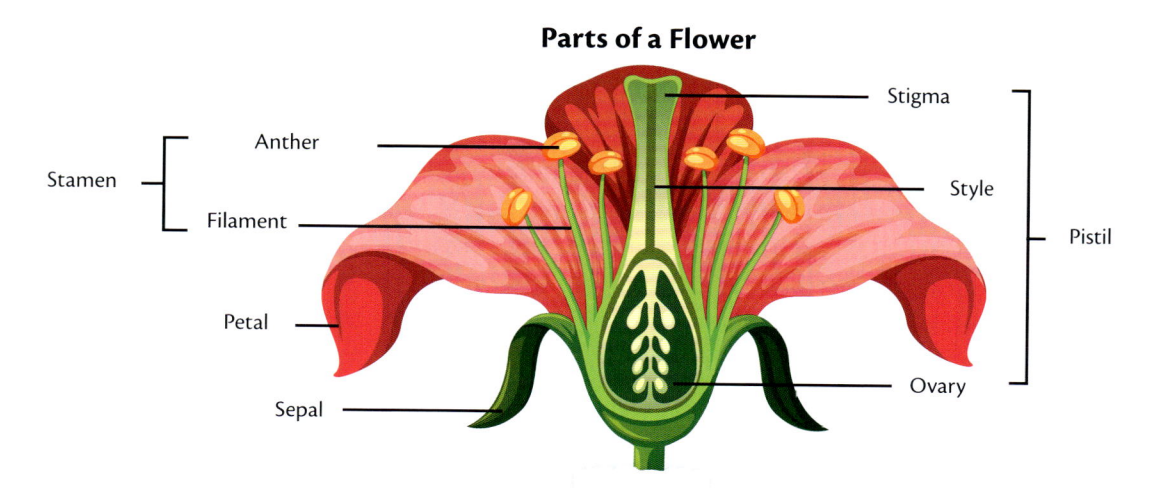

**Parts of a Flower**

Stamen
- Anther
- Filament

Petal

Sepal

Pistil
- Stigma
- Style
- Ovary

CHAPTER 5: GOD MADE PLANTS

2. The **petals** are the colorful parts of the flower. Not only do the petals attract the attention of humans, but bees and butterflies are attracted to the petals as well. Bees and butterflies are pollinators whose job it is to spread the male gametes (pollen), so they will meet up with the female gametes (eggs).

3. The **stamens** stick up near the middle of the flower. They contain thousands of pollen grains, which are the male gametes. A stamen is made up of a filament with an anther at the top holding the pollen.

4. The **pistil** contains the female gametes, also located towards the center of the flower. It is made up of the **stigma**, the **style**, and the **ovary**. The ovary contains the female gametes or **eggs**.

Our Creator God has ordered a very thoughtful and beautiful process for the creation of a seed. Here is how it works.

A bee crawls past the anther of a stamen. She gets the sticky pollen all over her feet, fuzzy body, and wings. She's interested in the flower nectar produced by special glands down at the base of the flower. The bee isn't thinking about the pollen grains sticking to her feet.

As she flits from flower to flower, the bee rubs her feet and wings up against the flower's stigma (at the top of the pistil). The stigma is pretty sticky, so it attracts the pollen carried by the bee.

When the pollen grain (male gamete) lands on the stigma, it makes its way down the pistil by growing a long tube through the style to the **ovary**. The seeds grow inside the ovary. Once inside the ovary, the pollen's DNA joins up with the DNA of the eggs (female gametes). The eggs can now develop into seeds. Each seed is made up of the zygote or embryo, some food, and a shell to protect the little embryo. The food is stored inside the seed in little containers called **cotyledons**.

What happens next? Imagine an apple blossom on an apple tree. The tree is full of beautiful blossoms. Then the blossom begins to shrivel, and the ovary grows, closing the seeds inside. All that's left of the flower is the dried-up sepal at the bottom of the apple. Most importantly for humans and animals,

133

# GOD MADE LIFE

is that the ovary becomes the apple or the **fruit**.

But why does God grow the ovary (fruit) with the seeds tucked inside of it? For one thing, He wanted to provide humans and animals with food to eat. Remember what God told Adam in the garden of Eden. "Of every tree of the garden you may freely eat; but of the tree of the knowledge of good and evil you shall not eat, for in the day that you eat of it you shall surely die." (Genesis 2:16-17). Another reason why seeds are wrapped in fruit is so that animals would distribute the seeds. Sometimes they spit out the seeds like we do. And sometimes the seeds drop out onto the ground when they defecate. Either way, seeds are scattered around — which is exactly what the Creator planned. They sprout here and there and more trees will grow and produce more seeds and more fruit.

The biggest fruits are just overgrown ovaries. A Minnesota farmer produced a pumpkin weighing 2,350 pounds (1066 kg) and the largest watermelon in the world was 350 pounds (159 kg).

The best fruit of all, however, is the spiritual fruit God brings about in our lives. What a blessing, when we begin to see the fruit of the Spirit sprout up — love, joy, peace, patience, kindness, gentleness, goodness, self-control, and faith! Sometimes this fruit increases thirty fold, sometimes at sixty fold, and sometimes at a hundred fold. All of it comes by faith, and by God's grace.

Blessed is the man who trusts in the LORD,
And whose hope is the LORD
For he shall be like a tree planted by the waters,
Which spreads out its roots by the river,
And will not fear when heat comes;
But its leaf will be green,
And will not be anxious in the year of drought,
Nor will cease from yielding fruit.
(Jeremiah 17:7-8)

## Other Kinds of Reproduction

Not all plants will reproduce using seeds and spores. Incredibly, God can use just a leaf or a stem to make a new plant! What versatility! Imagine if we could grow another person out of a thumb or a big toe!

The potato is a large, fat tuber that grows off an underground stem. If you let potatoes sit around long enough, they will grow little buds called "eyes." Plant these buds in the ground and they will grow into new plants. Believe it or not, you can plant a leaf of an African violet and get another flowering plant!

## How Plants Grow

We witness an extraordinary work of God Almighty when we watch a tree grow! One tiny seed grows into a gigantic oak tree. It is a truly wondrous thing. Where does all this organic matter come from? Where does the 1,600,000 pounds (730,000 kg) of wood in

CHAPTER 5: GOD MADE PLANTS

a redwood tree come from? Why doesn't a towering tree create a huge crater in the dirt? It turns out, the matter collected in a tree doesn't come from the dirt. The tree is mainly made up of carbon. The carbon comes from carbon dioxide, and carbon dioxide comes from the atmosphere.

Children grow by eating three solid meals a day. But, trees grow by **photosynthesis**. Here is the process. Tiny little pores on the underside of the leaves (called **stomata**) take in carbon dioxide from the air. Water is carried up to the plant leaves from the soil. God uses a green chemical called **chlorophyll** to turn sun energy into chemical energy. This chemical is contained in little cell organs called **chloroplasts**. These organs absorb energy from the sun, and this energy is used to make food by a special chemical reaction. A water molecule is made up of 2 hydrogen atoms and 1 oxygen atom, and the chemical reaction separates the oxygen and hydrogen atoms. The hydrogen and carbon dioxide combine to make carbohydrates or sugar (glucose). That's the plant's food. To complete the process, the plant will release

**Process of Photosynthesis**

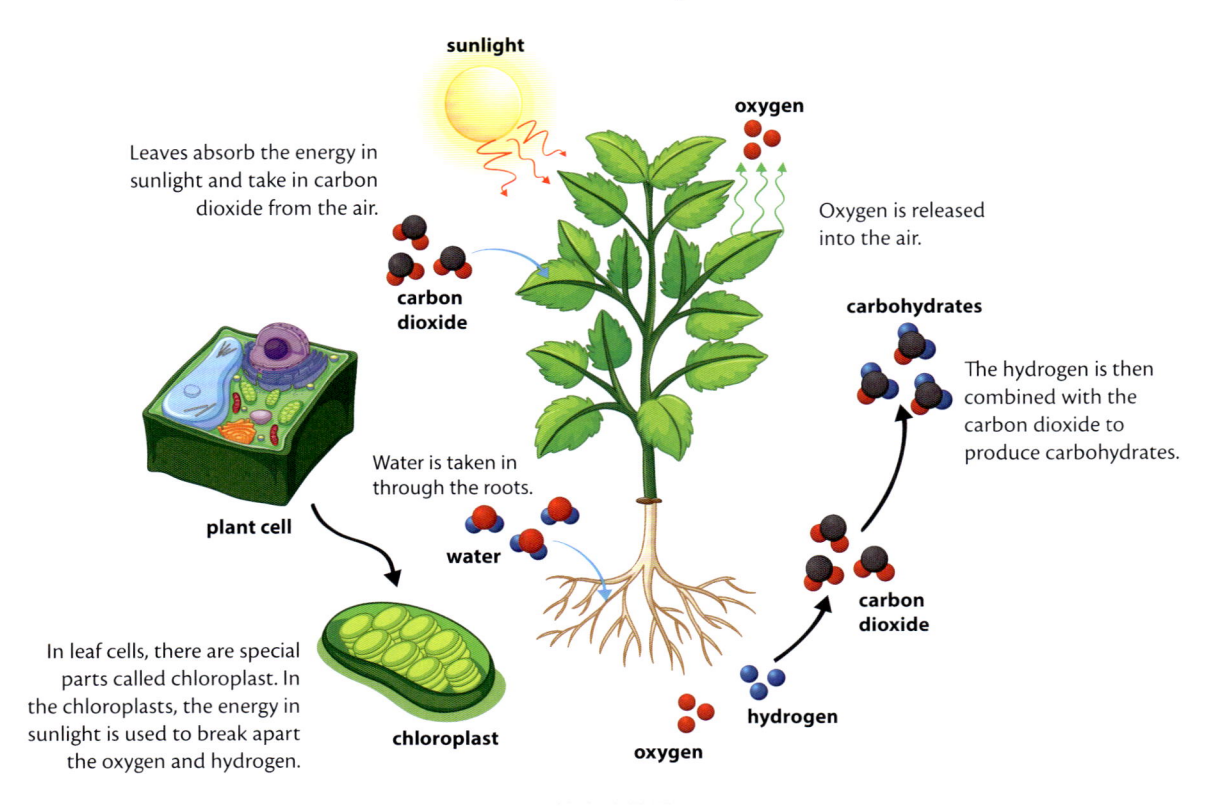

sunlight

oxygen

Leaves absorb the energy in sunlight and take in carbon dioxide from the air.

carbon dioxide

Oxygen is released into the air.

carbohydrates

The hydrogen is then combined with the carbon dioxide to produce carbohydrates.

plant cell

Water is taken in through the roots.

water

carbon dioxide

In leaf cells, there are special parts called chloroplast. In the chloroplasts, the energy in sunlight is used to break apart the oxygen and hydrogen.

chloroplast

hydrogen

oxygen

135

Cobra plant

the oxygen through the stomata in the form of gas.

As it turns out, we are dependent on plants for our breath, as well as for our food. About 30% of the oxygen we breathe comes from forests and fields, and the other 70% comes from marine plants in the oceans. The amount of oxygen required by each person to survive day by day is equivalent to the oxygen produced by 49 trees. Thankfully, there are 400 billion trees in the world and a lot of marine plants to keep us all well stocked in oxygen!

## The Cobra Plant

Plants make their own food sugars through the process of photosynthesis. But plants also need nitrogen in order to thrive. Most plants obtain nitrogen from good soil. When soils are depleted, people often apply man-made or natural fertilizers to provide a nitrogen boost for plants.

**Carnivorous plants**, however, are able to thrive in very poor soils. We call these amazing plants carnivorous (meat-eating) plants because God has given them ways to capture nitrogen-rich insects, digest them, and turn them into fertilizer!

The **cobra plant** is just one variety of the carnivorous plants. It can be found growing in Oregon and northern California bogs (in the US). Its leaves are wonderfully designed in the shape of hollow tubes. The opening of each leaf tube is covered by a large hood. This gives the plant a snake-like appearance of a cobra ready to strike. Insects are lured to the leaf's opening by nectar-covered, fang-like appendages. The opening itself is also saturated with nectar, smooth and slippery. Landing on this tempting edge, the unsuspecting insect soon finds itself sliding down the tube. When the bug tries to fly up and out, it gets confused by the many transparent "windows" on the plant's hood. The poor insect then keeps bumping into the hood, missing the opening. Then it will slide down the slippery tube once more.

Near the bottom of the tube, the insect runs into a lining of hairs all pointing downward into a pool of digestive juices. Try as it may, the little bug cannot escape by climbing over the hairs. Finally, in exhaustion the victim will collapse into the pool of toxic juices.

God has made the cobra plant very complex. On the one hand, the plant makes its own food by photosynthesis. But it also produces two more substances — the sweet nectar to attract insects, and the toxic liquid to digest them.

There is a spiritual lesson to take away from the cobra plant. We all face temptation to sin, but this turns out to be the way of death (Proverbs 14:12):

There is a way that seems right to a man,
But its end is the way of death.

Let us praise God for His creative wisdom in making the amazing and complex, carnivorous cobra plant!

## Fruits vs. Vegetables

Have you ever wondered about the difference between a fruit and a vegetable? As already covered, the fruit forms around a seed. It is part of the reproductive system of a plant. The vegetable is any other edible part of a plant. Technically, the following are fruits, not vegetables:

- Tomatoes
- Avocados
- Cucumbers
- Peppers
- Eggplants
- Olives
- Pumpkins
- Pea pods
- Zucchini

Besides the fruit, we eat stems, leaves, and roots of plants, and the following table provides a breakdown of common vegetables:

| | |
|---|---|
| **Roots** | beet, carrot, parsnip, radish, rutabaga, sweet potato, turnip |
| **Stems** | onion, celery, rhubarb, asparagus, potato |
| **Seeds** | lima beans, peas, corn, rice, wheat, pumpkin seeds |
| **Leaves** | lettuce, parsley, cabbage, spinach, collards, cilantro |
| **Flowers** | cauliflower, broccoli |

Of course, not every plant is edible. Some are very poisonous. Some plants are toxic to some animals, while not so much to others.

# GOD MADE LIFE

For instance, onions, garlic, and chives are toxic for cats; citrus fruits are toxic for dogs and cats; and grapes and raisins are toxic for dogs.

The most poisonous tree in the world is the **manchineel**, which grows in Florida and the Caribbean. Sap from the tree can cause an immediate breakout of blisters. If the sap comes in contact with your eyes, you could go blind immediately. A single bite of the tree's fruit could be fatal. Should someone burn the wood, smoke from the fire could cause blindness.

The most dangerous plant seed in the world is the Abrus precatorius, also called the **rosary pea**. It has a tough exterior shell, but if it is scratched and ingested, just 3 micrograms could kill an adult. That's one tenth of one drop of fluid.

Abrus precatorius

If you happen to get lost in the woods you should only eat things you know to be safe — like dandelions, clover, lamb's quarter, chickweed, wild amaranth, and curly dock.

Manchineel tree

As a rule of thumb, avoid all wild mushrooms, and yellow or white berries. Also, you should avoid eating plants with shiny leaves, bitter or soapy-tasting plants, and anything with an almond smell.

## Perennials, Annuals, and Biennials

God has designed some plants to grow back every year without the need of planting a new batch of seeds. What a convenient blessing for gardeners! The **perennial plants** will wither and die in the fall and winter. But then, in the springtime the rootstock grows new stems, leaves, and flowers. Many fruits and vegetables are perennials, producing more food for man and animal, year after year. Thanks be to God for His gift of perennials!

Annuals and biennials produce flowers

| Perennial Flowers | Perennial Fruits |
|---|---|
| Hibiscus | Apples |
| Lavender | Blueberries |
| Asters | Blackberries |
| Thyme | Cherries |
| Dahlia | Peaches |
| Most Roses | Pears |
| Some Geraniums | Olives |
| **Perennial Vegetables** | Oranges |
| Artichoke | Lemons |
| Asparagus | Limes |
| Broccoli (some varieties) | Grapes |
| Rhubarb | Plums |
| Spinach (some varieties) | Raspberries |
| Sweet Potato | Strawberries |
| Yams | |
| Potato | |

Orange tree (a perennial)

# GOD MADE LIFE

and fruit only once. Annuals come up from seed and produce more seeds in the same year. Then they die. Their roots cannot survive the winter. Biennials also sprout from seeds, but they don't flower and make more seeds until the second year. Their roots survive one winter but then, they will die the next winter after making seeds. Here you can see that God always wants to make sure the plant produces seeds for another generation before it dies. He creates life and sustains life, year after year, generation after generation. Praise Him for this wonderful creation!

| Annuals | Biennials |
| --- | --- |
| Corn | Pineapple |
| Wheat | Parsley |
| Rice | Leeks |
| Lettuce | Black-eyed Susan |
| Peas | Some cabbages |
| Watermelon | |
| Beans | |
| Zinnia | |
| Marigold | |
| Most tulips | |
| Most carnations | |

Will there be fruit-bearing plants in heaven? Revelation 22 speaks of a tree that bears fruit every month, and a different fruit for each month as well! While this earth is wonderfully fruitful, we may look forward to much more fruit from our bountiful God when we get to heaven!

And he showed me a pure river of water of life, clear as crystal, proceeding from the throne of God and of the Lamb. In the middle of its street, and on either side of the river, was the tree of life, which bore twelve fruits, each tree yielding its fruit every month. The leaves of the tree were for the healing of the nations. (Revelation 22:1-2)

Marigold

## Grasses

[The Lord] causes the grass to grow for the cattle,
And vegetation for the service of man,
That he may bring forth food from the earth,
And wine that makes glad the heart of man,
Oil to make his face shine,
And bread which strengthens man's heart.
(Psalm 104:14-15)

The most common food for animal and man is grass and grass seed. Wheat and rice are important grasses for humans, feeding billions around the world every year. About half of the world's nutrition for humans comes from grass! Grazing animals usually eat grass. Cows, for example, consume an average of 24 pounds (11 kg) a day. There are about a billion cows in the world, and that means that these cows are going to be eating 12 million tons a day, or 4.5 billion tons a year. That's more than humans consume in corn, rice, wheat, and sugar cane in a year. If you include grazing animals, close to 75% of the world's food comes from grasses.

There are 12,000 different kinds of grass, and these plants can be identified by their hollow stems. **Bamboo** is gigantic grass. In fact, the tallest bamboo will grow as high as 100 feet (30 m). And it grows fast! Some types of bamboo can grow an inch (2.5 cm) in 40 minutes, or a foot (30 cm) in a day. The adjoining picture of bamboo gives you a good view of the cross-section of this grass.

Bamboo

Cross-section of bamboo

The stems are hollow. Yet, you can still see a solid center at the nodes.

Bamboo has many practical uses. It's not wood, but it can be used to make roofing, flooring, and furniture. For thousands of years, bamboo has been used to make musical instruments. Elephants and panda bears, as well as humans, eat the bamboo

GOD MADE LIFE

shoots. Also, folks from Asia will make diapers, rugs, clothes, and utensils out of this versatile grass.

## Survivor Plants

Like penguins surviving in Antarctica, our all-wise Creator has also made plants to survive in the harshest places. In deserts, you will find cacti and other **succulents**. These plants store up as much water as possible when it rains. Succulent stems are designed with ridges to catch water, so as to absorb as much as possible. The spines on the cactus were designed to discourage animals from eating the plant. Nobody really wants a cactus needle stuck in their throat. Also, cactus roots spread out just below the surface of the ground to absorb all the rainwater they can get. Incredibly, some cacti can go for two years without water!

## Trees

The **tree** is another kind of plant. But it is different from other plants in that it has a hard, woody trunk with fewer branches and leaves on its lower section. Other plants usually feature lots of stems and many leaves growing out of the stems closer to the ground. Also, the "non-tree" plants don't grow much larger than ten feet tall, and perennial plants rarely live longer than ten to twenty years. However, the average tree will live for 100-300 years, and most trees will grow very tall. Trees continue to grow in width by adding an additional ring to their girth. But, most trees will stop growing in height after 150 years.

Here are a few definitions for various kinds of trees created by God:

**Deciduous trees** lose their leaves during the fall and winter months. The oak tree, the elm tree, and the maple tree are examples.

Evergreen trees retain their "green" needles all year round. The needles may be thought of as super narrow leaves.

**Conifers** or coniferous trees are usually evergreen. They drop their seeds (tucked into cones) on the surrounding ground.

**Ginkgoes** and **Cycads** are the broad-

Maple tree

## CHAPTER 5: GOD MADE PLANTS

leafed evergreen trees. Like conifers, these are gymnosperms. Examples of these include the stangeria, the Australian bowenia, and the maidenhair tree.

Trees live longer than humans, and some have survived since Noah's flood (about 4,300 years ago). The Methuselah tree, still alive in the Bristlecone Pine Forest in California, is probably about 4,300 years old. This long-living character of a tree is compared with man's short life in Job 14:7.

"For there is hope for a tree,
If it is cut down, that it will sprout again,
And that its tender shoots will not cease.
Though its root may grow old in the earth,
And its stump may die in the ground,
Yet at the scent of water it will bud
And bring forth branches like a plant.
But man dies and is laid away;
Indeed he breathes his last
And where is he?
(Job 14:7-11)

The promise of the new covenant provides us with the expectation that we will live a long time — even longer than a tree! This is the message of Isaiah 65:

For behold, I create new heavens and a new earth;
And the former shall not be remembered or come to mind
But be glad and rejoice forever in what I create;
For behold, I create Jerusalem as a rejoicing,
And her people a joy. . .
They shall not build and another inhabit;
They shall not plant and another eat;
For as the days of a tree, so shall be the days of My people,
And My elect shall long enjoy the work of their hands.
(Isaiah 65:17, 18, 22)

Methuselah Tree

## How Plants and Trees Get Water and

# GOD MADE LIFE

## Minerals

Water is as important to a plant as blood is necessary for the human body to function. The plant's **vascular system** has several jobs to do. It moves water from the roots up into the rest of the plant. Since the roots need nutrition as well (and they can't process food), the vascular system brings glucose down to feed the roots.

Tiny root hairs absorb water and nutrients from the soil. Using a microscope, you could count 250,000 root hairs in every square inch of root. At the end of the root is a root cap shaped like a torpedo. The Lord designed a special slimy liquid for the tip of the root. This makes it easy for a tiny root to push through tightly packed soil.

There are two kinds of piping or vascular plumbing in the plant — the xylem and the phloem. The **phloem** hugs the outside of the stem, and the **xylem** runs closer to the center of the stem. They are two transportation highways. The xylem carries minerals and waters absorbed from the roots up to the rest of the plant. And the phloem distrib-

**Xylem and Phloem**

XYLEM
- One-way flow
- Water and minerals
- No end walls between cells
- Stiffened with lignin

PHLOEM
- Two-way flow
- Water and food
- End walls with perforations

Photosynthesis products

Water and minerals

144

Deciduous trees in the four seasons

utes the sugar made in photosynthesis to the roots and throughout the plant system.

The largest redwood trees require as much as 500 gallons (1,900 l) of water a day. This must be absorbed into the roots and pumped up as high as 380 feet (115 m) in the air. So how does God arrange for 500 gallons (1,900 l) a day to seep up the trunk into the branches and leaves? Once again, we return to the sun's energy. Trees lose water from needles (or leaves) through a process called **transpiration**. As the sun heats the water droplets on the needles, the water evaporates. As the droplet disappears, this puts a slight tug on an adjacent water molecule. This creates a suction that carries water up through the xylem in the tree.

Now you can see that the Creator designed so much of nature to take care of itself. God brings sufficient rain to the forests all over the world, and the sun's energy helps to pull all that water up into the tree branches and leaves to keep the trees healthy, growing, and strong.

## Why Trees Change Colors

Chlorophyll is not only necessary for photosynthesis, but this chemical also provides the **pigment** to paint leaves a beautiful green during the spring and summer months.

## GOD MADE LIFE

When autumn brings shorter days, the chlorophyll begins to break down. The tree is getting less energy from the sun, and so the leaves begin to die. Since there is much less pigment coming from the chlorophyll, the other pigments contained in the leaves stand out more. These colors are usually yellow or red. Some leaves turn brown because they have no other pigments. All leaves turn brown after they fall because they lose all their pigments as they dry up.

When the tree detects that the leaves are dying, a special group of cells develops whose job it is to cut the leaves off the branches. Similar to the clotting of a cut on the human body, the tree branch will seal up the tear as the leaves tear off the branches. This process was carefully designed and put in place by the Creator at the beginning.

## Adventitious Roots

Our very-creative Creator has designed some plants to grow in some of the strangest ways, even when there is no farmer around to plant the seeds in the soil. These are called **adventitious roots**. One example of these resourceful roots is the mangrove tree. This tropical tree has **prop roots** that

Mangrove tree

## CHAPTER 5: GOD MADE PLANTS

offer a little extra support for the tree, which usually grows in water or super wet soil. Another example is the **aerial root**, a root that grows out of the side of a tree. Banyan trees produce seeds that fall on their own branches where they will eventually germinate. These new aerial roots develop and grow downwards into the soil. As these trees grow together, the banyan spreads and its leaves serve as wonderful shade for people who want to meet outside on a hot summer day. The largest banyan tree in the world is spread over 4 acres (16,187 m2). This incredible creation is found in India. It is made up of over 3,000 separate trees growing together! That would be enough shade for a crowd of 30,000 people sitting down.

## God's Design Seen in the Amazing Order of Fibonacci

God has produced an extraordinary order in creation, often using the **Fibonacci sequence**. The concept behind the Fibonacci sequence is simple if you have learned to add. Just add the two numbers before each subsequent number.

0
1
0 +1 = 1
1 +1 = 2
1 + 2 = 3
2 + 3 = 5
3 + 5 = 8
5 + 8 = 13

This yields the following sequence: 0, 1, 1, 2, 3, 5, 8, 13, 21, 34, 55, 89. . .

In nature, we find this pattern providing the spiral shape you find in seashells, flowers, pine cones, fern leaves, tornadoes, and galaxies.

For example, a lily has 3 petals,
the yellow violet has 5 petals,
the delphinium has 8 petals,
the mayweed has 13 petals,
the aster has 21 petals,

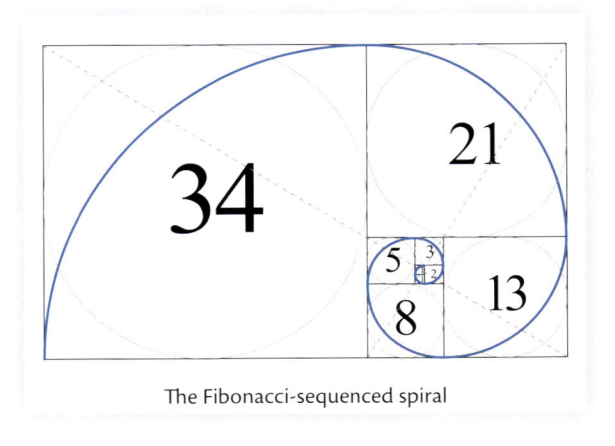

The Fibonacci-sequenced spiral

The Nautilus shell designed according to the Fibonacci pattern

# GOD MADE LIFE

The pyrethrum has 34 petals,
The helenium has 55 petals,
and the Michaelmas daisy has 89 petals.
Do you recognize these numbers?

You will find that smaller pine cones will spiral to the left with three scales, followed by a spiral to the right using five scales. Then, the larger pine cones will spiral with a 5-8 or 8-5. A few very big pine cones will spiral left and right with an 8-13. Do you recognize these numbers? The scales on pineapples will spiral as high as 13-21, and sunflowers have been known to pack into spiraling patterns as high as 89-55 or 144-89. Do you recognize these numbers? All of these quantities are Fibonacci numbers. When these are used for pine cones, pineapples, and sunflowers, they produce spirals of similar shape. This too points to a Creator, not random chance. God is a person, with a commitment to beauty, order, and patterns in His creative work. Thus, we see His preferred patterns in the creation and we understand something of His Personality.

None of this creation was left to chance. All of the plants and trees in the world reflect His beauty, His wisdom, and His glory.

Let them praise the name of the Lord,
For He commanded and they were created.
He also established them forever and ever;
He made a decree which shall not pass away.
Praise the Lord from the earth,
You great sea creatures and all the depths;
Fire and hail, snow and clouds;
Stormy wind, fulfilling His word;
Mountains and all hills;
Fruitful trees and all cedars... (Psalm 148:5-9)

God made pinecones using the Fibonacci pattern.

God made sunflowers using the Fibonacci pattern.

## Pray

- Praise God for the awesome beauty of His creation, particularly seen in flowering plants and trees. Praise the Creator for the huge variety of trees and plants in the world (between 400,000 and 500,000 different kinds).
- Thank the Lord for 20,000 different edible plants and the 9 billion tons of food grown by farmers in the world each year.
- Thank the Lord for 3 billion tons of grass grown each year for cows, sheep, and goats grazing in the fields.
- Thank the Lord for 1 billion tons of wood harvested each year for building homes and furniture, as well as all the cotton and wood used for our clothing.
- Thank our Creator for providing oxygen through plants for the people and creatures to breathe.
- Ascribe to God all the honor and infinite wisdom for His design of plant reproduction. Praise God for bees and butterflies that pollinate the plants. Give Him the glory for the resourcefulness and bountifulness in plant reproduction. Thank Him for all the reproduction in His creation that happens automatically without help from man.

## Sing

Having seen the marvelous work of God in this spectacular creation, an appropriate response is always worship and praise. If the student is unfamiliar with the hymn or psalm, some version of it is available on the internet and may be accessed (with supervision) for singing along.

*For the Beauty of the Earth*
For the beauty of the earth,
For the beauty of the skies,
For the Love which from our birth
Over and around us lies:

Chorus:
Christ, our Lord, to Thee we raise
This our hymn of grateful praise.

For the beauty of each hour
Of the day and of the night,
Hill and vale, and tree and flower,
Sun and moon and stars of light:

For the joy of ear and eye,
For the heart and mind's delight,
For the mystic harmony
Linking sense to sound and sight:

For the joy of human love,
Brother, sister, parent, child,
Friends on earth, and friends above;
For all gentle thoughts and mild:

For each perfect Gift of Thine
To our race so freely given,
Graces human and Divine,
Flowers of earth, and buds of Heaven:

## Do

Choose at least one of the following activities and apply the lessons you have learned in this chapter in real-life.

1. Plant a tree in your yard or somewhere else, with permission from the owner of the property. Here are the best steps to take for planting a tree.

    a. Find out which trees grow best in your area. Identify the best time of year to plant. This may be in early spring or the fall.

    b. If you want to grow a tree from a seed or acorn, follow these instructions (or another set of instructions that may better apply to your conditions):
    - Obtain a flower pot, put some stones at the bottom. Then fill the rest of the pot with compost.
    - Plant the seed(s) about an inch (2 cm) in the compost. Press the compost down, and pour water into the pot until you see it running out the bottom.
    - Place the pot under some shade outdoors. You probably want to cover it with wire mesh to keep animals from getting to the plant.
    - Check the soil every day to make sure it is still moist. Don't over-water it. Protect the plant from freezing temperatures.
    - Monitor the growth of the plant. You may need to transfer the tree to a bigger pot. Once the tree has grown to a height of 16 inches (40 cm), it should be ready to plant in the ground.

    c. Dig a hole for the sapling, about three times wider than the root mass. Watch the depth of the hole. You don't want any of the trunk flare to be buried underground.

    d. Plant the tree high such that about 25% of the root ball is above ground level. Then, you can taper the soil up over the roots. You want to make sure the entire root mass is covered by soil. Examine the roots. If they are cramped and pressing against the sides of the container (or pot), spread the roots out. Break up the circular pattern of the roots. Otherwise, the tree will probably die.

    e. Backfill with the native soil. If you add organic material, the tree roots probably won't grow into the more compacted native soil. Also, you don't want any air pockets in the soil, for close root to soil contact. You can tamp the soil down firmly. Or, you can apply a fast stream of water from a hose into the hole when you're half way done with the back fill. Then, apply more water when the back fill is completed.

## GOD MADE LIFE

- f. Add additional organic material as mulch on top of the backfill. Keep the mulch from contacting the trunk of the tree. Allow about a two inch separation of the mulch from the tree. You could use bark or shredded leaves. The mulch keeps the ground moist and provides needed nutrients for the young sapling.

- g. Keep your tree well-watered for several months, or perhaps up to a year. The best way to water the tree is by drip or soaker hoses. You don't want to pour a gallon or two of water over the tree all at once. When the tree is established, water at the drip line of the tree. The drip line is the circle on the ground underneath the outer branches where rain would drip off the leaves. This is where roots are actively growing and taking up water. Water the tree every day for the first week. Then, for the next two weeks water it every other day. It's easy to over-water trees and plants. Keep the ground moist, but not soaked and soggy. Keep an eye out for how the tree responds to the water.

2. Grow edible plants for your family. Choose a common plant that would grow well in a sunny window sill, such as basil, chives, parsley, mint, rosemary, or thyme. You want a window facing to the south, where the plant would receive over six hours of sunshine a day.

    - a. Choose your pot for growing plants. The pot should be 6-12 inches (15-30 cm) deep, and it should fit on a window sill or a table near a south-facing window. If you are planting seeds, you can use a smaller container and transfer them into the larger pot later. Your pot should have drainage holes for draining out excess water.

    - b. Fill the pot with potting mix.

    - c. If you are planting seedlings, fill the pot with a minimum of two to three inches (5-7 cm) of potting mix. Gently spread out the roots, and fill around the plant with more potting mix. Lightly tamp down the soil around the plant. Leave about an inch (2.5 cm) of space at the top of the pot for watering. If you are planting seeds, fill the pot almost to the top with potting mix. Place three to five seeds on top of the soil, and then cover them with a little bit of soil. You can place a plastic bag over the pot to provide a more humid environment. The soil should be moist, but not soaked with water. Once the seeds have germinated and the seedlings are visible, remove the plastic.

    - d. Don't over-water herb plants. You can poke your finger about an inch (2.5 cm) into the soil to check for moisture. If the soil is dry add more water.

    - e. Add fertilizer as recommended for herb plants.

f. Depending on the herbs, you may be able to harvest a few leaves within four to six weeks. Don't be afraid to clip off a few inches from the tips of the branches. This will make room for new growth and more branching.

## Watch

To watch the recommended videos for this chapter, go to **generations.org/GodMadeLife** and scroll down until you find the video links for Chapter 5. Our editors have been careful to avoid films with references to evolution, however, we would still encourage parents or teachers to provide oversight for all use of the internet. These videos may not give God the glory for His amazing creative work, so the student and parent/teacher should respond to these insights with prayer and praise.

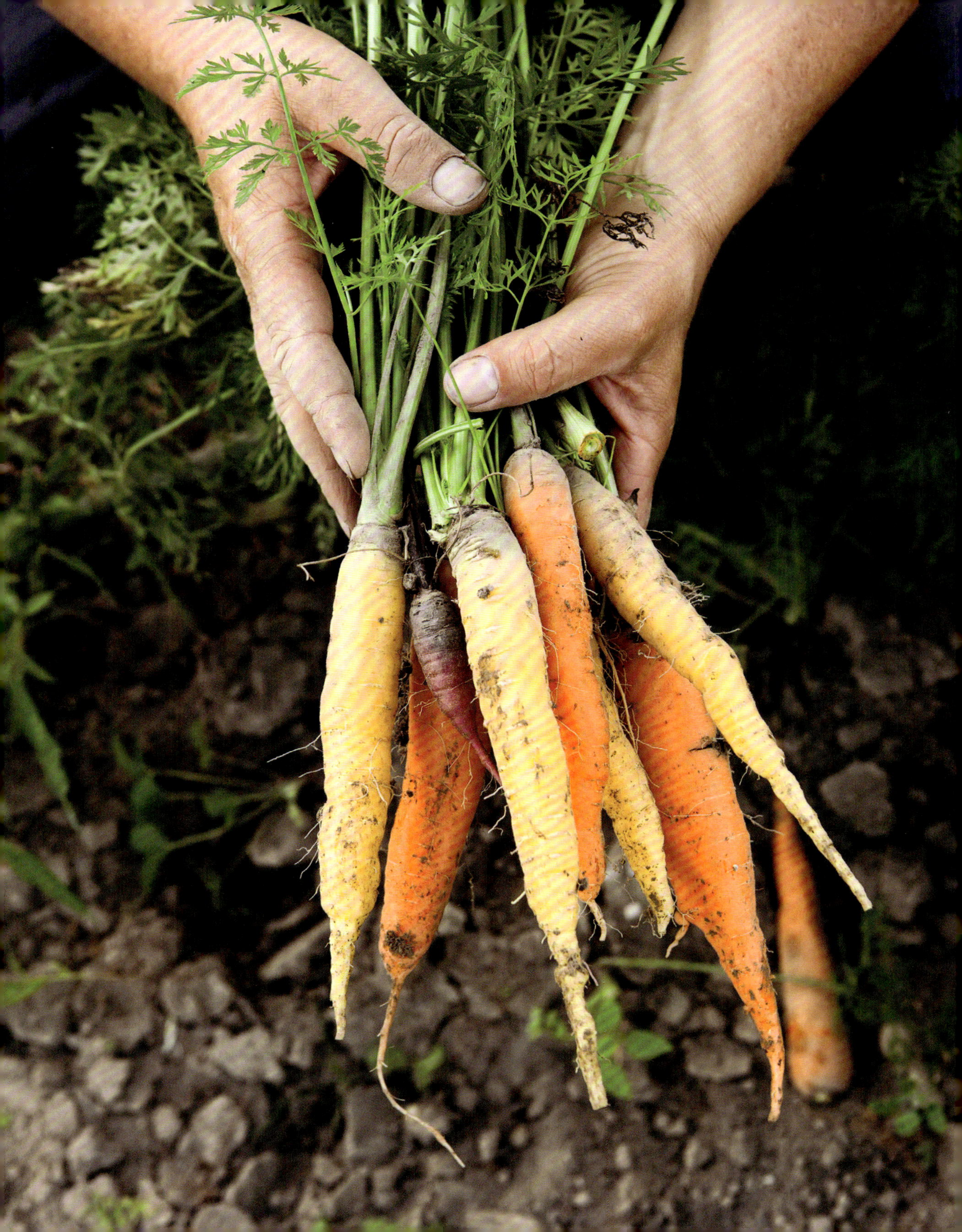

# CHAPTER 6
# GOD MADE FOOD

> If you walk in My statutes and keep My commandments, and perform them, then I will give you rain in its season, the land shall yield its produce, and the trees of the field shall yield their fruit. Your threshing shall last till the time of vintage, and the vintage shall last till the time of sowing; you shall eat your bread to the full, and dwell in your land safely. (Leviticus 26:3-5)

**S**curvy was one very dreaded disease throughout world history. It would often kill sailors on long trips across the ocean, especially in the 1600s and 1700s. During one British expedition across the Pacific Ocean in the 1740s, 1,300 out of the 2,000 men died of the terrible disease. Some Spanish galleons would be found floating out in the middle of the ocean. The whole crew was dead. Magellan returned from his voyage around the earth in 1520 with only 18 men. The other 212 sailors had died of scurvy.

The symptoms of scurvy are swollen, purplish, and bleeding gums, and loose teeth. The skin of the patient would sometimes start bleeding, and then, internal bleeding would leach into the joints and muscles.

The first to record the cure for the disease was a Christian sea captain by the name of Sir Richard Hawkins. Within his memoirs dated the year 1593, the Captain included these important words:

"There was great joy amongst my company and many with the sight of the oranges and lemons seemed to recover heart. This is a wonderful secret of the power and wisdom of God that hath hidden so great and unknown virtue in this fruit to be a certain remedy for this infirmity."

GOD MADE LIFE

Sir Richard Hawkins (1562-1622)

given the same diet and then provided with different supplements.

Two men receive one quart of cider a day.

Two men received several spoonfuls of vinegar a day.

Two men received 25 drops of medicine a day.

Two men received seawater each day.

Two men received garlic each day.

Two men received two oranges and one lemon every day.

Incredibly, the men who received the oranges and the lemons recovered in as little as six days. The men who received the cider were helped a little bit, and everybody else was as sick as ever. This was one of the most important experiments conducted in history.

Once again, nobody paid much attention to Dr. Lind's work for a long time. It wasn't until 42 years later that the British admiralty

But not very many paid attention to the Captain's recommendation. Millions of sailors continued dying of the terrible disease for 150 years. In 1747, a surgeon named James Lind, who worked for the Royal Navy, experimented with groups of sailors during a voyage of the HMS Salisbury. Taking 12 of the sailors, who had contracted scurvy, he broke them up into 6 groups of 2 each. All 12 were

James Lind (1716-1794)

**156**

**CHAPTER 6: GOD MADE FOOD**

ordered a provision of lemon juice for sailors on British ships.

What was it about oranges and lemons that cured the sailors of scurvy? Nobody really understood what the Creator had tucked into these citrus fruits for curing diseases until 1911. Three scientists teamed up to figure it out: Christiaan Eijkman (1858–1930), Gerrit Grijns (1865–1944)—Eijkman's assistant in Java—and Casimir Funk (1884–1967). These were the men who discovered vitamins.

While doing medical research in Indonesia in 1883, Dr. Eijkman noticed when the natives fed their chickens polished brown rice, the animals would get the disease called **beriberi**. But whenever the chickens ate unpolished brown rice, the chickens would recover. What was it in the whole grain that kept chickens healthy? Beriberi was common among Asians, especially for those who lived on rice. At first, nobody could figure out what was in the whole grain that

Brown rice

would protect them from the disease. Later, a Polish-American named Casimir Funk called this mysterious substance "vitamine," or what we call **vitamins**. And now we know that God gave us vitamin B1 (or **Thiamine**) to prevent beriberi, and vitamin C to prevent scurvy. When rice is polished into white kernels, the nutrient-rich, brown covering is removed (including the vitamin B1).

GOD MADE LIFE

By His wise and gracious provision, the Lord has given us all the nutrients we need in the food we eat. It was only after polishing the rice and removing the vitamins that the Asian peoples began to suffer from beriberi. If they had eaten the whole grains, nobody would have died of the disease.

The Lord God put vitamins and minerals into the food He has designed for man. The human body needs just the right amount of these nutrients. Getting too little or too much of them can be bad for you. A well-balanced diet of God's grains, fruits,

Salmon

| Vitamin | The Foods Containing the Vitamin | What the Vitamin Does for You |
| --- | --- | --- |
| Vitamin A | Cantaloupe, carrots, dairy, eggs, lettuce | Helps vision and growth |
| Vitamin B1 | Beans, peas, whole grains, nuts, pork | Converts food into energy |
| Vitamin B6 | Chickpeas, fruits, potatoes, salmon, tuna | Helps the nervous system and digestion |
| Vitamin B12 | Dairy, eggs, cereals, meat, poultry, seafood | Converts food into energy and helps nerves |
| Vitamin C | Citrus fruits, broccoli, peppers, tomatoes Rosa | Helps wounds to heal and aids the immune function |
| Vitamin D | Eggs, fish, dairy, mushrooms | Regulates blood pressure, bone health, and immunity |
| Vitamin E | Green vegetables, nuts, seeds, peanuts | Helps form blood vessels and boosts the immune system |
| Vitamin K | Green vegetables, Swiss chard, turnips | Helps blood clotting and bone health |

**CHAPTER 6: GOD MADE FOOD**

vegetables, and animal products is usually all you need. As sailors traveled across the high seas over a period of 200 days without any fruits or vegetables in their diet, they would get sick. Without nutrition that includes certain vitamins, any of us would certainly get sick and die. Everybody needs at least 30 different vitamins and minerals, none of which the body can produce on its own. These supplements help the body with hundreds of different functions. They convert food into energy, heal wounds, strengthen bones, and repair damage to the cells.

The tables on the left and following page list some of the important vitamins and minerals and a few of the foods containing them.

Vitamins are made partly of carbon. Minerals are non-carbon-based chemical elements God made for body function and other purposes. These nutrients don't give you energy like carbohydrates, protein, and

Chickpeas

fats, but they help your body perform all its necessary functions.

**Metabolism** is the process by which the body makes energy from food so all the body functions keep working. Metabolism is what keeps you alive and moving. Your **metabolic rate** is the sum total of the energy your body

| Mineral | The Foods Containing the Mineral | What the Mineral Does for You |
| --- | --- | --- |
| Calcium | Dairy, kale, broccoli, collard greens | Helps blood clotting and promotes strong bones and teeth |
| Chloride | Olive, rye, seaweed, salt, celery, lettuce, tomatoes | Helps digestion and acid-base balance |
| Copper | Chocolate, lentils, nuts, seeds, whole grains | Helps bone formation and energy production |
| Iron | Beans, eggs, fruit, meat, nuts, peas, poultry, seafood | Helps energy production, body growth, the healing of wounds, and immune function |
| Magnesium | Avocados, beans, peas, dairy, bananas, spinach | Helps blood pressure, bone health, and muscles |
| Manganese | Beans, nuts, pineapples, spinach, whole grains | Helps metabolism, bones, and the healing of wounds |
| Phosphorous | Beans, peas, dairy, meat, nuts, poultry, seafood | Helps bone formation and energy production |
| Potassium | Beans, dairy, bananas, vegetable juice, seafood | Helps blood pressure, fluid balance, and muscles |
| Sodium | Prepared foods, cheese, poultry | Helps blood pressure, fluid balance, and muscles |
| Zinc | Beans, peas, dairy, nuts, poultry, whole grains | Helps growth, the immune system, taste, the nervous system, and reproduction |

is consuming at any point in time. When you climb mountains and play sports, your metabolic rate is much higher than when you are lying on the sofa. Metabolism involves food digestion, heart and lung function, and cells converting the food into energy.

## Processed Foods

It wasn't just the Asians who polished the rice and lost valuable nutrients (vitamin B1). Every society likes to process foods, which usually removes some or most of the vitamins and minerals.

Drying foods will remove about 50% of the vitamins and minerals. Cooking vegetables takes out about 25-40% of the nutrients, but cooking and draining the fluids will remove 40-75% of the nutrients.[9] Frozen vegetables usually retain most of the nutrients, and fresh vegetables can lose nutrients if they are stored for a lengthy period. Bell peppers lose the most nutrients by cooking, so it would be best to eat them raw. On the other hand, carrots, celery, and green beans increase in usable nutritional value for your body when they are cooked.

Some food processing can also protect food from harmful bacteria. And, canned foods can provide access to a large variety of fruits and vegetables year-round.

Heavily processed foods like soda and potato chips lack nutritional value and can be detrimental to your health. Here are a few things to consider as you think about whether to eat a bag of chips or an apple[10]:

Soda

1. Processed foods can be highly addictive. As Christians, we want to resist the temptation to idolatry or excessive dependence on a substance. The Apostle Paul put it this way in 1 Corinthians 6:12-13: "All things are lawful for me, but all things are not helpful. All things are lawful for me, but I will not be brought under the power of any. Foods for the stomach and the stomach for foods..."

2. Some additives used in processed foods can cause weak bones, kidney problems, obesity, and aging. The high levels of phosphates in sodas can destroy your bones and internal organs.

3. Pound for pound, fresh food is usually cheaper than processed food. Instead of buying a pound of potato chips for $4.00, you could purchase a pound of apples for $1.30 or a pound of potatoes for $0.75.

4. Some processed foods can cause chronic inflammation, probably due to the use of refined sugars, excess salt, and processed flours.

5. Junk food can also ruin your digestive system. Remember, the same Creator who made your digestive system also made the food to go into it. Without natural fibers, enzymes, and vitamins, the body will have a hard time processing the substances.

6. Certain processed foods can also wreak havoc on your mind. People complain of brain fog or difficulty concentrating. Often, this is linked to interruptions in the digestive system.

# GOD MADE LIFE

## Reading a Food Label

Many countries require labels on food packages. A sample is provided here, and the following explains what you can learn from these labels:

**Sample Label for Macaroni & Cheese**

| Nutrition Facts | |
|---|---|
| Serving Size 1 cup (228g) | |
| Servings Per Container 2 | |

**Amount Per Serving**

| Calories 250 | Calories from Fat 110 |
|---|---|

| | % Daily Value* |
|---|---|
| **Total Fat** 12g | 18% |
| Saturated Fat 3g | 15% |
| *Trans* Fat 3g | |
| **Cholesterol** 30mg | 10% |
| **Sodium** 470mg | 20% |
| **Total Carbohydrate** 31g | 10% |
| Dietary Fiber 0g | 0% |
| Sugars 5g | |
| **Protein** 5g | |

| Vitamin A | 4% |
|---|---|
| Vitamin C | 2% |
| Calcium | 20% |
| Iron | 4% |

\* Percent Daily Values are based on a 2,000 calorie diet. Your Daily Values may be higher or lower depending on your calorie needs.

| | Calories | 2,000 | 2,500 |
|---|---|---|---|
| Total Fat | Less than | 65g | 80g |
| Sat Fat | Less than | 20g | 25g |
| Cholesterol | Less than | 300mg | 300mg |
| Sodium | Less than | 2,400mg | 2,400mg |
| Total Carbohydrate | | 300g | 375g |
| Dietary Fiber | | 25g | 30g |

1. The first part provides serving size, which in this case is one cup of Mac and Cheese.

2. The second part provides how many servings of this size are contained in the package. There are two servings in the package.

3. The third section provides the number of **calories** per serving. If the average pre-teen boy should receive 2,200 calories per day, and the average pre-teen girl should receive 1,800 calories per day, what is the maximum number of cups of Mac and Cheese the boy or girl could eat in a day?

4. The next section presents the weight of certain critical nutrients in each serving using grams and milligrams. Be careful not to eat too much of these substances: fat, cholesterol, sugars, and sodium. If you limit your fat intake to 30% of your daily calorie intake, how much Mac and Cheese could you eat in a day? The average pre-teen boy's fat intake should not exceed 660 calories. And, the average pre-teen girl's fat intake should not exceed 540 calories. The USDA recommends no more than 2,400 mg of sodium in a day. How much Mac and Cheese would one have to eat to exceed that number?

5. The fifth section lists the number of carbohydrates, sugars, and proteins contained in the food product.

6. The sixth section on the label lists the amounts of vitamins and minerals in the product.

CHAPTER 6: GOD MADE FOOD

You don't have to read every label on the food you buy. But you should get to know which foods are more helpful for your body's needs. God wants us to take good care of our bodies because the body is the temple of the Holy Spirit.

Or do you not know that your body is the temple of the Holy Spirit who is in you, whom you have from God, and you are not your own? For you were bought at a price; therefore glorify God in your body and in your spirit, which are God's. (1 Corinthians 6:19-20)

Fats are actually beneficial for you as long as you eat the right kinds in the right amounts. They are very helpful for the brain. Children especially need healthy fats in their diets.

Fatty foods to watch out for include things like bacon, ground beef that is not lean, breakfast sausage, hot dogs, processed meat, butter, ice cream, French fries, and cold cuts. You want to limit your intake of these fatty foods. Foods that contain healthier fats include avocados, nuts, olive oil, and peanut butter. The healthiest fats are polyunsaturated fats. These can be found in salmon, trout, walnuts, canola oil, tofu, and sesame oil. Meat and dairy products from cattle grazed on open range or grasslands also contain healthy fats.

The table on the next page provides the amount of fat in common foods.

If you were limiting yourself to 660 fat calories in a day, and you had already eaten two pancakes for breakfast and a salami sandwich for lunch, how many fat calories would you allow yourself for dinner?

Hear, my son, and be wise;
And guide your heart in the way.
Do not mix with winebibbers,
Or with gluttonous eaters of meat;
For the drunkard and the glutton will come to poverty,
And drowsiness will clothe a man with rags. (Proverbs 23:19-21)

The main concern about food found in

Nuts contain healthy fats.

Greenbeans

163

## GOD MADE LIFE

| Food | Fat Weight | Fat Calories |
|---|---|---|
| Three slices of salami | 40 grams | 360 |
| Cheesecake | 40 grams | 360 |
| Two hot dogs | 25 grams | 225 |
| One slice of fruit pie | 24 grams | 216 |
| Two pancakes | 24 grams | 216 |
| Two fried eggs | 24 grams | 216 |
| Steak (4 oz) | 21 grams | 189 |
| Milkshake (medium) | 20 grams | 180 |
| Two pieces of bacon | 19 grams | 171 |
| Two slices of cheddar cheese | 18 grams | 162 |
| One chicken breast | 17 grams | 153 |
| Hamburger (4 oz) | 17 grams | 153 |
| One doughnut | 16 grams | 144 |
| Twenty almonds | 16 grams | 144 |
| Whole milk (8 oz) | 10 grams | 90 |
| Yogurt (6 oz) | 8 grams | 72 |
| One scoop of ice cream | 7 grams | 64 |
| One tablespoon of peanut butter | 7 grams | 64 |
| Butter on one slice of toast | 5 grams | 45 |
| Two slices of bread | 1 gram | 9 |
| Cocoa powder | 1 gram | 9 |

CHAPTER 6: GOD MADE FOOD

High-fat foods

Low-fat foods

Scripture is the problem of overeating. Both the Old Testament and the New Testament allow for eating meat. Proverbs 23:19-21 only warns us not to give way to gluttony.

## Carcinogens

Researchers have looked into the things that increase the risk of cancer. These **carcinogens** either damage the DNA in your body or cause cells to reproduce too quickly. This can result in mutations of bad cells which multiply and spread through the body. Most recently, scientists say about one third of cancers are caused, at least in part, by what people eat. Overeating and taking in high amounts of fat are the things that increase the risk of cancer.[11]

The greatest amounts of carcinogens entering the human body come from meats barbecued at temperatures exceeding 446°F (230°C). Marinate the meat to reduce the risks of carcinogens developing during grilling. Also, let the flames die down before placing the meat on the grill, and keep the grill fairly high above the coals. Trimming excess fat is also a good idea.

We want to be careful not to set a standard for every person, man, woman, and child. Each person's body is a little different, and varying levels of

165

activity call for different nutritional needs. That said, the Cancer Society recommends a maximum of one pound (454 g) per week of red meat. Avoid eating burnt or charred meat, and avoid processed meats.

## Watch Your Sugar Intake

**Sugars** are not all bad for you. In fact, the body needs sugar to survive. The main problem with sugar is when people eat too much of it. Overeating contributes to obesity and Type 2 diabetes. This has become a major health crisis in many countries around the world. The American Heart Association suggests a limit of 100 calories per day of sugar for women, and 150 calories a day for men. That's only 9 teaspoons of sugar for men and 6 teaspoons for women. If one can of soda contains 11 teaspoons of sugar, how many cans of soda should a grown man drink in a day according to the American Heart Association? The table on the right lists the approximate amounts of sugar contained in common foods.

If you were limiting your sugar intake to 9 teaspoons in a day, which foods would you choose?

| Common Foods | Amount of Sugar |
|---|---|
| Starbucks Large Festive Hot Drink | 23 teaspoons[21] |
| Can of Dr. Pepper | 10 teaspoons |
| Chocolate cake with frosting (1 piece) | 6 teaspoons |
| One Cinnabon Classic Cinnamon Roll | 6 teaspoons |
| One slice Hostess Apple Pie | 4 teaspoons |
| One SNICKERS® chocolate bar | 3 teaspoons |
| Glass of Orange Juice | 2.5 teaspoons |
| Yoplait Yogurt | 2 teaspoons |
| Apple | 2 teaspoons |
| Banana | 1.5 teaspoons |
| Four slices of bread | 1 teaspoon |
| One cup of rice | 0.01 teaspoons |

## Obesity and Type 2 Diabetes

Lately, **obesity**, or putting on a lot of extra weight, has become a major worldwide problem. A 5-foot tall girl is obese if she weighs over 155 pounds. She is extremely obese if she weighs over 205 pounds.

In 1975, only 7 out of 1,000 girls and 9 out of 1,000 boys were obese. Forty years later, the obesity rate was 60 out of 1,000 girls and 80 out of 1,000 boys.[12] On average, American children are more unhealthy than

CHAPTER 6: GOD MADE FOOD

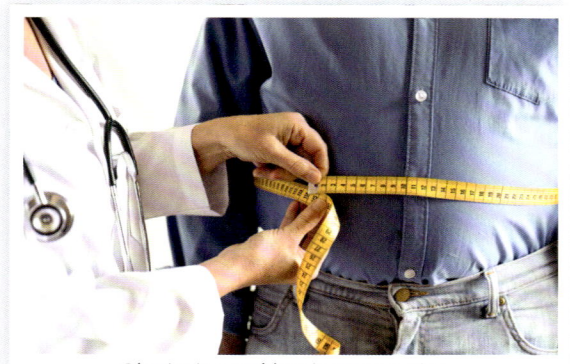

Obesity is a problem in many nations.

the rest of the world's children. Since 1988, the obesity rate for American kids has increased from 10% to 20% (or 20 in 100).[13] Because increased weight can contribute to **Type 2 diabetes**, there are many more people getting it now. As of 2020, about 1 in 10 Americans suffer from this disease, whereas only 1 in 100 people had it in 1960.

Why are so many people getting Type 2 diabetes? Scientific studies point to eating too many sugary products, meat, dairy products, and fatty foods as the problem. Obesity can also cause high blood pressure (hypertension), heart disease, strokes, gallbladder disease, and death.

In one medical study of 6,000 Americans, researchers found that obese people were 27% more likely to die at a younger than average age. An extremely obese person was twice as likely to die of some medical problem compared to the rest of the population.[14] Overweight conditions contribute to about 20% of American deaths.[15]

Not everyone who has extra weight has the problem of eating too much. But God's Word warns us several times about eating too much and drinking too much alcohol.

Hear, my son, and be wise;
And guide your heart in the way.
Do not mix with winebibbers,
Or with gluttonous eaters of meat;
For the drunkard and the glutton will come to poverty,
And drowsiness will clothe a man with rags. (Proverbs 23:19-21)

Have you found honey?
Eat only as much as you need,
Lest you be filled with it and vomit. (Proverbs 25:16)

## God's Plants Help to Heal

Not only are plants good for eating, but the Lord God has also purposefully made plants for medicine. Amazingly, scientists have explored the plant world and found medicinal uses for 35,000–70,000 different plants![16] Sometimes doctors will use the plant itself. At other times, companies will copy God's design from the plant and come up with their own medicines. Aspirin was developed in 1899 after a company called Bayer found the plant-based drug called salicin. This drug was discovered in the White Willow tree found in Europe. Most drugs prescribed by doctors don't come directly from plants. Synthetic drugs are processed in a laboratory, and they

## GOD MADE LIFE

are usually more powerful and designed to treat specific diseases or symptoms.

Herbal medicines are typically taken from the plants with very little processing on the part of companies that sell them. God has blessed us with some useful herbs from plants that have proven to help with some ailments.

What other helpful medicines will we uncover from the 400,000 plants growing in God's beautiful world? It is for us to discover new uses for His vast array of plants, all given to us for our benefit. Man brought disease and death into the world when he sinned against God. What tremendous mercy that He should still provide us with so many medicinal plants to help us in our pain and suffering!

And he showed me a pure river of water of life, clear as crystal, proceeding from the throne of God and of the Lamb. In the middle of its street, and on either side of the river, was the tree of life, which

| Herb | Plant | Usefulness |
|---|---|---|
| Echinacea | Coneflower | Used to treat wounds, burns, toothaches, sore throats, and upset stomachs, and to prevent colds. |
| Ginseng | Ginseng | Used to improve immunity and reduce inflammation. Helps brain function and energy levels. |
| Gingko Biloba | Maidenhair tree | Used to treat heart disease, dementia, and mental problems. |
| Elderberry | Elderberry tree | Used for headaches, nerve pain, toothaches, colds, and viral infections. |
| St. John's Wort | St. John's Wort | Helps wounds to heal, insomnia, depression, and various kidney and lung diseases. |
| Ginger | Ginger | Used to treat colds, migraines, nausea, and high blood pressure. |
| Turmeric | Turmeric | Used to treat inflammation, arthritis pain, metabolic syndrome, and anxiety. |

## CHAPTER 6: GOD MADE FOOD

Elderberry

bore twelve fruits, each tree yielding its fruit every month. The leaves of the tree were for the healing of the nations. (Revelation 22:1-2)

## God Made Tobacco and Drugs

God made tobacco, opium, and other drugs as well as edible fruits and vegetables. Some of God's plants are more poisonous than others for the human body. We are not supposed to eat or drink poison, but often humans will consume these substances anyway. Tobacco is one such substance. Christians did not smoke or chew tobacco until the 1500s. That is when the non-Christian tribes of South America introduced the habit of using tobacco to the Europeans. For many hundreds of years, the witchdoctors would use tobacco in their false religious worship. The European explorers tried the tobacco and found that it produced "a kind of peaceful drunkenness." Soon, the Europeans began cultivating fields of tobacco and sending the product back to Europe. It wasn't long before the whole world was addicted to the dangerous drug.

All things are lawful for me, but all things are not helpful. All things are lawful for me, but I will not be brought under the power of any. (1 Corinthians 6:12)

Tobacco is very addictive, which means that many people come under the control of it. They have to smoke throughout the whole day, every day. And they have a hard time quitting the habit.

About 13% of deaths each year are caused by smoking and tobacco use. Each year, about 7 million people die of smoking-related diseases. Compare that to the 1.4 million people who die in automobile accidents. Only 14% of adults smoke. That would mean smoking is about 30 times riskier than driving a car. The deadliest diseases which come from smoking are lung cancer, COPD, heart disease, and strokes.

Fresh turmeric

## GOD MADE LIFE

Cigarettes contain tobacco

The Word of God condemns all forms of murder. To hasten your death intentionally is to disobey the sixth commandment. If you knew that a certain poison contributes to the deaths of 20% of the people who die each year, and if it is 30 times more dangerous than driving a car, why would you put that poison into your body?

"You shall not murder." (Exodus 20:13)

Do not be overly wicked,
Nor be foolish:
Why should you die before your time?
(Ecclesiastes 7:17)

When people get very sick and are on the brink of death, they need something to reduce their pain. We learn from Proverbs 31:6 that God provided strong drink (alcohol) as a painkiller for the dying. In ancient Greece, **morphine** was taken from the opium plant and used by doctors to help patients in extreme pain. The opium plant also produces a less powerful drug called **codeine**. However, people will abuse these drugs and use them for recreational purposes. In 1905, the U.S. Congress banned the sale of opium. This resulted in the illegal sale of these drugs. Eastern countries (mostly Afghanistan and Myanmar) export about 2,500 tons of these illegal drugs each year.

The opium plant is not evil. God has created it for a purpose. Blenders are wonderful tools made for a good purpose. But, you wouldn't want to use a blender to clip your fingernails. Sticking your hand in a blender would result in a terrible tragedy. And so, God provided painkillers for people who are in severe pain or for those who are about to die. This is what we read in Proverbs 31:

It is not for kings, O Lemuel,
It is not for kings to drink wine,
Nor for princes intoxicating drink;
Lest they drink and forget the law,
And pervert the justice of all the afflicted.
Give strong drink to him who is perishing,
And wine to those who are bitter of heart.
(Proverbs 31:4-6)

When people use drugs or "strong drink" for other purposes, they will hurt themselves. They bring about their own deaths more quickly. They violate God's law by killing themselves. They become addicted to drugs. They open themselves up to evil spiritual influences. They also destroy family relationships. Sometimes they take up the life of crime and then die a terrible death. This is the warning of Proverbs 23:

Who has woe?
Who has sorrow?
Who has contentions?
Who has complaints?
Who has wounds without cause?
Who has redness of eyes?
Those who linger long at the wine,
Those who go in search of mixed wine.
Do not look on the wine when it is red,
When it sparkles in the cup,
When it swirls around smoothly;
At the last it bites like a serpent,
And stings like a viper. (Proverbs 23:29-32)

The problem of drug abuse is much more common today than it was in the 1950s. Deaths caused by drug overdose (when people take too much at one time) have increased from 1 in 100,000 persons in 1950 to 21 in 100,000 persons in 2020. [17]

When people turn to drugs for comfort and their own destruction, they have lost hope. Now more than ever, the world needs the Gospel of Jesus Christ. He has come to save us from sin and death. No matter how painful life can be, we can have hope that we are going to heaven. We know that all the forces of evil are overcome by our Lord Jesus Christ.

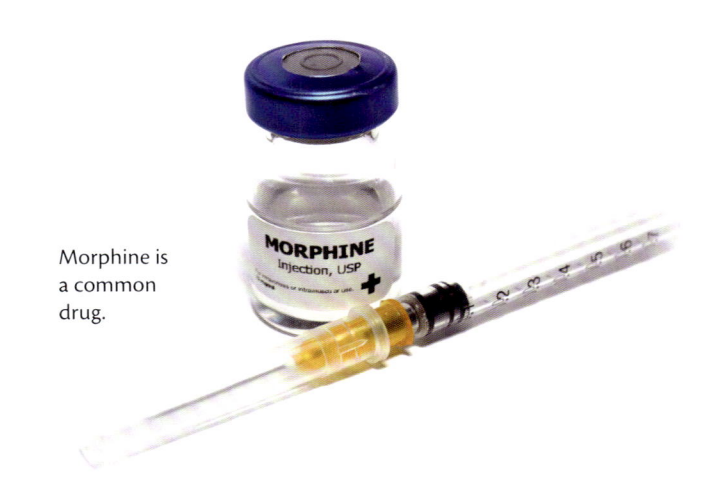

Morphine is a common drug.

## The Most Important Principles for Eating

Receive one who is weak in the faith, but not to disputes over doubtful things. For one believes he may eat all things, but he who is weak eats only vegetables. Let not him who eats despise him who does not eat, and let not him who does not eat judge him who eats; for God has received him. . . . He who eats, eats to the Lord, for he gives God thanks; and he who does not eat, to the Lord he does not eat, and gives God thanks. (Romans 14:1-3, 6)

Some people eat only vegetables. They sometimes refer to themselves as "vegetarians" or "vegans." Other people eat beef and poultry. Some people need more protein, and some need more fat. Some people have larger bodies, and they need to eat more. Some people burn their food more quickly

CHAPTER 6: GOD MADE FOOD

than others. Some people work in fields, requiring more physical exertion. Others work desk jobs where they type on computers all day. There is no one-size-fits-all diet for everybody. And so, we must be careful not to argue over diets in our church community.

The most important question to ask is this: "Are we giving God thanks?" Those who idolize food and those who are addicted to alcohol are not giving thanks to the true and living God. They are ignoring Him. They worship the gift more than the Giver.

Also, Romans 14 encourages us to eat in faith. As we eat the bacon, lettuce, and tomato sandwich, will we receive it in faith as a gift from God? Do we believe this will bless our bodies with health and strength?

> But he who doubts is condemned if he eats, because he does not eat from faith; for whatever is not from faith is sin. (Romans 14:23)

## A Balanced Diet

God created the body to use carbohydrates, proteins, and fats for energy. And, at the same time, He provided food containing these important components.

Carbohydrates are turned into glucose in your body's cells for an immediate source of energy. The only drawback with carbs is that the body can't store any more than it is able to use in a day or two. However, God

## CHAPTER 6: GOD MADE FOOD

made the body such that it can store up fats for future use. So, if you are stranded in the woods with nothing to eat, your body can live off the fat for a month or two. In fact, over half of the body's energy comes from fats. While proteins are used for energy, their first job is to make hormones, muscles, and other proteins the body needs.

*Oatmeal for breakfast*

Since God made your body and gave it to you to take care of, you should feed your body a balanced diet. Every day, you should try to eat a good mix of carbs, fats, and proteins. Although everybody's needs are slightly different, the table on the right provides a good guideline for children 9-13 years of age.

Keep in mind that foods containing high amounts of protein include eggs, almonds, meat, milk, Greek yogurt, broccoli, Brussel sprouts, and lentils. Foods containing high amounts of fat include avocados, cheese, butter, eggs, nuts, olive oil, regular yogurt,

## A Good Balanced Diet — Guidelines for Daily Food Intake

| Food | Girls | Boys |
|---|---|---|
| Proteins | 4-6 ounces | 5-6.5 ounces |
| Fruits (Carbs) | 1.5-2 cups | 1.5-2 cups |
| Vegetables (Carbs) | 1.5-3 cups | 2-3.5 cups |
| Grains (Carbs) | 5-7 ounces | 5-9 ounces |
| Dairy (Fats/Proteins) | 3 cups | 3 cups |

and meats. And foods containing a large number of carbohydrates include breads, sugary foods, fruits, grains, oatmeal, and sweet potatoes.

## Growing Food

*Then the LORD God took the man and put him in the garden of Eden to tend and keep it. (Genesis 2:15)*

When God first created man, He put him in a garden and gave him the charge to keep it. Of the 1.9 billion acres in America, people live on 133 million acres, and farms grow food on 349 million acres. That leaves about

# GOD MADE LIFE

1.4 billion acres of forests and rangeland, all undeveloped and mostly unfenced. Now, let's compare that to the country of Mexico. About 12% of Mexico's land is good for growing crops, but only about 1% is farmed. Another 33% of the land is forested. Mexico has to import about half its food from other countries.[18] The U.S. only imports about 15% of its food. Much of what grows in Mexico and other places around the world is just weeds and useless bushes. God grows His grasses and trees all over the earth. But He still wants us to tend the land, plant seeds, and harvest crops as He instructed Adam. There is much work yet to be done to make good use of the land.

## What Are You Doing with Your Yard?

God has given most of us a little bit of ground to take care of. It may be a small yard or large acreage. Some yards are meant to be pretty, where the owners grow grassy lawns and flower beds. But, some yards actually produce food for consumption. A man named Greg Peterson turned his little yard in Phoenix, Arizona, into a fruitful garden. He planted 70 fruit trees and grew all the herbs and vegetables he needed each year on a third of an acre in the desert city. And he used natural rainwater and gray water (bath water) from his house to water his garden.

Weeding

## The Things that Ruin Your Farm and Garden

Then to Adam [the LORD God] said, "Because you have heeded the voice of your wife, and have eaten from the tree of which I commanded you, saying, 'You shall not eat of it':
"Cursed is the ground for your sake;
In toil you shall eat of it
All the days of your life.
Both thorns and thistles it shall bring forth for you,
And you shall eat the herb of the field.
In the sweat of your face you shall eat bread
Till you return to the ground,
For out of it you were taken;
For dust you are,
And to dust you shall return."
(Genesis 3:17-19)

When man fell into sin, God cursed the dirt, and He told Adam he would have to sweat and toil to bring forth food from the ground. Many thorns and thistles would grow, which means Adam would have to pull a lot of weeds.

There are many challenges to gardening. Diseases and bugs will try to destroy the plants. Birds and other animals will try to eat from the garden. Weeds will choke out the plants. A **weed** is any unwanted plant that steals nutrition and water from the good plants.

A new batch of weeds can poke through the soil within 5-7 days, so you have to be consistent with weeding your garden. There are several ways to control weeds:

1. Pull the weeds. In some cases, you can pull the weed out by its roots. But you need to be careful not to sever the roots

# GOD MADE LIFE

of one of your garden plants. It may be safer to use a hoe to scrape the weed off at ground level. Weeding with a hoe also takes less time and is easier than bending over.

2. Pour boiling water over them if the weeds are far enough away from your garden plants.
3. Place breathable landscaping fabric around the plants so weeds can't see the light of day.
4. Overpower the weeds with good plants. If the weeds are shaded by thick leaves from the rest of the plants, there won't be much of an opportunity for weeds to grow.

---

What the chewing locust left,
the swarming locust has eaten;
What the swarming locust left, the crawling locust has eaten;
And what the crawling locust left, the consuming locust has eaten. . . .

He has laid waste My vine,
And ruined My fig tree;
He has stripped it bare and thrown it away;
Its branches are made white. . . .
The field is wasted,
The land mourns;
For the grain is ruined,
The new wine is dried up,
The oil fails.
Be ashamed, you farmers,
Wail, you vinedressers,
For the wheat and the barley;
Because the harvest of the field has perished.
The vine has dried up,
And the fig tree has withered;
The pomegranate tree,
The palm tree also,
And the apple tree—
All the trees of the field are withered;
Surely joy has withered away from the sons of men. (Joel 1:4, 7, 10-12)

Aphids

There are five different kinds of pests active in gardens and on farms:

1. Deer, birds, and other varmints.
2. Insects
3. Mites
4. Nematodes (roundworms or eelworms)
5. Mollusks (snails)

Bugs are a serious threat to fruit-bearing plants. The most common pests are locusts (grasshoppers), Japanese beetles, aphids, whiteflies, the Colorado potato beetle, stink bugs, the spider mite, the diamondback moth, the red flower beetle, the fall armyworm, and the brown planthopper.

Most farms use man-made fertilizers and pesticides to control pests and provide for a healthy crop. Sometimes, farmers will try to minimize the use of these chemicals by introducing predatory insects that hunt down and eat the bad insects. But plant diseases (mostly fungi) cause three times more damage than these pests for farms around the world. To take care of these issues, you can remove infected leaves and plants, and, thereby protect the rest of the crop. You can put them in the garbage, burn them, or bury them in a hole at least one foot deep. Also, some farmers use fungicides to kill the destructive fungi. Here are some homemade **fungicides** you could try:

- Mix a 50:50 solution of milk and water. Use a spray bottle to apply it to the plants in direct sunlight every 10 to 14 days.

- Mix 1 tablespoon of apple cider vinegar with 4 cups of water. Use a spray bottle to apply it to the plants during a cool time of day. Do not apply this solution in direct sunlight. Alternate between the milk solution and the apple cider vinegar solution to prevent the development of fungal resistance when treating powdery mildew.

- Mix 1 tablespoon baking soda, 1 tablespoon horticultural oil, and 1 to 2 drops of dishwashing liquid to make a natural fungicide.

The costs are pretty minimal for growing a garden. A pack of seeds will cost between $1-$5. Apple tree seedlings cost $12-$20. You may have to fence your garden to keep the deer and other animals out. Or you may need to protect the plants from birds using netting. The main thing you need to grow a successful garden is hard work. The farmer who plants his seeds and doesn't do anything else won't have any fruit at harvest time.

Harvesting fresh tomatoes

## GOD MADE LIFE

Plants need the right minerals in the soil, the right amount of sunshine, the right amount of water, and protection from bugs and diseases to thrive. Before planting a garden or fruit trees, you need to do some research. Every neighborhood and every region presents different challenges and advantages. There are usually successful farmers who have experimented with the soil and found the best way to grow crops in their area. Visit successful farmers and gardeners and ask for advice. Take careful notes and ask as many questions as possible. Here are some important things you need to know:

1. What fruits and vegetables grow best in this area?
2. What is the best way to till the soil?
3. Do you need to supplement the soil with peat moss or compost? Are wood chips helpful?
4. How do you protect your plants from bugs?
5. How much water do the plants need?
6. When is the best time to plant the seeds?

## Organic Food or Regular Food

Since the 1980s, people from all around the world have preferred food from **organic farms** over food from conventional farms. By 2018, organic foods made up about 6% of the food sold in U.S. grocery stores. Organic farms use "natural products" to control pests, and they will use natural fertilizer (like manure and compost mixtures) for the plants. Produce can officially be labeled organic if it has grown in soil that has not had man-made fertilizers and pesticides added to it for three years.

Organic produce

It is hard to know for sure if eating food grown on organic farms will actually result in better health and longer life. Hundreds of studies have been done on organic foods grown on organic farms. Organic fruits and vegetables sometimes contain a few more vitamins than the farms that use man-made pesticides. However, organic pesticides can be just as dangerous for your health as regular pesticides if consumed. So you must still wash all fruits and vegetables before eating them.

Pesticides are dangerous. Medical studies have confirmed that farmers, pesticide manufacturing workers, and golf course maintenance personnel are more likely to get certain kinds of cancer.[19] That's probably

## CHAPTER 6: GOD MADE FOOD

because they have regular contact with pesticides. About 60% of both man-made and natural pesticides have caused cancer in rats. But the laboratory rats in the experiments are forced to eat large amounts—more than a human would ever eat. The average person eats only about 0.2% of the allowable daily limit of these pesticides. [20]

## Steward the Soil

Soil

"Speak to the children of Israel, and say to them: 'When you come into the land which I give you, then the land shall keep a sabbath to the LORD. Six years you shall sow your field, and six years you shall prune your vineyard, and gather its fruit; but in the seventh year there shall be a sabbath of solemn rest for the land, a sabbath to the LORD. You shall neither sow your field nor prune your vineyard. What grows of its own accord of your harvest you shall not reap, nor gather the grapes of your untended vine, for it is a year of rest for the land.'" (Leviticus 25:2-5)

The **soil** is your most precious resource provided by God for producing crops. In the Old Testament, the Lord required His people to let the land rest every seven years. This was a good principle for stewarding the land. If farmers get too greedy, they will work the land too hard and try to make as much money as possible. But, the land needs rest too. Letting the land rest for a year can restore nutrients, improve the soil's ability to hold moisture, and return healthy microorganisms to the soil. The farmer that allows his fields to lie fallow once every 2-7 years will usually produce more crops during the years he plants.

**Crop rotation** is one more strategy for **sustainable farming**. This means that every 3-5 years, the farmer changes the types of plants he grows. He may grow corn for two years and then switch to soybeans for a year or two. This helps to reduce insect problems. Insects that like to eat corn will be disappointed when the farmer switches to soybeans, and they will die off or move somewhere else. Rotating crops also helps with weed control, improving soil health, and crop yields. Sometimes farmers switch from a shallow-rooted grass (like wheat) to a deep-rooted broadleaf (like sunflowers). This improves drainage for the soil and pulls nutrients up from deeper down in the soil. Farmers will also rotate heavy nitrogen feeders like corn with plants in the legume

# GOD MADE LIFE

family. These bean and pea plants are able to take nitrogen from the air and replace what the corn has removed from the soil.

Corn

## Seedless Fruits and Strange Experiments

"You shall not sow your vineyard with different kinds of seed, lest the yield of the seed which you have sown and the fruit of your vineyard be defiled." (Deuteronomy 22:9)

Beginning in the 1970s, scientists began to change the DNA of certain plants like tobacco, corn, soybeans, and potatoes. These researchers would take a gene from a different plant and insert it into the food crop—thus mixing different kinds of plants. These

Harvesting tons of wheat

actions seemed to violate God's command found in Deuteronomy 22:9. Could it be that God wanted His seeds to remain unmixed with other kinds for good reason?

Genetically modified seeds have increased yields quite a bit and brought down the price of some crops like corn by 30%. America, Canada, Argentina, and Brazil grow a lot of this GMO food, but other nations including Kenya, China, and the European Union have banned much of it.

## What Does This Mean for You and Me?

God wants us to rule over His creation, and our first job is to take good care of the fields and gardens. This doesn't mean that everybody should become a farmer when they grow up. Yet, we should take good care of the yards around our own homes. We all must be interested in how food is grown and how much food is grown. We all want to make sure we are eating healthy food. Foods can lose their nutrients within 24 hours of harvesting. That is why local foods are often better for you than foods transported from other countries. Every country should be concerned about food first. After that, nations can take care of their military, mining, and the manufacturing of cars and appliances.

Farming also teaches us to trust in God. He sends the rain. He provides the sun-

## GOD MADE LIFE

light. He controls the bugs. We plant seeds, "but God. . . gives the increase." We have no control over how many tomatoes will grow on the vine. There may be 4 tomatoes or there may be 40. God gives us the harvest. People living in the city sometimes think that food comes from the grocery store. So many people today get their money from the government or the big corporation, and they purchase food at the store. Then, they forget about God, and they don't give Him thanks. Farmers should know better. Ultimately, our food comes from God. Let us always remember that our food comes by God's blessings. Then we will be ready to give Him thanks each day with warm hearts of gratitude.

God is very interested in how we treat His plants and animals. He doesn't want us ruining the soil because He wants more plants to grow in that soil next year. The best ways to maintain healthy soil are:

- Minimize tilling (turning over the soil).
- Keep enough organic material mixed into the soil and limit the addition of inorganic material. Tilling dead plants that were free of disease back into the soil can improve it. Aged animal waste products are also usually very good for it.
- Provide enough water.

Collecting God's rich bounty of peaches

Harvesting God's gift of pumpkins

- Keep erosion down. Wind and water can remove the topsoil, leaving very little room for plants to grow deep roots.
- Provide mulch to help prevent erosion and to aid with shade and moisture retention. Mulch will keep the soil cool and moist.

Also, let us remember that the earth is the Lord's and the fullness thereof. We are His stewards tasked with taking care of His land. He is the Creator, the Designer, and the Owner of all this creation. By studying the soil, the plants already growing there, and the nature of the plants, we can copy His method of growing things. He planted the first Garden for Adam, and He wants us to plant more gardens. The goal is to create more beauty, fruitfulness, and sustainability. Our goal is to continue planting and producing each year. That's sustainability. We want a balance between getting a good yield out of the garden and meeting the needs of the land so that it can keep producing in the following years. Above all, we should pray to God for wisdom. How can we be good stewards of His land? How can we get a good crop this year and provide for sustainability next year? Farming and gardening require much wisdom and hard work—and always a spirit of gratitude to God for His good gifts.

183

## Pray

- Thank the Lord for His rich provision of food and the varieties He made to nourish our bodies.
- Thank God for the thousands of plants from which we can get medicines to help the body.
- Praise Him for the disease-preventing vitamins and minerals available in foods.
- Pray for wisdom and self-control when it comes to the foods you eat.
- Pray for wisdom to properly care for the yard and the garden God has given your family.

## Sing

Having again witnessed the marvelous work of God in this wonderful creation of plants, the appropriate response is more worship and praise. If the student is unfamiliar with the hymn or psalm, some version of it is available on the internet and may be accessed (with supervision) for singing along.

*Great Is Thy Faithfulness*
"Great is Thy faithfulness," O God my Father,
There is no shadow of turning with Thee;
Thou changest not, Thy compassions, they fail not;
As Thou hast been Thou forever wilt be.

*Refrain:*
"Great is Thy faithfulness!
Great is Thy faithfulness!"
Morning by morning new mercies I see.
All I have needed Thy hand hath provided—
"Great is Thy faithfulness," Lord, unto me!

Summer and winter, and springtime and harvest,
Sun, moon, and stars in their courses above,
Join with all nature in manifold witness
To Thy great faithfulness, mercy, and love.

Pardon for sin and a peace that endureth,
Thine own dear presence to cheer and to guide;
Strength for today and bright hope for tomorrow,
Blessings all mine, with ten thousand beside!

CHAPTER 6: GOD MADE FOOD

## Do

Choose at least one of the following activities and apply the lessons you have learned in this chapter.

1. Are you feeding your body a healthy diet? Examine your food intake for a day and compare it to the suggested balanced diet for 9-13-year-old children laid out in this chapter. You will need a scale and a measuring cup to record the amount of food you eat in a day.

   |  | Morning | Lunch | Supper/Evening |
   |---|---|---|---|
   | Proteins (ounces) | _____ | _____ | _____ |
   | Fruits (cups) | _____ | _____ | _____ |
   | Vegetables (cups) | _____ | _____ | _____ |
   | Grains (ounces) | _____ | _____ | _____ |
   | Dairy (cups) | _____ | _____ | _____ |

   How does your daily diet compare with the guidelines provided in the chapter?
   Based on your body and your lifestyle—such as the amount of activity you are involved in or the kinds of chores you do daily—do you tend to need more or less of certain nutrients, vitamins, or minerals compared to the guidelines provided? Are there any adjustments you might need to make in your eating habits to keep your body healthy?

2. Research your own country. What percentage of the food consumed is grown within your country? What are the foods imported into your country? What crops grow best in your country? Are there other crops which may be introduced to your country — crops which your country is now importing from other countries?

## Watch

To watch the recommended videos for this chapter, go to **generations.org/GodMadeLife** and scroll down until you find the video links for Chapter 6. Our editors have been careful to avoid films with references to evolution; however we would still encourage parents or teachers to provide oversight for all internet usage. These videos may not give God the glory for His amazing creative work, so the student and parent/teacher should respond to these insights with prayer and praise.

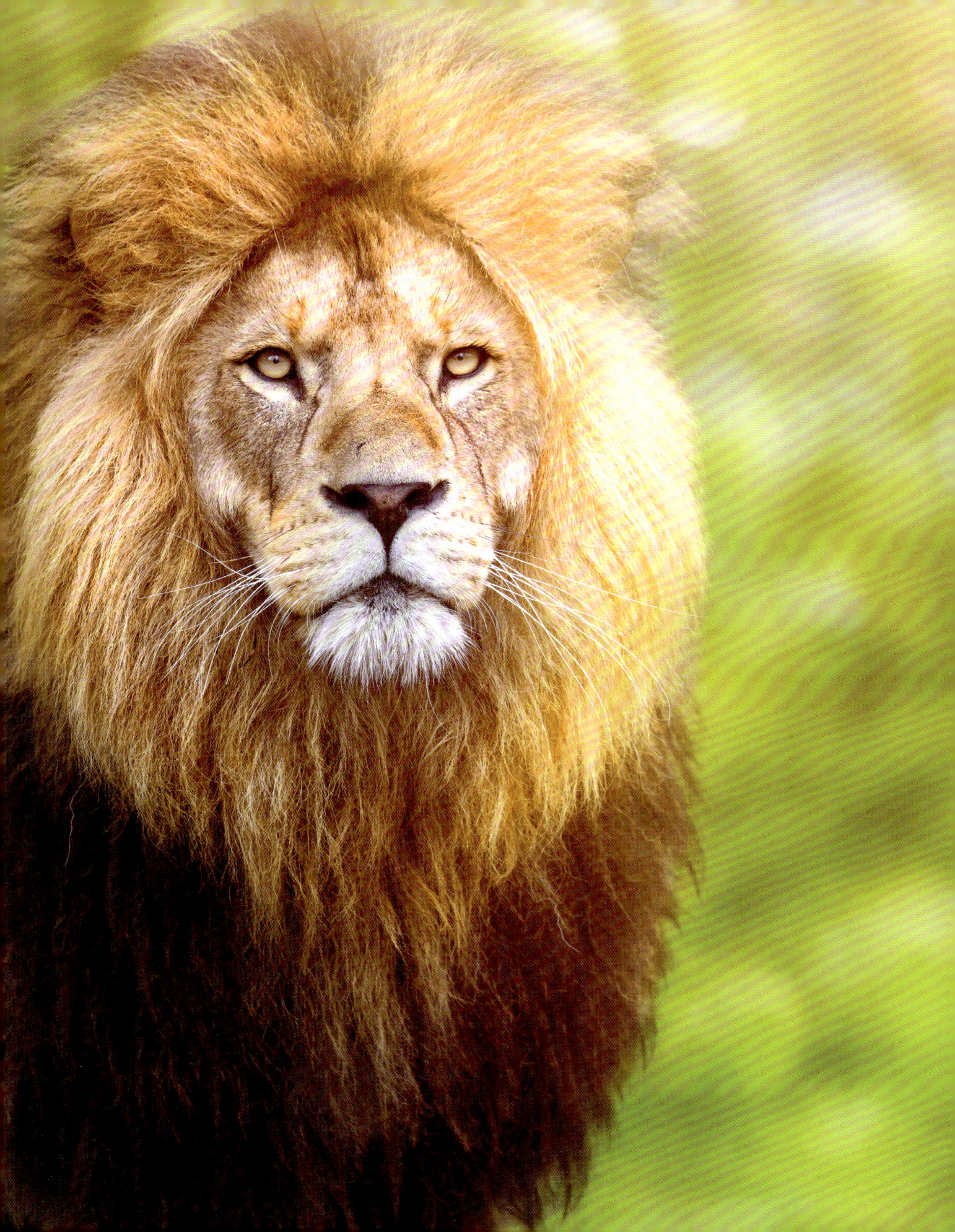

# CHAPTER 7
# GOD MADE ANIMALS

*"Are not two sparrows sold for a copper coin? And not one of them falls to the ground apart from your Father's will." (Matthew 10:29)*

Typically, biologists specify six kingdoms in their studies of life forms—animals, plants, protists, fungi, Archaebacteria, and Eubacteria. Thus far in this study, we have looked at four kingdoms.

We do not agree with the division of organisms into six kingdoms. From a Biblical perspective, it would be better to divide God's amazing creation of organisms into eight kingdoms. First, humans must be kept distinct from the animal creation because God made Adam in His image. And secondly, God's Word makes a distinction between organisms with the breath of life and those without it. In this study, we will separate those animals with the breath of life from those without the breath of life.

## Eight Kingdoms

- Humans
- Animals with the Breath of Life
- Animals without the Breath of Life
- Plants
- Protists
- Fungi
- Archaebacteria
- Eubacteria

Let us not forget that the most honorable living creatures are angels, who would make a ninth kingdom (Psalm 8:5). However, angels do not have material bodies. So they cannot be studied the way we can study the rest of creation.

God created the world with varying levels of complexity. As this study shifts from protists and plants to the animal creation, you will see more of His wisdom and genius. He created all this variety—from the simple to the complex—in six, 24-hour days. This is the clear teaching of Exodus 20:11:

> "For in six days the LORD made the heavens and the earth, the sea, and all that is in them, and rested the seventh day. Therefore the LORD blessed the Sabbath day and hallowed it."

Evolutionists think that simple animals turned into complicated animals over many generations. This came about by billions of mutations, they say. But there is no evidence of this gradual process. If a fish turns into a bird over a billion years, where are all the half-fish, half-bird creatures in the fossil record? If a worm turns into a fish over a billion years, where are the millions of half-worm, half-fish creatures in the fossil record? The fossil record is full of worms, sea urchins, shellfish, fish, insects, and birds, but no intermediate creatures. We still find live worms, sea urchins, shellfish, fish, insects, and birds all over the world today. The fossils we find are all fossilized creatures mostly killed and buried in rock layers during the worldwide flood. The evolutionary explanation for the appearance of all the different animals is a myth or a fairy tale. There is no scientific basis for it. But more importantly, God tells us He made the world in six days.

# CHAPTER 7: GOD MADE ANIMALS

This is the biggest reason we cannot believe the evolutionists.

Mammals give birth to babies, and they nurse them. These animals are more complex and thus are part of a higher life form. God made some mammals to live on land, and He made some mammals such as whales, seals, and dolphins to live in the sea. Evolutionists want to believe that land mammals evolved into sea mammals over a hundred million years. But how would a cow evolve into a sea creature? How could a half-cow, half-whale creature survive? If her tail had grown into large flukes, and her legs turned into flippers, how would she have survived on land? She still couldn't walk, but she wouldn't be able to swim very well either. How would a half-cow animal be able to nurse her babies under the water? How could she breathe unless her nostrils were replaced by some inlet for oxygen positioned at the top of her body? The whole idea is silly. From the beginning, God made whales capable of living in sea water, birthing baby whales, and nursing them in the sea. A cow or some other land mammal did not gradually turn into a whale by some strange process.

## Animals without the Breath of Life

Since insects do not inhale the breath of life using lungs, we do not include them in the same category as the rest of the animal cre-

Dolphin

## GOD MADE LIFE

ation. There are several categories (phyla) of animals that do not inhale oxygen using lungs.

- Phylum Porifera (sponges)
- Phylum Cnidaria (jellyfish and sea anemones)
- Phylum Echinodermata (sea stars)
- Phylum Platyhelminthes (flatworms and tapeworms)
- Phylum Mollusca (octopi, clams, and squids)
- Phylum Arthropoda (crabs, insects, spiders)
- Part of phylum Chordate (fish, sharks, and rays)

Another way to organize the animal creation is by separating those with backbones and those without backbones. Insects, crabs, fall worms, jellyfish, and sponges don't have backbones, and they are called **invertebrates**. With a few exceptions, all animals with hearts, lungs, and blood flow have backbones. These are called **vertebrates**. The exceptions to this rule are fish, sharks, and rays, all of which are vertebrates, but they do not have hearts and lungs. Of all the invertebrates, only the land snail has lungs and breathes using a pore.

About 90% of the animals crawling around the earth are invertebrates. We've identified at least 1.25 million different kinds of invertebrates on the earth. There could be as many as 30 million varieties. No doubt there are many more of God's creatures to discover in the years to come.

Most invertebrates have an **exoskeleton**—a hard, external casing that keeps the animal together. The cricket's crunchy skin and the snail's shell are examples of exoskeletons. God decided to make these exoskeletons out of different materials such as protein, glass, limestone, and water.

Jellyfish (example of an invertebrate) and giraffes (example of vertebrates)

## How Invertebrates Differ from Plants

*Then God said, "Let the waters abound with an abundance of living creatures, and let birds fly above the earth across the face of the firmament of the heavens." So God created great sea creatures and every living thing that moves, with which the waters abounded, according to their kind...* (Genesis 1:20-21)

The genius of our Creator God is seen in the **animate creation**. These are the creatures that move. While we don't find trees and plants walking around, insects and jellyfish are made to move. Their main goal is to find food, so they are always on the lookout for something to eat. The only invertebrates that cannot move are sponges and corals, yet they still find ways to collect food. A plant makes its own food, while animals have to collect their food from their environment.

Both plant and animal cells contain a nucleus, DNA, and mitochondria. However, some of their organelles are different. For example, plant cells have chloroplasts whose job it is to make food via photosynthesis. Also, plant cells are equipped with an additional cell wall that surrounds the cell membrane. So, if you look at plant cells under a microscope, they will look rectangular.

Plants can respond to light and touch, but animals are much more sensitive, responding very quickly to outside stimuli. With that said, we are still amazed at the wisdom of God's plant creation! For instance, cicadas cut small slits in trees in which to lay their eggs. To defend themselves, the trees grow calluses around the eggs which are supposed to crush the eggs before they hatch. Should something touch the *Mimosa pudica* plant, the leaves fold in on themselves and begin to droop. This would make the plant appear dead and unappetizing to the predator. Corn plants release volatile organic compounds when they are under attack. Some scientists think this is a distress call intended to attract wasps. When other plants or trees intrude on the space of the black walnut tree, the roots of this tree will emit a toxin called juglone to kill off the intruder. All of these are impressive design features in plants, but animals are even more awe-inspiring.

The most impressive features of animals without the breath of life are movement, sense and response mechanisms, feeding methods, defense mechanisms, and reproduction. Before an animal moves, a series of events need to happen. Our wonder-

Mimosa pudica plant

## GOD MADE LIFE

ful Creator equipped these organisms with systems that all work together to bring about movement and thriving life:

1. The animal picks up stimuli (or messages) using its senses.
2. The animal processes the stimuli using its brain. Some invertebrates do not have brains, but God gave them bundles of nerves called ganglia that send and receive messages in response to the stimuli.
3. The animal's nervous system signals the muscles to act.
4. The animal's muscles stimulate movement.

There are millions of designs to study in the animal kingdom. God is very creative. It would be impossible to look at all this amazing variety in a short study like this.

## Movement

But now ask the beasts, and they will teach you;
And the birds of the air, and they will tell you;
Or speak to the earth, and it will teach you;
And the fish of the sea will explain to you.
Who among all these does not know
That the hand of the Lord has done this,
In whose hand is the life of every living thing,
And the breath of all mankind?
Does not the ear test words
And the mouth taste its food?
Wisdom is with aged men,
And with length of days, understanding.
With [God] are wisdom and strength,
He has counsel and understanding.
(Job 12:7-13)

How does a grasshopper jump and fly around? Animals use muscles to move. A muscle is made up of long, thin cells bundled together. When the nerve signals the muscle, proteins and other chemicals release energy to either contract or relax the muscle.

Earthworm

God created **sponges** as one of the few animals that can't move, but they can pump water. A host of little flagella are attached to the central cavity of the sponge. This causes the water to flow through the sponge, allowing collar cells to catch and digest the bacteria and algae passing through. They also absorb oxygen out of the water and kick out waste materials. The sponge is a very efficient organism, constantly feeding and fil-

Octopus

tering out waste products.

The Creator equipped the **jellyfish** with muscles around its gastrovascular cavity (or stomach) near the opening of the jellyfish's bell. These muscles squeeze down on the cavity, which contains seawater. When the muscles contract, it pushes the water out and propels the jellyfish along in slow, jerky movements.

Then, the **earthworm** moves using circular muscles and long muscles which run the length of its body. How would you pull yourself through the soil if you were a worm? The Lord came up with a super-creative, four-step process for the earthworm to slither through the dirt. First, it digs tiny hooks called setae located on its rear end into the soil. Then, it contracts its inner circular muscles. This stretches the front end of the worm, making it longer and thinner. Releasing the rear setae, the worm digs into the soil using the setae on its front end and contracts its long outer muscles. This pulls the worm forward. Suffice it to say, it takes a lot of mechanical functions for an earthworm to make it through soil. Praise God for such an innovative design!

The **octopus** usually moves using its eight arms. But when it really needs to scoot, this creature will shoot water out through a muscular tube called the **siphon**.

The **clam** uses powerful muscles to keep its shell closed, protecting itself from predators. To move through the sand, the clam extends a muscular foot. Then the "toes" on the foot expand into the sand, anchoring it. The foot muscle contracts, pulling its body forward an inch (2.5 cm) or so at a time.

You might be able to open a clamshell using a hammer and a chisel. But the **starfish** can do it with his bare feet! Actually, this little creature has tube feet at the bottom of each leg or ray. He uses these to grab on to surfaces so he can pull himself forward. But

## GOD MADE LIFE

the starfish can also grip onto a clamshell with these feet. He tugs away as the clam resists the attack for dear life. The contest can go on for several hours. When the starfish finally cracks open the shell a tiny bit, he enjoys his lunch. He turns his stomach inside out and pushes it through the crack. Digestive enzymes pour into the clam and consume the clam innards. Finally, the starfish retracts his stomach and continues the hunt for more food.

The strongest animal in the world turns out to be the **Rhinoceros Beetle**. This impressive, horned beetle grows to 6 inches (15 cm) in length and can lift as much as 100 times its own weight. If a human could lift 100 times his own weight, he could lift a school bus off the ground. In a laboratory experiment, one rhinoceros beetle carried 30 times its weight for over a mile (1.6 km). That would be akin to a human carrying a Cadillac on his back for a whole mile without getting tired! Scientists concluded that the incredible metabolic efficiency and strength of these little animals simply "cannot be explained" by typical biological theory.

Among arthropods, you will find a huge variety of moving mechanisms:

- **Insects** walk on six legs. And there are more than 1,000,000 different kinds of these creatures walking around the earth.
- **Arachnids** walk on eight legs. There are about 100,000 different kinds of these creatures, including spiders, scorpions, ticks, and mites.
- Centipedes and millipedes are made with 15-200 pairs of legs. These are called **myriapods**. We have identified 16,000 different kinds of these creatures so far.
- Crustaceans called **decapods** move around in the water or on land using 10 legs. There are 15,000 types of these creatures in the world, including lobsters, shrimp, crabs, and crayfish. Shrimp also have three pairs of feeding legs used to grab things when eating. Crustaceans like woodlice (pill bugs) have 7-18 pairs of legs. Usually, each segment of the body is equipped with a pair of legs.

## Flight

Does the hawk fly by your wisdom,
And spread its wings toward the south?
Does the eagle mount up at your command,
And make its nest on high? (Job 39:26-27)

*Rhinoceros beetle*

The technology required for flying is even more complex than that used when crawling, walking, or swimming. It took humans 5,900 years for him to figure out the Creator's flight technology. The Wright brothers would never have successfully pulled off the first self-propelled flight without studying birds and copying God's original design.

Besides birds, grasshoppers, cockroaches, bees, wasps, dragonflies, true bugs, butterflies, moths, and other insects are created to fly. And the various designs are awe-inspiring, to say the least.

The **Painted Lady butterfly** has been known to migrate a distance of 4,000 miles (6,000 km) by flight! Some butterflies fly at altitudes of 20,000 feet (6,100 m) above sea level.

Originally, God made the **dragonfly** with a humongous wingspan of more than two feet (61 cm). After the worldwide flood, these insects were much smaller. With an astounding speed of 35 mph (50 km/h), the dragonfly beats out the Hummingbird Moth as the fastest flying insect. No man-made airplane can do what the dragonfly can do. They can abruptly change direction in midair and can fly backwards or upside down. They can turn on a dime at sharp 90-degree angles. The Lord designed this flying machine with a long, tapered body to minimize drag from the wind. The delicate wings of the insect are also incredibly light but tough. Most insects have to move both pairs of wings together in concert, but the dragonfly's wings function independently. This acrobatic flyer can move each wing up or down, backward or forward, to fly wherever he wants to go. And, the dragonfly usually gets his food with his feet. One university study found that dragonflies successfully capture 90% of the fruit flies they try to catch.

Dragonfly

## Those Amazing God-given Senses

*The Lord is good to all,*
*And His tender mercies are over all His works.*
*All Your works shall praise You, O Lord,*
*And Your saints shall bless You.*
*(Psalm 145:9-10)*

Our God of all goodness has also equipped animals with ingenious methods of sensing their immediate surroundings. Humans have five well-tuned senses—sight, hearing, touch, smell, and taste. To this may be added some amazing, special senses the Lord decided to distribute to His animal creatures.

A bumblebee uses an electric charge to

find flowers in the garden. As the bee flies through the air, its body gains a small positive charge. Using what scientists think are **mechano-sensory hairs** on its legs, the bee can sense a negative electric charge contained in the flower.

The Creator also invented a high-tech device to help sharks find their food. An electricity-conducting jelly called Lorenzini is stuffed into pores on a shark's face. As the shark swims closer to its food, the animal begins to sense the electrically-charged body of its prey.

Roundworms can figure out which way is up to the surface of the earth by sensing the earth's magnetic field. Honeybees are also sensitive to the earth's magnetic field, but nobody really knows how they detect it at this point.

Earthworms can't see or hear, but their sensory receptors pick up light, certain chemicals in the soil, and temperature. If your full-time job is eating dirt and breaking up soil underground, there is no real need for eyesight. There is nothing much to see underground. The Creator knew what He was doing when He equipped each of His creatures for the job assigned to them.

## The Miracle of Sight

The hearing ear and the seeing eye,
The LORD has made them both.
(Proverbs 20:12)

Vision is the most versatile and helpful sense created by God. Using eyesight, we can see fine detail in color, shape, size, and distances all at once. We can pick up on very slight differences between the distances and appearances of objects and people. Sight helps us to quickly assess the safety of a situation, and sight helps to coordinate our movements. Think about how hard it would be to eat a meal while sitting in the dark.

If you look at a house fly under a microscope, you'll see a thousand eyes looking back at you. These are called **compound eyes**. When the bug looks at you, he sees your image on a thousand split screens in front of him. Since each lens is angled in a different direction, the fly has an extremely wide field of vision and can see something move in practically any direction. That's one reason it's so hard to kill a fly with your bare hands.

The **planarian** (flatworm) has two eyes, but it cannot see very well. Its eyes are only sensitive to light, which means the flatworm

Bumblebee

CHAPTER 7: GOD MADE ANIMALS

would notice if you turned on the lights in a dark room. So, the worm generally feeds at night.

The dragonfly has tremendous eyesight capability. Practically the whole head of this insect is made up of the eye, and he can see everything going on around him at the same time. Although this insect has six legs, he cannot walk. But, be aware that he can bite. The dragonfly can see ultraviolet light—something the human eye cannot see. Similar to the fly, the dragonfly sees a single image through 30,000 receptors. He can pick up any movement very quickly and respond to it—making this insect hard to catch.

The Lord our Creator gave the **mantis shrimp** eyes mounted on stalks, each one with the capability of moving independently. Each eye is given three segments, effectively giving the shrimp a view of the world through **trinoculars**. Their eyes can pick up 12 different wavelengths, whereas humans can only detect 3 (red, green, and blue). Also, this is the only creature on Earth that can detect both linear and circular polarized light underwater. Whereas the human brain processes a 3-D image from what the eyes pick up, the mantis shrimp's eyes can pick up the 3D image right away without having to process it. Don't forget that God is the author of these extraordinary designs. Such stunning creations are very much praiseworthy.

The **trilobite** has evolutionists stumped. This invertebrate is extinct, probably since the worldwide flood. But this creation of God possessed eyes more advanced than any

Trilobite—created with phenomenal eyesight

animal living today. Of course, the animal was not a lower creature that went extinct millions of years ago to make way for more advanced designs. Scientists have studied the lenses of the trilobites that have been preserved for thousands of years in the rocks. They found this animal that walked around on the seafloor had compound eyes like a dragonfly and special lenses. These lenses were made of calcite, maximizing light in

Mantis shrimp

197

# GOD MADE LIFE

the dark waters. No other creature has been so blessed with such clear sight. Each eye could see in three dimensions because the lenses bulged out and overlapped each other. The trilobite had no trouble focusing. The Creator gave him a double lens for each eye so he could see underwater without distortion. His ability to see things close up and far away was phenomenal. One evolutionist had a hard time believing this animal was 500 million years old. He admitted that, "With the 'space-age' geometric symmetry presented by each lens, these eyes seem to defy their incredibly ancient ancestry."

A few things need to happen if an insect or any other creature is going to discern images using sight:

1. There has to be some light shining on the objects to be seen. Nobody can see in the pitch dark. Some animals like the cat (especially big cats like lions, tigers, leopards, and jaguars) can see better in the dark. When you're riding in a car at night, sometimes you see two little cat eyes shining back at you. Cats have a reflective membrane designed to maximize the light entering the eye, helping them to see better at night. God made light so that we can see the world around us. What would life be like without light?

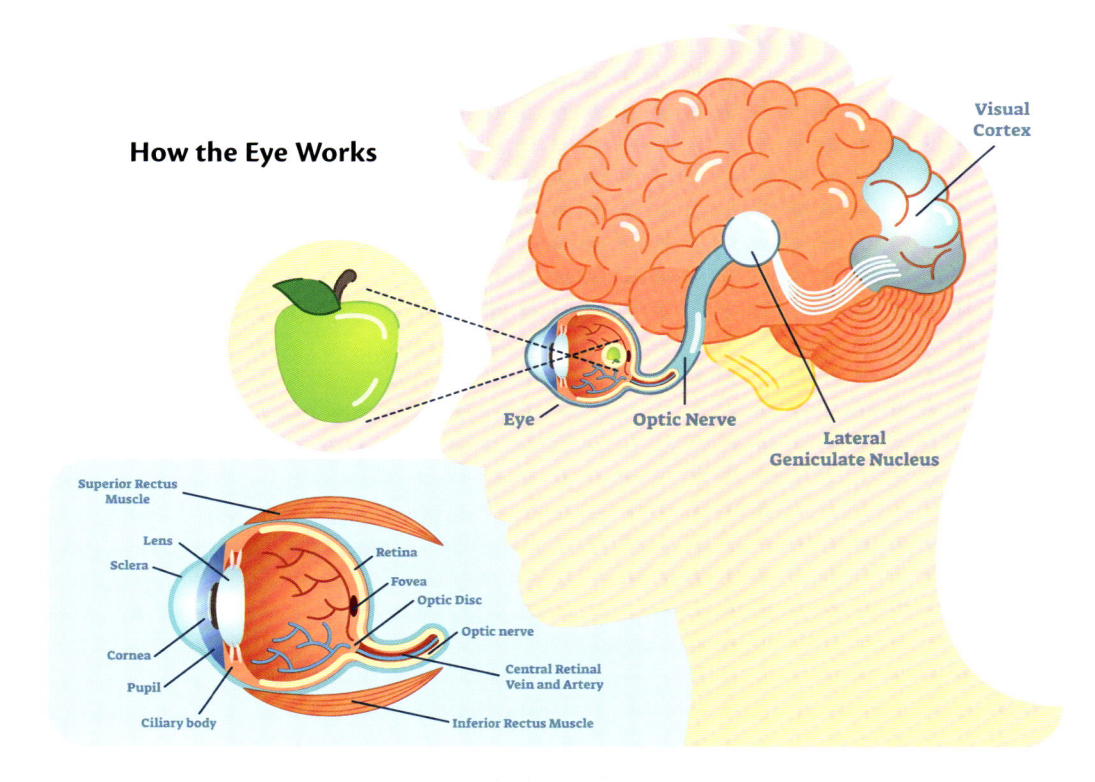

**How the Eye Works**

## CHAPTER 7: GOD MADE ANIMALS

> God is light and in Him there is no darkness at all. (1 John 1:5)

2. Many light rays will bounce off those things the eye is looking at and will enter the eye. First, the light passes through a lens in the front of the eye. Next, the lens focuses the light rays on the **retina** at the back of the eye. The retina is made up of millions of light-sensitive cells called **photoreceptors**. Nobody knows for sure how the photoreceptors work. Their job is to collect the data which will be transmitted to the brain (or the ganglia in some invertebrates).

3. This sensory information collected by the photoreceptors is sent to the ganglia (for insects) or the brain. For humans, the message is communicated through the **optic nerve** to the **occipital lobes** of the brain.

Once again these are examples of "irreducible complexity." These systems were created all at once. It would have been impossible for sight to have developed without everything working at the same time. The evolutionary theory that these complicated systems appeared over a long time by gradual processes is far-fetched, foolish, and completely unbelievable. God created the miracle of sight and gave this gift to animals and man alike.

The miracle of sight was witnessed again as Jesus healed blind men when He walked the earth 2,000 years ago.

> Then [Jesus] came to Bethsaida; and they brought a blind man to Him, and begged Him to touch him. So He took the blind man by the hand and led him out of the town. And when He had spit on his eyes and put His hands on him, He asked him if he saw anything.
> And he looked up and said, "I see men like trees, walking."
> Then He put His hands on his eyes again and made him look up. And he was restored and saw everyone clearly. (Mark 8:22-25)

## The Wonderful Sense of Hearing

> "But blessed are your eyes for they see, and your ears for they hear." (Matthew 13:16)

God uses different kinds of technology for hearing, but the human ear is a good example. Sound comes through the air in the form of waves. These waves hit the **eardrum** (or the **tympanic membrane**). This membrane starts to vibrate. The Lord attached a **hammer** on the other side of the membrane, and the vibration causes the hammer to move back and forth. Like a complicated machine, the hammer moves another bone called the **anvil**, which moves a third bone called the **stirrup**. Now, the stirrup presses up against the **cochlea** (KOHK-lee-uh), a tubular device curled up like a snail shell. It's filled up with fluid and sealed up with

## The Human Ear

**The Human Ear** diagram with labels: Malleus, Incus, Staples (attached to oval window), Vestibular Nerve, Cochlear Nerve, Cochlea, Round Window, Eustachian Tube, Tympanic Cavity, Tympanic Membrane, External Auditory Canal.

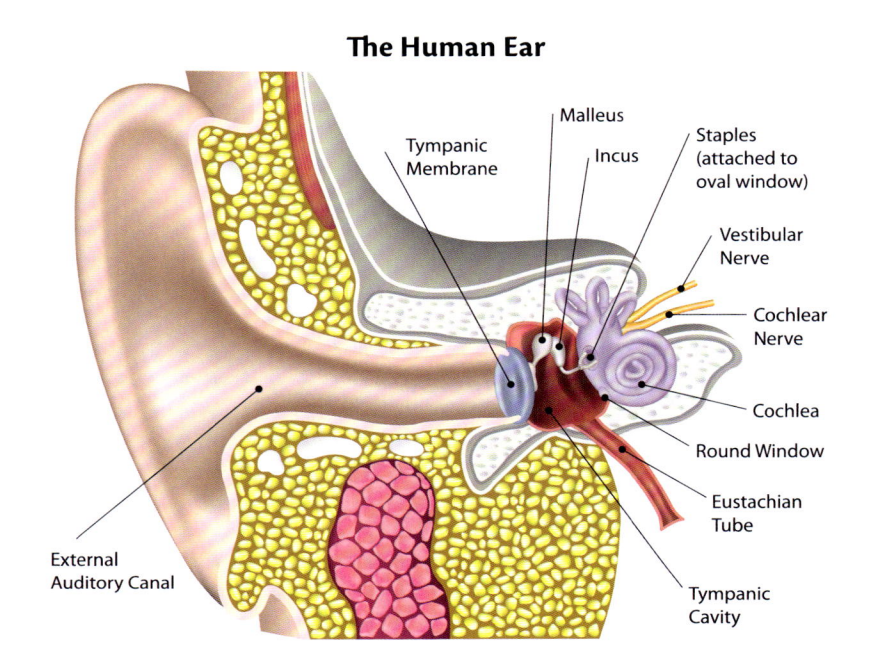

a membrane. When the stirrup vibrates against the membrane, waves ripple through the cochlea's fluid. These waves are picked up by little hair cells which bend one way or another as the fluid passes over them. They release neurotransmitters (little packages of electrical energy) which transmit messages to the brain through the neural pathways. This translates the sound into an electrical signal that's passed on to the brain. Finally, the brain interprets the message, and the human or animal reacts to the message. Is this a danger signal? Is that a predator over there, or is it prey? Is that the voice of a friend or foe?

As you can see, this is a very high-tech process designed by God Himself! This is all quite amazing. Have humans created any technology so finely tuned and complicated as the hearing ear? For some creatures, these ears continue to function over a lifespan of 100 years!

There is more to the wonderful creation of the ear than just sound. Our genius Creator attached **semicircular canals** to the cochlea which are also filled with fluid. He gave us these devices to equip us with a sense of balance. You can maintain your balance when walking in the dark because the fluids in your right and left ears slosh up against tiny hairy neuron cells. These cells are made to transfer information through the body. The neurons in the semicircular canals communicate the message about your body's position to the brain.

Have you ever been dizzy after riding

a merry-go-round? Even if your eyes were closed on the ride, you probably experienced dizziness. As you spin around on the merry-go-round, those fluids in your inner ear slosh against tiny strands of neurons. When you get off the merry-go-round, the fluid is still jiggling around. The neurons continue firing messages off to the brain, telling it that the body is still in motion. The brain is telling you that you are moving or that your surroundings are moving. You might even lose your balance and fall because your brain is misinterpreting the message from the juices sloshing around in the semicircular canals.

Every animal was given the ability to hear a certain range of **frequencies**. Some animals can't hear very high screams (high frequencies) or very low drumbeats (low frequencies).

Besides the five human senses, the Lord added a few more interesting senses for some of His creatures. The bat and the dolphin use **echolocation** to figure out where they are going. As the dolphin swims through the water, he emits squeaking sounds. These sound waves vibrate through the water and then bounce off surfaces surrounding the dolphin. Amazingly, this animal picks up the returning sound vibrations using its teeth. They can pick up all kinds of sound vibrations using their teeth. Once again, the neurons behind the teeth send the message to the dolphin's brain, which will interpret the data.

The dolphin can hear screams and squeaks at frequencies that are seven times higher than those that humans can hear. They can hear sounds as low as 20 Hz and as high as 150 kHz. Compare that to humans, who can hear at a range from 20 Hz to 20 kHz. Dogs can hear sounds as high as 45 kHz, a range about double that of humans. But rats can hear sounds up to 90 kHz, and mother rats talk to their young at frequencies around 50 kHz. So humans could never hear these moms "talking" to their kids.

Amazingly, elephants can "hear" with their feet and trunks by picking up vibrations made by other elephants. They don't need telephones to communicate because their friends can hear them stomping on the ground from as far as 10 miles (16 km) away. Somehow the elephant can tell if the messenger is a friend or stranger, and he can figure out the messenger's location by comparing the vibrations sensed in all four of his feet. Elephants also have very good ears, picking up sounds from as far as 3 miles (5 km) away.

While the outside of the human ear is

Elephant herd

Pigeons

controlled by just three muscles, the Creator gave the cat 36 muscles so that it can rotate its ears 180 degrees and listen to what's going on behind it.

Believe it or not, the lowly moth may have the best hearing of any animal. Scientists estimate that this insect's hearing is 150 times better than a human's. The moth can pick up sounds as high as 300 kHz.

Incredibly, the pigeon can hear sounds as low as 0.5 Hz. Because they can hear the rumbling noise of earthquakes, volcanoes, and storms before other creatures, these birds are the first to make their escape.

Snakes don't have ears. Instead, the Creator installed a delicate little bone in the jaw of the animal. The bone vibrates as sounds come from the ground or the air.

## Smelling and Tasting

*Oh, taste and see that the LORD is good! (Psalm 34:8)*

Imagine what life would be like without the ability to smell and taste. How dull life would be without the ability to smell and taste food! Smelling can also help humans and animals sense invisible dangers. We might smell smoke and realize the building is on fire. We can sniff out gas leaks and take proper action.

Animals rely on smell for hunting, escaping predators, and communication. Actually, compared to many animals, the human nose is not very good at smelling. Here's how the olfactory system works for humans:

1. Inside the nasal cavity is a membrane with special nerve endings. As different chemicals in gas form float through the air, they dissolve in the mucus on the membrane. These nerve endings react to the chemicals.

2. Millions of **olfactory receptor neurons** are sitting under the mucus. These neurons activate if they "recognize" the chemical or a similar chemical.

3. These neurons pass the information on to the **olfactory bulbs** located at the back of the nose.

4. The information is then processed in the brain where memory and emotions are active. The brain interprets bad smells such that the individual responds with negative emotion. Memories of a father's cologne or a mother's perfume might bring back positive emotions.

Dogs are much better sniffers than humans. The human nose contains 5-10 million of these olfactory receptor neurons, but dogs have 300 million of them. If a human can detect a teaspoon of sugar in a cup of coffee, a dog could detect a teaspoon of sugar mixed into two Olympic-sized swimming pools. The dog could find one rotten apple mixed into two million barrels of apples.

The human nose inhales air and then exhales it right away. But dogs tuck away about 12% of the air they inhale. This works its way into a maze of bony caverns. This gives the dog's nose a chance to sort out all the chemicals (or smells) mixed with the air.

Certainly, the Creator designed the dog to assist man by this powerful ability to smell. Dogs have saved lives by sniffing out bombs. They have hunted down scores of lost children. Now, these animals are being used to detect cancer in patients. People with diabetes can go into shock if they have low blood

Golden retriever nose

sugar, but dogs can be trained to sniff out this problem.

The world record for a tracking dog was set in 1925 by a Doberman Pinscher. A dog named Sauer tracked down a thief on foot over 100 miles (160 km) across the Great Karoo desert in South Africa. One bloodhound followed a trail of skin cells that was 3 years old to find a body buried 10 feet under the soil. How does the dog follow a trail? He sniffs out dead skin cells flaking off a person's body. (You lose about 40,000 cells a minute.)

God made an extraordinary variety of

smell and taste arrangements for the animal and human kingdoms. For example, the octopus smells using 10,000 neurons located on each sucker on his eight arms.

Since your nose and mouth are connected inside, you can smell your food as you chew it. Smelling your food this way helps you enjoy it. You have probably noticed that when your nose is stuffy, food doesn't taste very good. That's because the smell of the food you are chewing can't get to the patch of smelling cells.

## How Taste Works

Tasting is more than just smelling the food you are chewing. You also have special cells that send other information to your brain about what you are eating. These tasting cells are found in the little bumps on your tongue. These little bumps are called **tastebuds**. Tasting cells are also found on the roof of your mouth and in your throat. There are five different kinds of tasting cells that taste five different tastes: sweet, salty, bitter, sour, or savory. Tasting cells are not neurons, but they give their information to neurons that send it to your brain.

When we want to feel something, we usually use our fingers to touch it. The skin on our fingers has a lot of nerves in it. We have nerves in the skin all over our bodies. We have nerves for feeling inside our bodies too. Your stomach has nerves to tell you when you're hungry. If you stub your toe, you feel pain because of nerves inside your toe.

You have four kinds of nerves in your

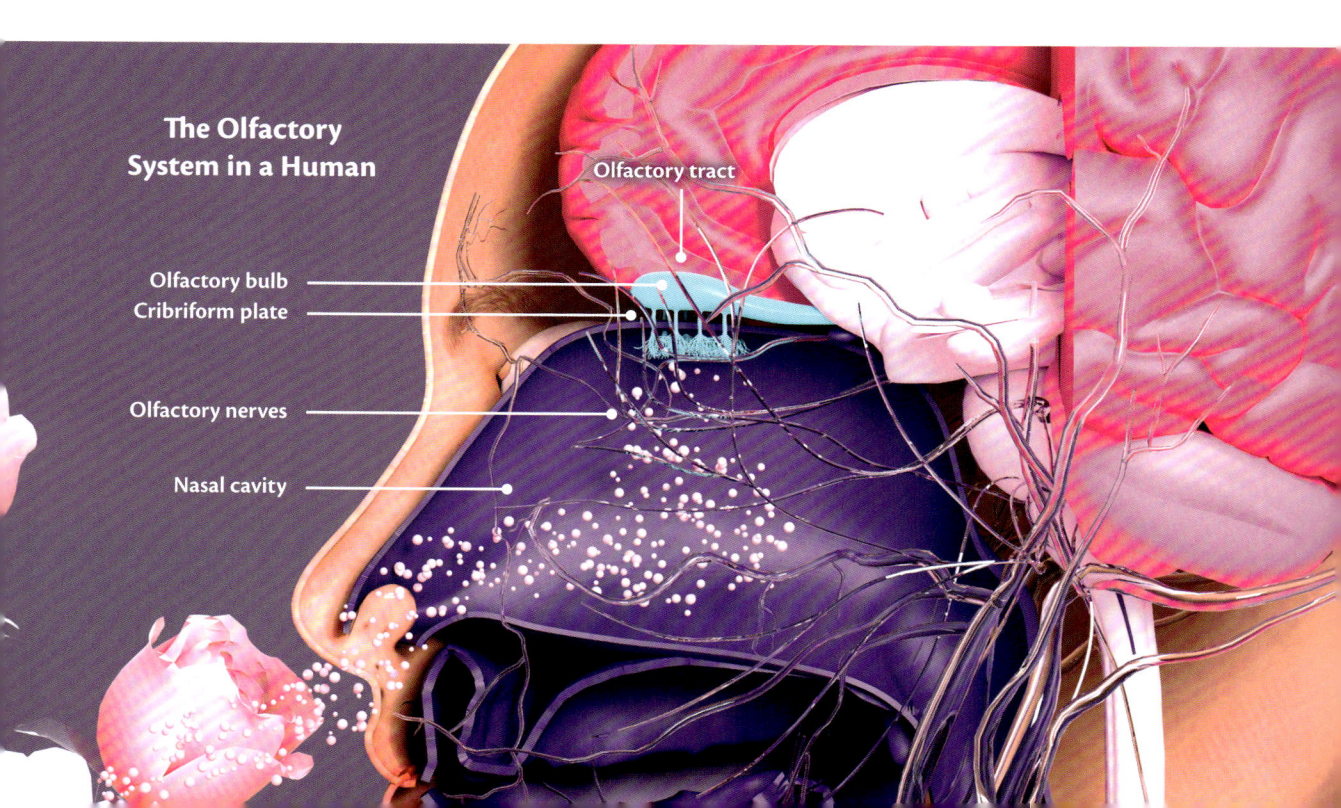

**The Olfactory System in a Human**

Olfactory tract

Olfactory bulb
Cribriform plate

Olfactory nerves

Nasal cavity

## CHAPTER 7: GOD MADE ANIMALS

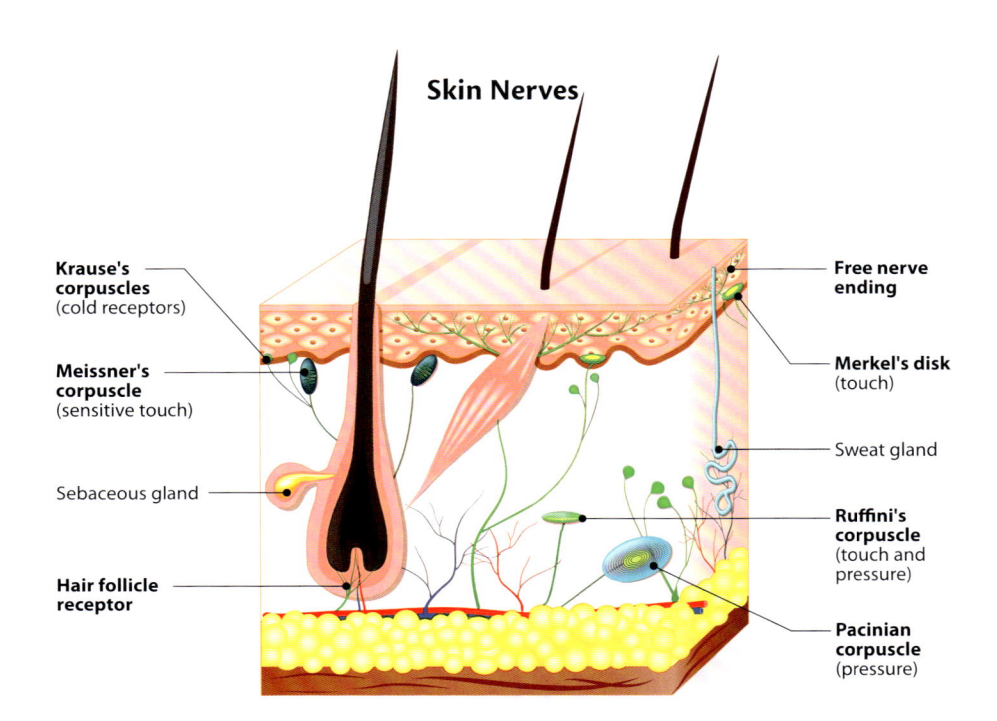

**Skin Nerves**

- Krause's corpuscles (cold receptors)
- Meissner's corpuscle (sensitive touch)
- Sebaceous gland
- Hair follicle receptor
- Free nerve ending
- Merkel's disk (touch)
- Sweat gland
- Ruffini's corpuscle (touch and pressure)
- Pacinian corpuscle (pressure)

skin. These nerves can sense four kinds of touch. They send their information to the brain. The brain helps us understand it.

*Understand, you senseless among the people;
And you fools, when will you be wise?
He who planted the ear, shall He not hear?
He who formed the eye, shall He not see?
He who instructs the nations, shall He not correct,
He who teaches man knowledge?
The LORD knows the thoughts of man,
That they are futile. (Psalm 94:8-11)*

## Ganglia, Brains, and Nerves

Some animals are "smarter" than others. In His infinite wisdom, the Creator assigned a level of intelligence to each animal and equipped them with what they need for their jobs. Typically, invertebrates are not as intelligent as mammals, and they are given a simple brain called a ganglion. The jellyfish's movements are controlled by a nerve net, which is just a series of interconnected nerves throughout its body. They are effectively brainless. However, it would be fair to say that a jellyfish is 100 times more intelligent than a plant. A roundworm has a pair of ganglia and is 100 times more intelligent than a jellyfish. A dog or an ape is 100 times more intelligent than a roundworm. But, a human is a million times more intelligent than a dog or an ape. God made these

205

GOD MADE LIFE

| Animals | Means of God-given Intelligence Controlling Animal's Movement |
|---|---|
| Jellyfish | Nerve net surrounding the jellyfish's body |
| Sea anemone | Nerve net surrounding the anemone's body |
| Sea star | Nerve ring around its mouth and nerves in each leg |
| Flatworm | Nerve net connected to ganglia (simple brains) |
| Clams | Three ganglia—one for the clam's foot, one for its shell muscles, and one for the esophagus (or throat and mouth) |
| Snails | Six ganglia |
| Lobster | Brain, stomach ganglion, and a pair of ganglia for each body segment |
| Octopus | Nine brains |
| Insects | Brain, stomach/mouth ganglia, and a pair of ganglia for each body segment |
| Mammals | Brain |

creatures with varying levels of intelligence. Nonetheless, all of these creations demonstrate God's limitless genius and command our praise.

The nerve's job is to communicate the messages to the brain or ganglion. Then the brain interprets the message and orders the animal to respond in some way. While the human brain functions as the intelligence control center for the whole body, the ganglion is more specialized. An earthworm has a ganglion assigned to each of its 100-150 segments. Each ganglion then controls the movements for each segment. Yet, the earthworm has a simple brain made up of a ganglion inside its head, which controls the other ganglia throughout its body. If this brain is removed, the worm's body will move continuously like it is out of control.

Insects like grasshoppers can jump and fly without the use of their brains. The local ganglion at each segment can keep the insect moving, while the brain helps to coordinate

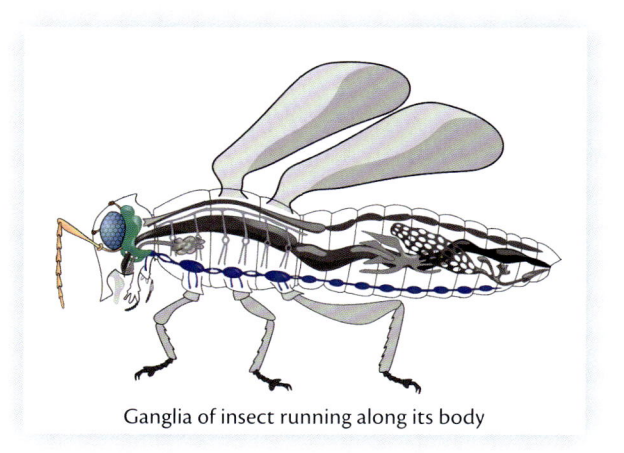

Ganglia of insect running along its body

## CHAPTER 7: GOD MADE ANIMALS

movements.

The octopus is probably the smartest invertebrate created by God. He's equipped with 500 million neurons, most of which are in his arms. He has only 150 million in his brain. That's about the same number of neurons as dogs have, but it is only 0.5% of the number of neurons found in the human body. Each octopus arm is equipped with a brain, while a centralized brain located in the foot or head controls the rest of the body.

The intelligence of the octopus is no secret. While living in aquariums, some of these amazing creatures have learned how to turn off the lights in the room by squirting water at the bulbs and short-circuiting the outlets. The octopus can twist a jar lid open—from the inside!

As we study the different senses and the ways all these animals process information, something remarkable appears to us. Compare the octopus to the grasshopper, or the flatworm with the dragonfly, and what do you see? All these designs are very different. The one cannot be related to the other. There are a few common patterns where it is obvious the Lord used the same design for different creatures. But the differences are even more remarkable. What wondrous creativity shows up in the wide variety of shapes, sizes, intelligence, and capabilities found in the world of moving organisms! None of these sensory organs, ganglia, or brains could have come about by unintelligent chance. How can unintelligence create intelligence? How could a rock create a bug? How could a bug create a human? How could a world of dirt and rocks produce the first bug with the intelligence to escape a predator and search for food? The evolutionary theory is ultimate irrationality (or craziness).

## Defense Mechanisms

The LORD by wisdom founded the earth;
By understanding He established the heavens;
By His knowledge the depths were broken up,
And clouds drop down the dew.
(Proverbs 3:19-20)

Besides brains, ganglia, and wonderful senses, the Creator perfected His work with a set of tools for His animals. He made a huge array of defense mechanisms and hunting tools to help His creation to survive and thrive. Once again, there are far too many of these to list here. The variety of creative ideas God worked into His creation is astounding. You could discover a new and different design in this marvelous creation every day. By studying His creative work, you would find enough reasons to praise Him in excited wonder and grateful devotion for the rest of your life.

Although they are usually very tiny creatures, God has outfitted the invertebrates with quite the armor. Hard shells, camouflage, venom, inky spray, stingers, pincers, and spines are all included in God's armory for bugs. Many animals that use camouflage include octopuses, squids, and chameleons.

Baby opossum

Some animals can make themselves look big and scary to startle a predator. This is called **deimatic behavior**. Some moths have wings designed to look like snakes. Other moth wings are decked out with large spots that look like owl's eyes. A predator backs away when the moth spreads its wings in its face. The peacock butterfly can make itself look like a leaf floating in the wind. The arctic moth can make a clicking noise, mimicking a bat. The little frill-necked lizard spreads out his frills around his head in an attempt to look very ferocious. The rattlesnake's rattle warns everybody to stay away!

Some animals like **opossums** will play dead when they feel threatened. They will even give off an odor of rotting carrion to discourage predators. Many lizards, rodents, sharks, spiders, and birds also play dead.

To protect its young, the **Mourning Dove** will walk away from the nest, dragging one of her wings over the ground. She acts like an easy target for the predator. But when the bird has distracted the predator from her young lying in the nest, she quickly recovers from her injury and flies away. All of these are intelligent designs invented by an all-wise Creator God.

Mourning dove

## CHAPTER 7: GOD MADE ANIMALS

Bombardier beetle

Perhaps the most amazing defense mechanism is that designed for the **Bombardier beetle**.

This insect's defense system could be the most amazing and the most complicated in all of creation. Using a mixture of chemicals this beetle keeps inside his body, he lights up an explosion in a chamber contained in his abdomen. This blows out his rear and fries any small animal that might be trying to attack him. Like a slow-motion machine gun, this little guy can repeat the explosions 15-20 times in just a few minutes.

The Bombardier Beetle keeps two chemicals in different sacs tucked away in his body—a watered-down hydrogen peroxide and hydroquinone. If these chemicals leaked out and mixed in his body, they might react and blow him up. So God provided the beetle with another chemical inside of him that acts as an inhibitor, preventing a chemical reaction. When the beetle squirts the hydrogen peroxide and hydroquinone into the chamber towards his rear end, the chemicals mix with two enzymes: catalase and peroxidase. This causes a reaction that increases the pressure and temperature inside the chamber, resulting in a small explosion and the emission of burning toxic gases into the face of the predator. Also, the combustion chamber inside the body of the beetle has to be able to sustain very high temperatures over 212 °F (or 100 °C ).

All of this is very complicated, and a twelve-year-old child might have a hard time understanding how these chemical reactions work. But that's okay. Scientists have had to study this process for a long time to figure out how the catalyst (the enzymes) works in the chemical reaction. What can we learn from all of this? Obviously, a brilliant Designer, our awesomely wise God, must have come up with this idea. The important thing to remember about Mr. Bombardier's defense system is that *nothing works until everything works*. The beetle needs both active chemicals, the catalyst, and the inhibitor for the whole thing to work.

Unbelieving scientists try to tell us that all of this came about over millions of years by evolutionary process. That would be utterly impossible. When humans try to create controlled explosions and chemical reactions, they have to work at it for a long time. It is very hard to engineer a safe and reliable chemical process. Researchers have to be very careful not to blow themselves up while they are doing their experiments. Wonder of wonders, God's design of this little beetle's defense system works thousands of times

# GOD MADE LIFE

without blowing the little guy up. And it continues to work over six thousand years after the beetle was first created. That is amazing!

*Because the foolishness of God is wiser than men, and the weakness of God is stronger than men. . . . God has chosen the foolish things of the world to put to shame the wise, and God has chosen the weak things of the world to put to shame the things which are mighty; and the base things of the world and the things which are despised God has chosen, and the things which are not, to bring to nothing the things that are. (1 Corinthians 1:25, 27-28)*

## Invertebrate Digestion

*The eyes of all look expectantly to You, And You give them their food in due season. You open Your hand And satisfy the desire of every living thing. (Psalm 145:15-16)*

Invertebrates must eat stuff they find in the sea or on the land. Unlike plants, they use a simple digestive system. For example, the earthworm eats dirt, fungi, decayed grass, and leaves using its mouth. As the food enters the **pharynx**, liquid is produced by glands and mixed with it.. This helps lubricate the food as it slides down the esophagus to the **crop** (a storage area). From the crop, the food moves into the **gizzard** where the

Digestive system of earthworm

food is broken down. Special enzymes break up the food. But the gizzard is also outfitted with muscles that grind up the food into tiny particles. Then, the food passes into the worm's **intestines**. The usable nutrition passes through the intestinal walls into blood vessels. The unusable part is passed out as waste. Some invertebrates digest their food like this.

## Invertebrate Respiration and Nutrient Transportation

Insects such as grasshoppers are equipped with **hemolymph** made up of mostly water and nutrients. Hemolymph is different from blood in that it contains no red blood cells. This fluid isn't colored red. Sometimes it looks like a clear fluid, and sometimes it is lime green. Both blood and hemolymph carry water, food, and other important things around the creatures' bodies. In humans and vertebrates, the blood carries oxygen through the body. Invertebrates like insects get their oxygen through a system of tubes accessing the open air through the skin.

Grasshoppers have a heart-like organ that pumps the hemolymph past the gizzard and the intestines from which it picks up food. This fluid then moves throughout the insect's inner cavity delivering the food to hungry cells. It also picks up waste materials which are filtered out through **excretory tubules** and delivered back into the intestines.

Crustaceans are a large group of invertebrates that don't get their oxygen through their skin. God gave them gills for that purpose. Also, spiders get their oxygen using book lungs.

## Invertebrate Reproduction

"Do you know the time when the
wild mountain goats bear young?
Or can you mark when the deer gives birth?
Can you number the months that they fulfill?
Or do you know the time when they bear young?
They bow down,
They bring forth their young,
They deliver their offspring.
Their young ones are healthy,
They grow strong with grain;
They depart and do not return to them....
Shall the one who contends with the
Almighty correct Him?
He who rebukes God, let him answer it."
(Job 39:1-4, 40:2)

Whiptail lizard reproduces by parthenogenesis

God's various systems of reproduction very well might be the most impressive part of His physical creation. How life produces new life is a mystery and a wonder. Scientists and doctors have studied this complex process for thousands of years. With each generation, they learn more about this amazing creative work of almighty God.

Sometimes it takes two organisms to reproduce, and sometimes only one. Earthworms and planarians reproduce either way. Because the Creator provided each worm with both the male cell and the female cell, a single earthworm can reproduce baby worms by itself. This is called **parthenogenesis**.

The planarian can reproduce itself into two worms when it is cut in two pieces. Incredibly, even small pieces of the worm can grow back into another full-sized worm. This is called reproduction by regeneration.

With these few exceptions, most insects and other invertebrates reproduce with a male and a female animal each contributing a cell to conceive new life. Most arthropods will lay eggs which hatch anywhere from two days to two months later. However, the

# GOD MADE LIFE

**Process of Metamorphosis**

Lord makes exceptions for scorpions and a few other invertebrates. Their eggs actually hatch inside the mother, and the live little babies crawl out of her.

Some invertebrates will come out of their eggs looking and functioning like adults. However, when baby insects hatch, many of them look completely different from their mature forms—think of grubs and caterpillars. The Lord has designed a very wonderful and miraculous process of change called **metamorphosis**. This is how the immature life form of the insect is transformed into a mature adult. But how does this work? How does an ugly caterpillar tun into a beautiful butterfly inside the mysterious chrysalis? Such things are too wonderful for us to comprehend.

What can we learn from this is that God loves variety, creativity, complexity, and beauty. This is what marks His creation! There are many unique features built into His creatures. So, there are many little exceptions to the rules as biologists try to categorize millions of organisms. Our Creator has designed many special abilities and behaviors for each little organism in this vast creation.

Monarch butterfly

> For since the creation of the world His invisible attributes are clearly seen, being understood by the things that are made, even His eternal power and Godhead. . . (Romans 1:20)

## CHAPTER 7: GOD MADE ANIMALS

## Taking Dominion of Animals

O LORD, our Lord,
How excellent is Your name in all the earth,
Who have set Your glory above the heavens! . . .
What is man that You are mindful of him,
And the son of man that You visit him?
For You have made him a little lower than the angels,
And You have crowned him with glory and honor.
You have made him to have dominion over the works of Your hands;
You have put all things under his feet,
All sheep and oxen—
Even of the beasts of the field . . .
(Psalm 8:1, 4-7)

In accord with God's wise plans for our world, some animals are pests. They hurt people. They cause health problems. They destroy our crops and cause billions of dollars of damage each year. This presents quite a challenge for us. Yet, the Lord still commands us to take dominion over this part of the creation. First, we are to identify the pest, and then we must find some way to get rid of it.

A **parasite** is an animal living off another organism, either animal or human. Most flatworms are parasitic, but not all of them are. Dogs and cats can pick up tapeworms (one kind of flatworm). When deworming a pet, you are trying to kill either **tapeworms** or

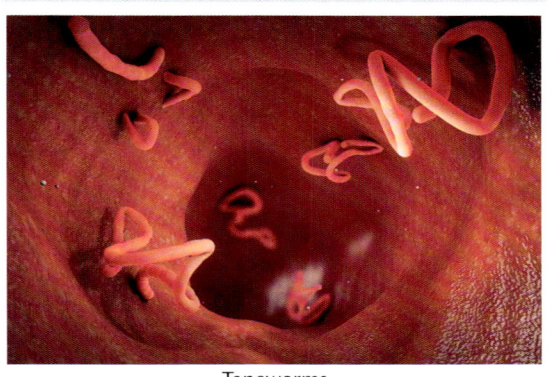

Tapeworms

**roundworms**. When the tapeworm makes its way into the digestive track of a puppy, it grabs on to the track's surface with hooks or suckers. It doesn't want to get flushed out by fluids and wastes. The worm grows longer by feeding on the food coming through the host. Some tapeworms have grown to 66 feet (20 m) in length. When the tapeworm releases its eggs, they show up in the feces of the dog. When animals feed on feces or rotting meat, they will pick up more of these worms. That's why you want to clean out cages and whelping boxes where puppies are kept. And, don't forget to deworm your dog from time to time. Roundworms (Ascaris) are more likely to infect humans.

The mosquito is the most dangerous animal in the world because it spreads the malaria-carrying protozoan parasite in its saliva when it bites. This pesky insect will also spread the West Nile virus and yellow fever. When animals pass diseases around like this, they are called vectors. Ticks also pass around the Lyme disease. Rats have

# GOD MADE LIFE

Most dangerous animal: the mosquito

carried about fleas which spread the bubonic plague.

America and other nations were saved from many of these mosquito-borne diseases by using a chemical called DDT in the 1940s and 1950s. Christians have taken dominion over this terrible insect and eliminated diseases like typhoid fever, yellow fever, and malaria. This chemical worked well for "insect control in crop and livestock production, institutions, homes, and gardens."

When nations worship the creature more than the Creator, they will not obey God. They refuse to take dominion over the mosquito and other parts of creation. In 1972, the USA and other countries stopped using DDT. This has resulted in about 50 million human lives lost to these mosquito-borne diseases in Asia and Africa.

While some animals are especially bothersome and even dangerous to humans, that's rarely the case. Most animals are harmless, and many are even useful to us. Cows and goats give us milk. Earthworms dig up the ground and fertilize it with their droppings. And bees pollinate our fields, blessing us with honey.

## Edible Animals

So God blessed Noah and his sons, and said to them: "Be fruitful and multiply, and fill the earth. And the fear of you and the dread of you shall be on every beast of the earth, on every bird of the air, on all that move on the earth, and on all the fish of the sea. They are given into your hand. Every moving thing that lives shall be food for you. I have given you all things, even as the green herbs. But you shall not eat flesh with its life, that is, its blood." (Genesis 9:1-4)

Since the time of the flood, the Lord has given us every living thing to eat. But He warned Noah not to eat the blood of the animal. The Apostles repeated this rule Acts 15:29. To eat the blood is to disrespect the life God created.

For a while, under the Mosaic covenant, God did not allow His people to eat "unclean meats." This included camels, horses, dogs, cats, rabbits, pigs, lobsters, shrimp, carrion birds (vultures), owls, bats, and most insects except for grasshoppers. There is disagreement among sincere Christians on whether or not to eat these animals. Some Christians do not eat bacon, shrimp, horse meat, vultures, and cockroaches.

Puffer fish

In the New Testament, the Apostle Peter had a dream about unclean animals, and a voice from heaven told him, "Kill and eat… What God has cleansed you must not call common" (Acts 10:13,15). This was a strong message to Christians that they have much liberty in what they can eat.

## Are There Any Inedible Animals?

Generally, those animals which God allowed His people to eat in the Old Testament were vegetarian, such as cows, sheep, and grasshoppers. Some animals are **omnivores**, which means they will eat just about anything. And some are **carnivores**, which means they eat only meat. Some carnivores like carrion birds will eat rotting meat. Some sea creatures such as shrimp, lobsters, squid, shellfish, crabs, and octopuses are known as **bottom feeders**. This means they eat algae, plants, and dead animals that settle at the bottom of the sea. Similar to disagreements among Christians on these things,

Poison dart frog

health experts disagree on whether to eat the bottom feeders. Most are agreed on this, however: The health benefits of eating a cow, a chicken, a pig, or a lobster depend on what the animal was feeding on.

Some animals are more risky to eat than others. For instance, the liver and reproductive organs of the puffer fish are extremely poisonous—a stronger poison than cyanide. An average of two people die each year from consuming puffer fish in Japan. There are also highly poisonous salamanders, as well as the poison dart frog and the Hawksbill sea turtles which are too dangerous to eat.

You may be "grossed out" by certain kinds of strange foods. And you should be cautious when it comes to certain substances known to be poisonous. Don't eat foods known to be very bad for you. But, we are not defiled spiritually when we step in cow manure. And, we are not terribly defiled when we eat a shrimp that has been crawling over a dead fish or feeding on poisonous chemicals. Some food may be fatal to our bodies. However, the Lord Jesus Christ reminded His disciples of the ultimate dirty defilement. And that is the sin in our hearts. Only the blood of Christ can clean our hearts of this terrible filth.

So Jesus said, "Are you also still without understanding? Do you not yet understand that whatever enters the mouth goes into the stomach and is eliminated? But those things which proceed out of the mouth come from the heart, and they defile a man. For out of the heart proceed evil thoughts, murders, adulteries, fornications, thefts, false witness, blasphemies. These are the things which defile a man, but to eat with unwashed hands does not defile a man." (Matthew 15:16-20)

God made animals beautiful, cute, and majestic!

## Pray

- Take a moment and praise God for His animal creation. Praise Him for the great variety of invertebrates He has created.
- Praise God for His wisdom and power seen in the creation of the senses of sight, hearing, and smell.
- Praise God for the brain, for the ability of animals to process information, and for their hundreds of defense mechanisms.
- Thank the Lord for the helpful animals like worms and bees.
- Thank God for the blessing of meat from the animals He has made.

## Sing

Having once again witnessed the marvelous work of God in this wonderful creation of animals, the appropriate response is more worship and praise. If the student is unfamiliar with the hymn or psalm, some version of it is available on the internet and may be accessed (with supervision) for singing along.

*Psalm 148—Hallelujah Praise Jehovah*
Hallelujah, praise Jehovah,
From the heavens praise His Name;
Praise Jehovah in the highest,
All His angels, praise proclaim.
All His hosts, together praise Him,
Sun and moon and stars on high;
Praise Him, O ye heav'ns of heavens,
And ye floods above the sky.

Refrain:
Let them praises give Jehovah,
For His Name alone is high,
And His glory is exalted,
And His glory is exalted,
And His glory is exalted
Far above the earth and sky.

Let them praises give Jehovah,
They were made at His command;
Them for ever He established,
His decree shall ever stand.
From the earth, O praise Jehovah,
All ye seas, ye monsters all,
Fire and hail and snow and vapors,
Stormy winds that hear His call.

All ye fruitful trees and cedars,
All ye hills and mountains high,
Creeping things and beasts and cattle,
Birds that in the heavens fly,
Kings of earth, and all ye people,
Princes great, earth's judges all;
Praise His Name, young men and maidens,
Aged men, and children small.

## Do

Choose at least one of the following activities and apply the lessons you have learned in this chapter.

1. One research study found that 36% of dogs are infected with worms at any point in time. That's about one in three dogs. Here is what the researchers found to be the most common worm infestations:

   - The hookworm: 19% of infected dogs
   - The roundworm: 15% of infected dogs
   - The whipworm: 14% of infected dogs

   Deworm your dog, or you may help friends deworm their dog. Veterinarians recommend that dogs be dewormed every three months. Puppies should be dewormed every two weeks until they are twelve weeks old. After that, they should be dewormed once a month until they are six months old. Go with the vet's advice for the best anti-parasitic product to use and the appropriate dose for the dog. Follow the instructions on the packaging.

2. Go fishing (or hunting). Clean the fish and cook the meat for eating. Drain the blood from the animal. To drain the blood of a fish, sever the main artery behind the gills. Dunk the fish in cold water for 30 seconds to remove the blood. Usually removing the blood from an animal helps to preserve the meat. And, removing the blood can minimize the "fishy" smell (or taste) of the meat.

3. Protect your pets and yourself from worms. Your dog can pick up worms and then share them with you and your family. Dogs get worms from rolling around in contaminated soil and feces. They can get worms from eating dead rodents and birds. So, you can take action now. Clean up feces and rotting plants and animals strewn about the yard or any area where the dog runs and plays.

## Watch

To watch the recommended videos for this chapter, go to **generations.org/GodMadeLife** and scroll down until you find the video links for Chapter 7. Our editors have been careful to avoid films with references to evolution; however, we would still encourage parents or teachers to provide oversight for all internet usage. These videos may not give God the glory for His amazing creative work, so the student and parent/teacher should respond to these insights with prayer and praise.

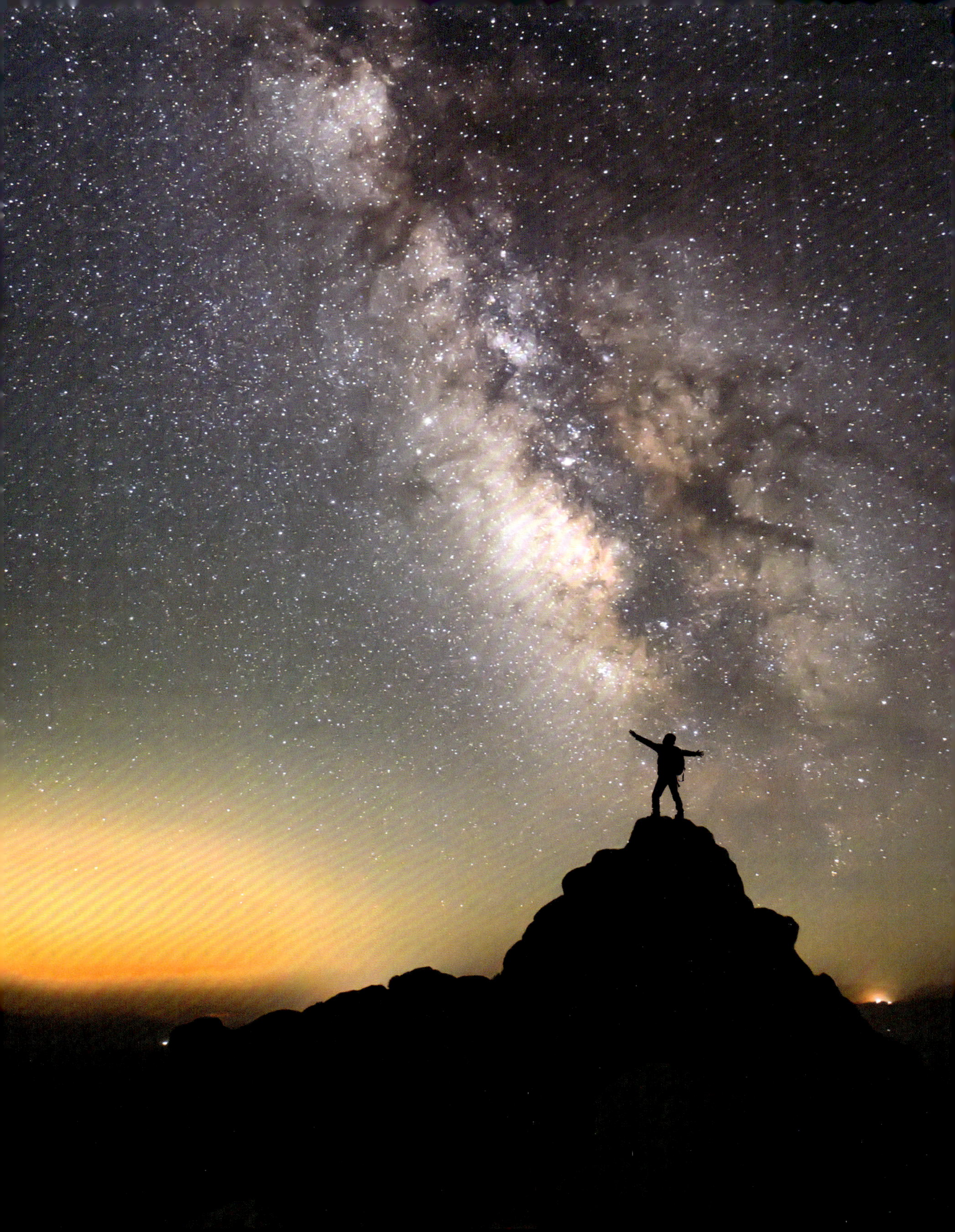

# CHAPTER 8
# GOD MADE MAN

> And the LORD God formed man of the dust of the ground, and breathed into his nostrils the breath of life; and man became a living being. (Genesis 2:7)

Working from the simple creatures to the more complex, this study has covered the microbes and invertebrates first. The second major, and more complex, category of animals is vertebrates. Some within this category have hearts and lungs, while others such as fish do not. Those receiving the breath of life and blood are the higher creatures. Now, there is a higher creature, more advanced than dogs and cats, to which the Creator gave the breath of life. Man is a vertebrate with heart and lungs, and yet he was created a little lower than the angels. Man is the highest creature on Earth.

During the worldwide flood, the Lord destroyed both man and animal with the breath of life according to Genesis 6:17.

> "And behold, I Myself am bringing floodwaters on the earth, to destroy from under heaven all flesh in which is the breath of life; everything that is on the earth shall die." (Genesis 6:17)

The five most common classes of vertebrates include fish, amphibians, reptiles, birds, and mammals. Typically, zoologists add two more classes to this list. The class **Chondrichthyes** includes sharks, rays, and chimeras, and the class **Agnatha** is made up of jawless fish. There are 120 species of agnathous fish, including lampreys and hagfish. In total, there are between 40,000 and 50,000 kinds of vertebrates.

Vertebrates are more complex than in-

# GOD MADE LIFE

vertebrates. They have better sensory organs. And instead of ganglia, these advanced creatures are equipped with brains. So, they are much smarter than invertebrates (with the exception of the octopus). But, above and beyond all animals, both vertebrates and invertebrates, is the highest of God's material creation—the **human being**. Because the human is by far the most complex and the most wonderful of God's biological creation, we will study this vertebrate for the remainder of this survey.

## The Differences Between Man and Animal

*Then the Lord God took the man and put him in the garden of Eden to tend and keep it. And the Lord God commanded the man, saying, "Of every tree of the garden you may freely eat; but of the tree of the knowledge of good and evil you shall not eat, for in the day that you eat of it you shall surely die." (Genesis 2:15-17)*

From the beginning, God communicated with man, and He intends for man to communicate using language. This is something animals cannot do. Most dogs can learn the meaning of 100 words. Exceptionally "smart" dogs can learn the meaning of 250 words. But, these are still simple words that dogs can associate with something they like, such as "food" and "lunch." Humans, on the other hand, learn about 30,000 words, many of which are conceptual. They can learn about "history" and "faith" and "forgiveness." They can put sentences together with these words—all of which make humans millions of times smarter than animals.

Humans are created in the image of God, with the capacity to think in a similar way to the way God thinks.

Men and monkeys are very different.

Skyscrapers

> **You have put off the old man with his deeds, and have put on the new man who is renewed in knowledge according to the image of Him who created him. (Colossians 3:9b-10)**

Humans can take care of creation. They can solve complex problems, treat medical problems, and build skyscrapers. They can use technology to convey information and move people from one place to another very quickly. Rats and monkeys don't manufacture cars.

Humans are moral creatures. That means they have a moral conscience. They feel guilt when they disobey God's laws. Therefore, humans are concerned about atonement. They need to know that their sins have been paid for, that they are reconciled to God, and that they are in a right relationship with God.

Humans live longer than most other creatures. Except for the Greenland shark, the bowhead whale, and the giant tortoise, the human being lives longer than any other member of the animal kingdom. The Lord designed the human body to live a very long time despite the effects of the fall. Your body can withstand a lot of troubles and still live for 80 to 90 years. Not many automobiles can continue running every day for 90 years. The average car lasts only 10 to 20 years.

## Evolutionists Believe Animals Evolved into Humans

Most colleges today teach that animals evolved into humans over millions of years. They do not believe that God created Adam out of the dust of the earth.

The evolutionists claim the world has been around for 4 billion years. And, they say there were only fish and no land animals on Earth 400 million years ago. Somehow a fish was supposed to have turned into a four-legged, amphibian-like creature. There is no evidence that a half-fish, half-amphibian creature ever existed. But, how would this half-fish, half-amphibian survive? This creature couldn't swim with half of a fin, and

Car assembly

he couldn't walk with half-developed legs. At the same time, this creature would have to develop lungs because land vertebrates wouldn't be able to survive if they used only gills to receive oxygen. Gills are designed, not for breathing, but for taking in oxygen gas dissolved in water. But how could the fish survive in the water if his gills had shrunk to make room for developing lungs?

Evolutionists have tried to find fossils of half-ape, half-human creatures. For a long time, they thought the **Neanderthal man** was a half-human creature. When they dug up their skeletons, evolutionists discovered these creatures were short, stocky, and had sloped foreheads. But, they used tools and buried their dead like most humans do today. When scientists tested their DNA, they found them to be as human as you and me.

While digging around in Tanzania 50 years ago, evolutionists found about one quarter of an ape's skeleton. They called her "Lucy" and introduced her as a half-human, half-ape creature. After further research, however, they discovered her toes were curved like most tree-dwelling apes. Her wrists were bent like knuckle-walking monkeys and apes. And, her shoulders looked exactly like those of modern apes. Lucy was just another ape. But, most evolutionists still want to believe this animal was related to one of their ancestors.

Evolution is not science. The process of one animal changing into another has never

CHAPTER 8: GOD MADE MAN

been observed. Evolution is an imagination game played by grown men who don't want to believe in God.

Just take a look at your body. It should be obvious that you are fearfully and wonderfully made.

## The Wonder of the Human Cells

*I will praise You, for I am fearfully and wonderfully made;*
*Marvelous are Your works,*
*And that my soul knows very well.*
*(Psalm 139:14)*

A typical car is made of 30,000 parts. It takes about 200 moving parts to get the car to propel itself down the road. Compare that to the human body created by God. This has 30,000,000,000,000 separate parts. Within each of these cells, there are thousands of things going on at the same time. Every cell is full of activity — lots of moving parts. Some parts of the cell are making energy, some are making new proteins, and some are involved in transportation and communication.

The cell is the smallest unit of life. It is

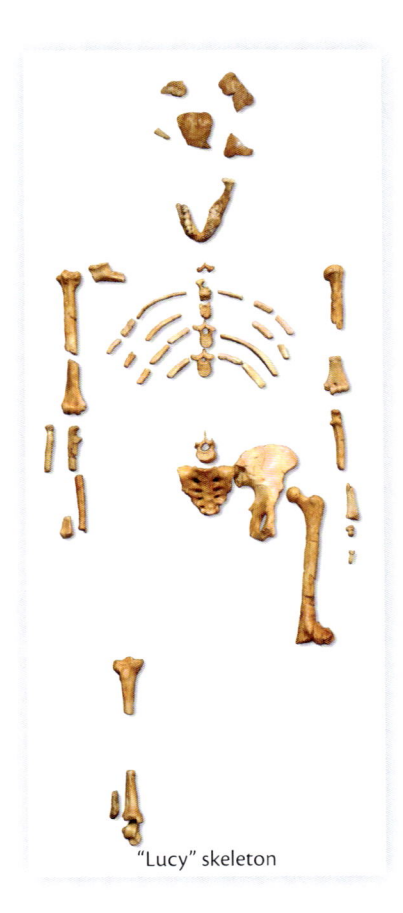

"Lucy" skeleton

the basic building block for all of life, both for plants and animals. The same can be said for human cells. There are 200 types of cells used to put together the human body. Here are a few important cells:

**Red blood cells** are very critical to the life of the body. Called **erythrocytes**, these guys are tasked with carrying oxygen around the body. **Liver cells** are much more complicated than red blood cells. If the liver is damaged, it can quickly rejuvenate itself by producing more liver cells to replace the damaged cells. The liver cells are made for digesting fats and getting rid of toxins.

The **muscle cells** are

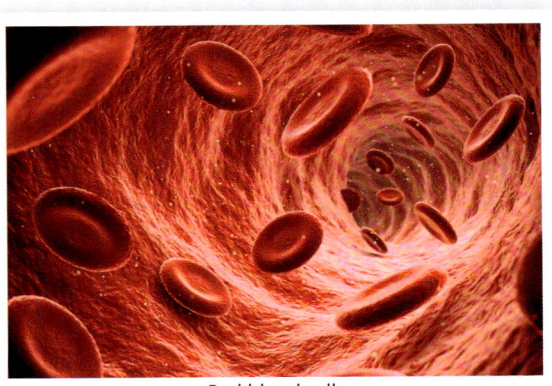

Red blood cells

225

designed to contract (to get shorter and fatter). When the body calls for a muscle to go to work, millions of muscle cells contract together as a team. As these muscle cells contract and the muscle shortens, the bones connected to the muscle will move with it.

The **pancreas cells** produce insulin, which regulates sugar in the body.

The **nerve cells** are made up of stuff that gets excited by an electrical signal. They carry these electrical signals to and from the brain.

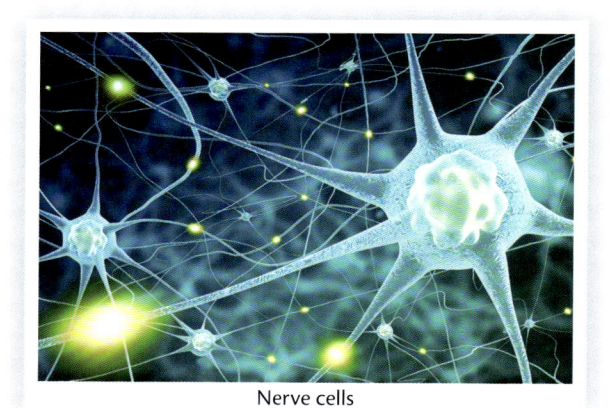

Nerve cells

As already covered in the previous chapter on cells, the **eukaryotic cells** in animals and humans have special organelles. These are the parts swimming around in the cell, each with a special job to do. There is the nucleus containing the DNA, the mitochondria that produce energy, and the endoplasmic reticulum which produces new proteins. Some cells need a lot of a certain kind of organelle. For example, muscle cells need a lot of mitochondria. These cells convert food into energy, and muscles need a lot of energy!

Some specialized organelles show up in certain kinds of cells. For example, the intestines are lined with cells whose membranes have a lot of **microvilli** sticking out. These finger-like protrusions provide an increased amount of surface area that can pick up useful food swimming down through the intestines. Pancreas cells get quite a few **storage granule organelles**. These are like storage barrels used to hold insulin, which can be released at any time when the body needs it.

## The Wonder of Human Tissue

The hand of the LORD came upon me and brought me out in the Spirit of the LORD, and set me down in the midst of the valley; and it was full of bones. Then He caused me to pass by them all around, and behold, there were very many in the open valley; and indeed they were very dry. And He said to me, "Son of man, can these bones live?"
So I answered, "O Lord GOD, You know." Again He said to me, "Prophesy to these bones, and say to them, 'O dry bones, hear the word of the LORD! Thus says the Lord GOD to these bones: "Surely I will cause breath to enter into you, and you shall live. I will put sinews on you and bring flesh upon you, cover you with skin and put breath in you; and you

**CHAPTER 8: GOD MADE MAN**

shall live. Then you shall know that I am the LORD." ' "
So I prophesied as I was commanded; and as I prophesied, there was a noise, and suddenly a rattling; and the bones came together, bone to bone. Indeed, as I looked, the sinews and the flesh came upon them, and the skin covered them over; but there was no breath in them.
Also He said to me, "Prophesy to the breath, prophesy, son of man, and say to the breath, 'Thus says the Lord GOD: "Come from the four winds, O breath, and breathe on these slain, that they may live." ' " So I prophesied as He commanded me, and breath came into them, and they lived, and stood upon their feet, an exceedingly great army. (Ezekiel 37:1-10)

As the basic building blocks for the body, similar human cells are grouped together to make up **tissue**. Common cells team up and connect to each other so the whole body can hold together. There are four kinds of tissue in the human body:

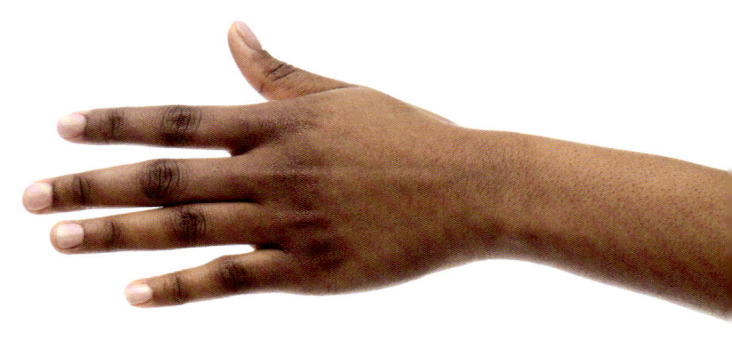

| Tissues in the Human Body | Where the Tissue is Found |
|---|---|
| Epithelial Tissue | Skin, hair follicles, and outside covering of bodily organs |
| Muscle Tissue | Muscles |
| Connective Tissue | Ligaments, tendons, skin, bones, fat, and blood |
| Nervous Tissue | Nerves, brain, and spinal cord |

When you eat fried chicken, remember that you are eating muscle tissue. The skin of the chicken is epithelial tissue. The chicken's ligaments and tendons are usually chewy and tough. These are connective tissues. The thicker dermis under the outer skin of the chicken is also connective tissue. If you clean a chicken for cooking, you see the sciatic

Fried chicken

227

# GOD MADE LIFE

nerve (tissue) running alongside the thigh.

The **epithelial tissue**'s main job is to confine the body and its organs and to protect them. But different epithelial tissues have additional special jobs assigned to them:

## What Epithelial Tissue is Designed to Do

| Location in Body | The Function of the Tissue |
|---|---|
| Skin | Protects the body against injury |
| Skin hair follicles | Grows hair |
| Esophagus and inner mouth | Protects against injury |
| Heart lining, lung air sacs, and blood vessels | Filtrates materials, secretes lubricants |
| Digestive tract and bronchi | Secretes mucous and enzymes |
| Bladder and urethra | Allows the urinary system to expand and contract |
| Glands and kidneys | Secretes and absorbs fluids |
| Trachea and throat | Secretes and moves mucous |

## Muscle Tissue

**Muscle tissue** runs through your whole body. Your stomach uses muscles to grind up your food. It takes muscular action to keep your lungs functioning and your heart pumping. The three kinds of muscle tissue are skeletal, smooth, and cardiac.

**Cardiac muscles** are special tissues God made just for the functioning of the heart. **Smooth muscles** are used in the organs of your body, such as the digestive tract, the urinary tract, and the respiratory tract (the system used for breathing). Smooth muscles and cardiac muscles are involuntary muscles. That means you don't have to think about exercising these muscles. You don't think about breathing, or digesting food, or processing urine and body wastes. Your body automatically takes care of these functions for you. Actually, the body is very busy taking care of these things throughout the day and night. God has designed much of your body to function automatically. If you had to remember to take care of these functions all day long and all night long, you would never survive.

**Skeletal muscles** are connected to the bones in your skeleton. You can move your fingers, toes, hands, and feet when you contract your skeletal muscles. These are **voluntary muscles**. This means that when you want to move your finger, your brain signals the muscles in your finger to contract. These skeletal muscles, in turn, move the finger bones in the direction you want the hand to move.

There are between 650 and 840 skeletal muscles in the human body. God gave these to you for maximum range of motion, dexterity, and flexibility. He gave them to you to perform a million different tasks. No other creature was given such a variety of volun-

228

**CHAPTER 8: GOD MADE MAN**

**Three Types of Muscle Tissue**

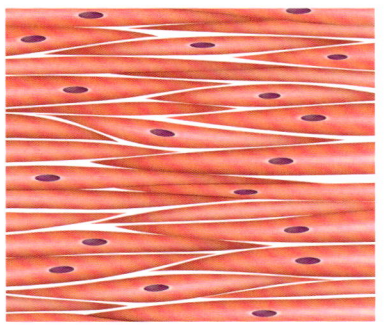

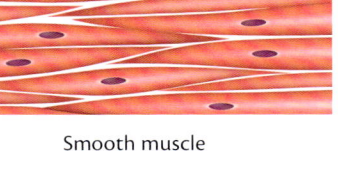

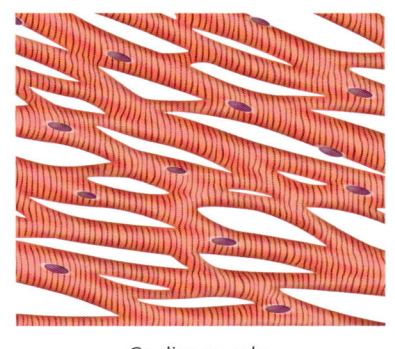

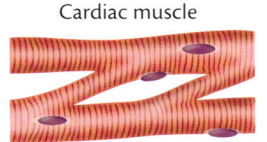

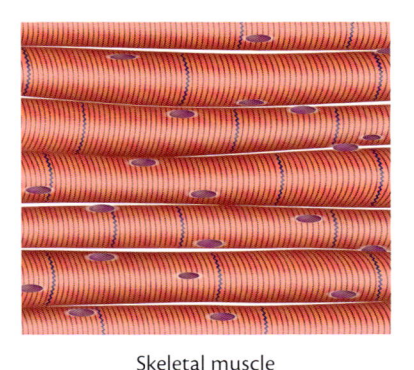

Smooth muscle

Cardiac muscle

Skeletal muscle

tary movements. Your eyes move 10,000 times an hour as you read a book. Your heart is pumping 2,500 gallons of blood every 24 hours. Pound for pound of muscle, a jaw muscle called the **masseter** is the strongest muscle in your body. This muscle enables your molars to crunch down at 200 pounds of force.

Human facial muscles

Have you not known?
Have you not heard?
The everlasting God, the LORD,
The Creator of the ends of the earth,
Neither faints nor is weary.
His understanding is unsearchable.
He gives power to the weak,
And to those who have no might He increases strength. (Isaiah 40:28-29)

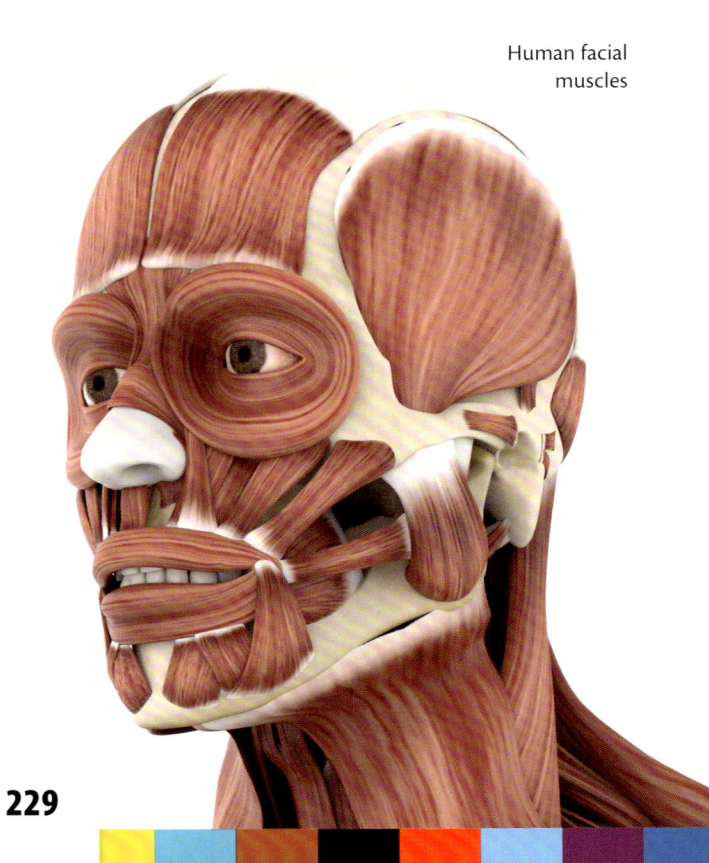

GOD MADE LIFE

## Connective Tissue

Connective tissue holds the body together. It connects organs and transports stuff around the body. Although blood flows like a liquid, it is also a form of connective tissue. Different kinds of cells make up connective tissue. The following table lists connective tissue cells with their various functions as designed by God.

There is no need for children to know the names of so many different cells and all that they do. What you do need to know is this—the human body is a very complicated machine. Two trillion white blood cells run through your body, fighting diseases all day long. Billions of mast cells are always busy balancing the body's immune system, reducing pains, and alleviating headaches and inflammations. They can even counter the effects of venomous snake bites. Fibroblasts rush in to heal a wound when you fall and skin up your legs. Did you know all of this is going on in your body right now? Indeed, you are fearfully and wonderfully made! In His kindness and mercy, God has given your body the means to take care of itself. He has created ways for the body to heal itself and to defend itself from dangerous germs.

| Connective Tissue Cells | The Function of the Cell |
| --- | --- |
| Adipose cells | Fat cells that store and release energy as needed |
| Lymphocytes | White blood cells that fight disease (bacteria, viruses, etc.) |
| Macrophages | White blood cells that detect and destroy bacteria and other bad organisms |
| Mast cells | White blood cells that reduce inflammation and regulate immune responses |
| Fibroblasts | Connective tissue cells that heal wounds and produce collagen fibers |
| Collagen fibers | Structural proteins in connective tissue that provide strength in tendons and bones |
| Elastic fibers | Long, thin, elastic cells that give flexibility to the skin |
| Reticular fibers | Delicate fibers that support gland cells and blood cells |
| Capillaries | The smallest type of blood vessel that conveys blood through body tissue |

Make a joyful shout to the LORD, all you lands!
Serve the LORD with gladness;
Come before His presence with singing.
Know that the LORD, He is God;
It is He who has made us, and not we

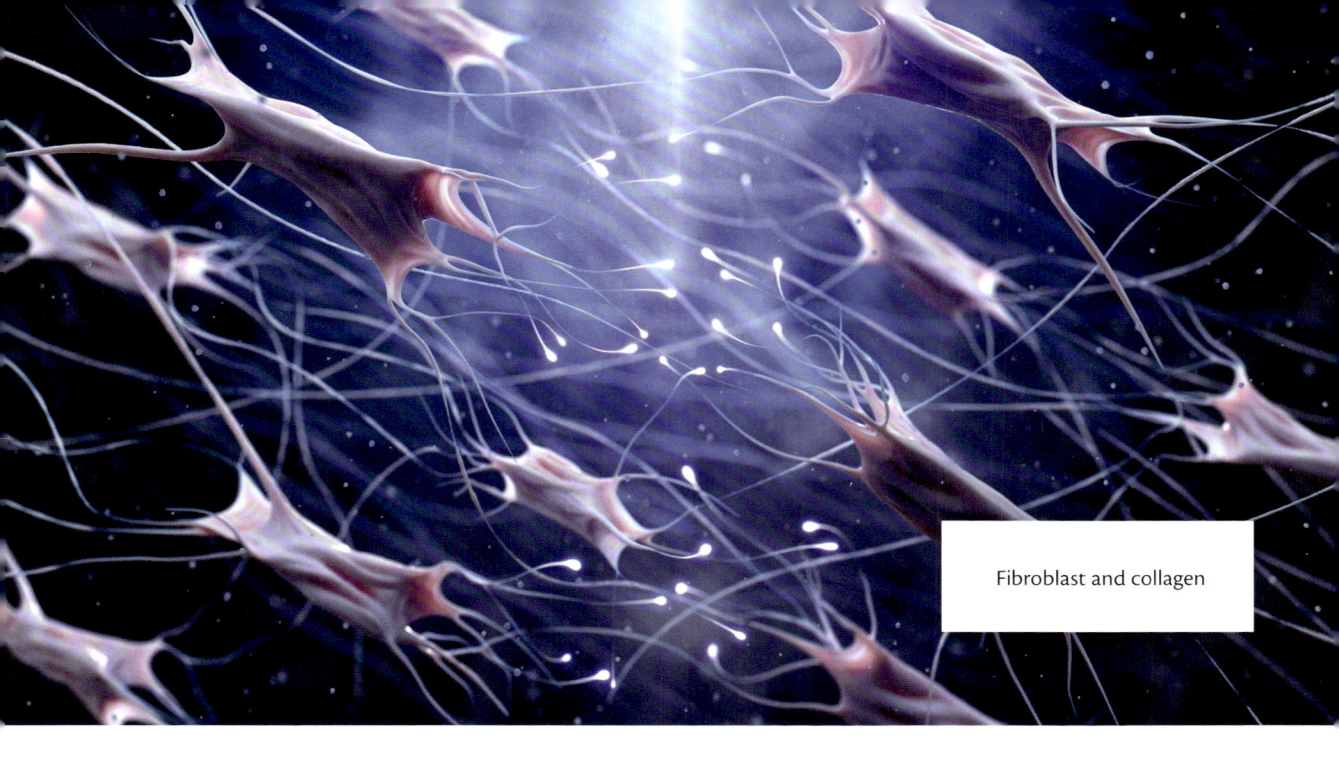

Fibroblast and collagen

ourselves;
We are His people and the sheep of His pasture.
Enter into His gates with thanksgiving,
And into His courts with praise.
Be thankful to Him, and bless His name.
For the LORD is good;
His mercy is everlasting,
And His truth endures to all generations.
(Psalm 100)

## The Wonder of Skin

There are three layers of **skin**: The **epidermis** (the part you see), the dermis underneath it, and the **deepest subcutaneous layer**. What you see of the epidermis are dead cells. But, just underneath it, there are living cells busily producing new cells for the epidermis layer. If you wear dark socks, you will see flakes of skin on the inside of your socks when you remove them. That's because your skin is constantly shedding. You lose about 40,000 cells a minute. In fact, scientists tell us that the human body sheds the whole outer layer of skin every six weeks or so.

Skin is a wonderful protective covering for the body. Imagine what it would feel like to walk around without any skin covering your body! Does that sound painful?

Skin blocks bacteria and dangerous chemicals from getting inside your body. When you go swimming or climb into a bathtub, your skin keeps water from soaking your muscles and organs. Skin makes for a good waterproof raincoat!

Your skin can take a lot of abuse. When you fall, the rest of your body is protected from injury.

God also provided you with toenails and

# GOD MADE LIFE

**Structure of the Skin**

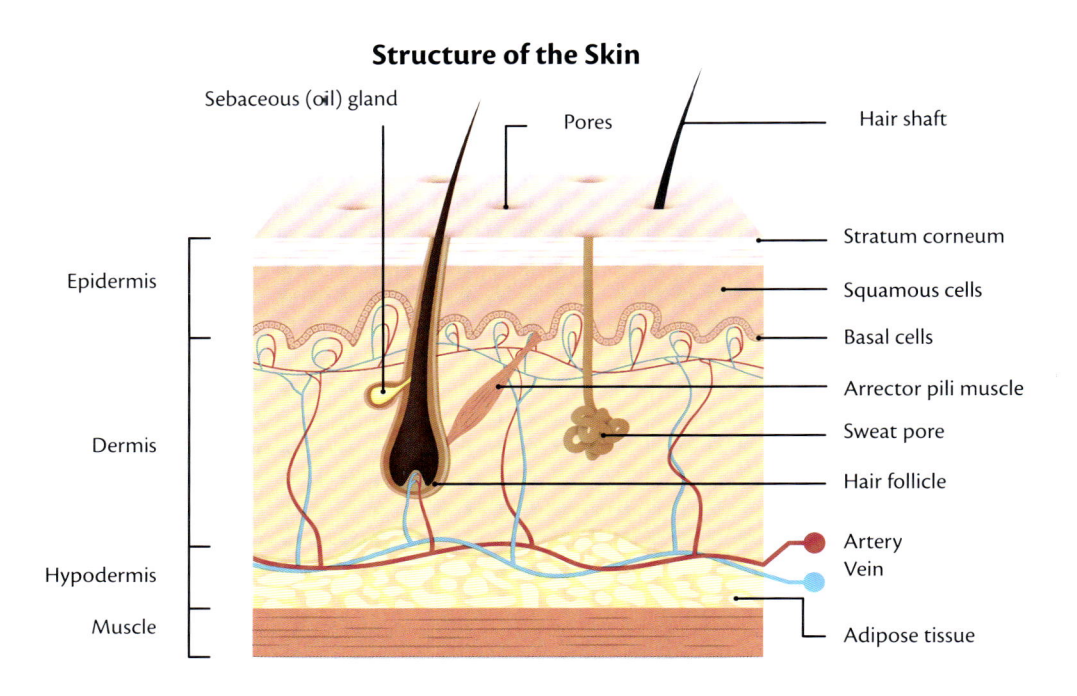

- Sebaceous (oil) gland
- Pores
- Hair shaft
- Stratum corneum
- Squamous cells
- Basal cells
- Arrector pili muscle
- Sweat pore
- Hair follicle
- Artery
- Vein
- Adipose tissue
- Epidermis
- Dermis
- Hypodermis
- Muscle

fingernails. These are thicker layers of epidermis made of epithelial tissue. The Lord knew the fingers and toes would often come into contact with hard surfaces. That's why He provided extra protection.

Skin is tough. To weaken a piece of paper or plastic, all you need to do is cut a notch into it. But a tear in the skin does not weaken it. The collagen strands twist together to make the skin really strong. But if a tear happens in the skin, God made the collagen fibers so the skin won't easily tear any more. The collagen fibers separate to keep the stress from concentrating.

When you use your hands over and over again for heavy work, your body feels the extra pressure. Over time, the epidermal cells produce even more cells in that area, thickening the skin into a **callus**. This is one more way the skin protects the body from injury. Praise God for His wise provision for the body!

If the skin is subject to too much pressure, friction, or heat over a short time, a blister forms.

This can happen if you take a long hike in shoes that rub against your feet. The dermis separates from the epidermis, and fluid gets in between, causing the blister.

There are other parts of the body poking through the under layers of skin. Embedded in your skin are sweat glands, oil glands, wax glands, hair follicles, and nerve endings. Also, blood vessels lie just under the epidermis, which is only about 0.1 mm thick—

CHAPTER 8: GOD MADE MAN

about as thick as a sheet of paper. Even a tiny paper cut can pierce into the blood vessel and cause bleeding.

Sweat caused by heat while working outside

The **sweat glands** and blood vessels team up to regulate the body's temperature. When you are outside working on a hot summer day, your blood vessels get bigger. More blood flows up towards the outer layer of the skin, transferring the increased heat from your body into the air. Meanwhile, the sweat glands open and secrete drops made of 99% water and 1% salt onto the surface of your skin. As the sweat evaporates off your skin, you will immediately feel a lot cooler. This is God's automatic cooling system He has installed into every square inch of the skin that covers your body.

On cold days, when the inside of your body needs to conserve heat, the blood vessels in your skin will constrict. The skin's surface becomes cold and clammy. If the rest of the body is rather warm, but some part of your face is exposed to the cold, the blood vessels open up again near the surface of your face. Your face turns red as the blood vessels pump more blood to warm up the cold surface. Here again we find the goodness and wisdom of God providing for our body's needs in the heat and the cold!

Praise God for your body! As the psalmist said it, we are fearfully and wonderfully made!

## Skin Care

Like it or not, the human skin is layered with "microbial life" all the time. Many of these organisms actually defend your body against an invasion from bad organisms. Scientists have found 1,000 species of microbes living in the human skin.

God has made your body to defend itself from skin diseases like psoriasis, eczema, dermatitis, acne, skin ulcers, and dandruff, as well as yeast and fungal infections. So, you want to be careful you are not killing the good bacteria while still pursuing good

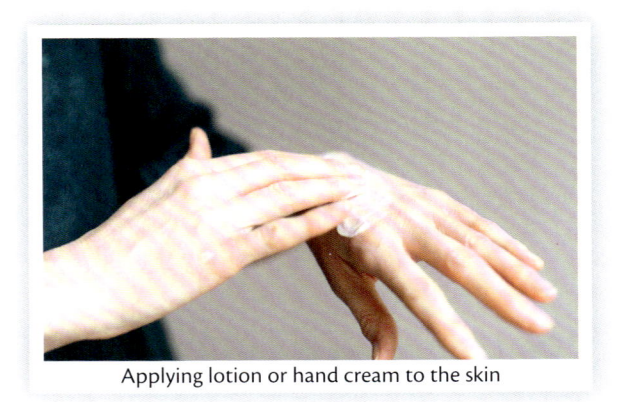

Applying lotion or hand cream to the skin

hygiene. The idea is to keep a balance of a wide variety of microbes, and that is one more reason diet is important. Yogurt, garlic, onions, and asparagus are good for healthy skin. And, you should drink plenty of water every day.

You want to be careful not to use too much antimicrobial hand sanitizer. Use gentle soaps for washing your hands. Working up a sweat and taking a warm shower helps to keep the body in healthy shape.

Too much time in a hot shower can dry out the skin. Not only does your skin become itchy, but cracks in the skin can allow bacteria to break through the skin barrier. This can result in skin infections and allergic reactions. Skin doctors recommend showering for 3-5 minutes, a couple of times a week. Obviously, if you are sweaty, grimy, or have been working in the dirt, you would want to clean yourself up.

Also, be aware that damp towels can grow a nice crop of yeasts, molds, viruses, and bacteria. After taking a bath, you could be spreading these bad organisms over your body as you are drying off. So it's a good idea to wash your towel at least once a week. If you live in a humid place, you may want to wash it even more often.

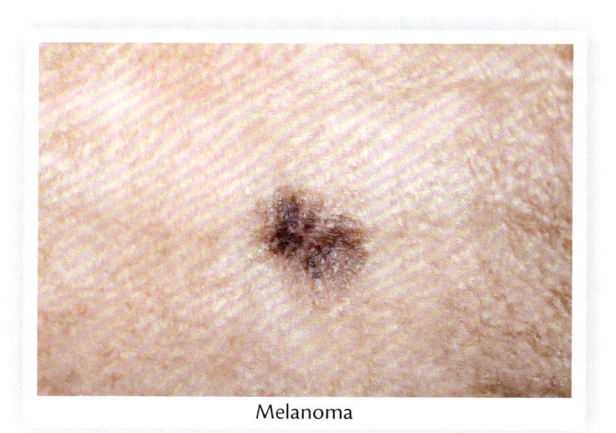

The sun is the most dangerous threat to your skin. Extended exposure to the sun can result in skin cancer. **Melanoma** is the most serious kind of skin cancer. Squamous cell carcinoma and Merkel cell carcinoma are not as serious. Children can increase risks of getting skin cancer later in life if they get sunburned over and over again. Serous sunburns causing severe blisters increases the possibility of contracting skin cancer.

Here are a few pointers for taking better care of your skin:

- Avoid picking your skin or scratching yourself too much. It's okay to remove dead skin with a light scrub. But don't scrub too vigorously. This can create tiny tears in the skin. Remember, your skin is a barrier to keep harmful bacteria from getting into your body.

- Use sunscreen when you are out in the hot sun. This protects your skin from burning and skin cancer. This same care for your skin is recommended when outdoors in the winter snow, especially if the sun is shining. The sun's rays reflecting off the snow can cause severe sunburn.

- Treat fungal infections like ringworm and athlete's foot. Because yeast loves sugar, it would be a good idea to reduce

Melanoma

Not applying sunscreen can result in sunburn

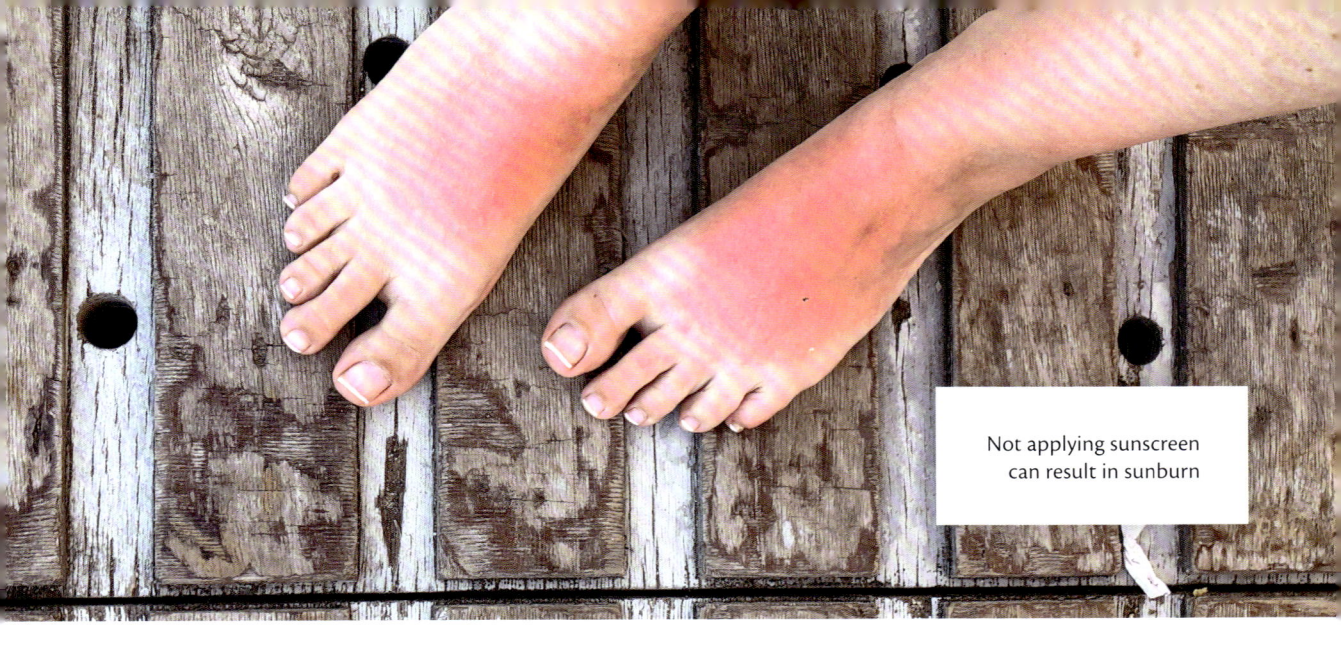

your sugar intake while fighting these infections.

## Treating Burns

Whether caused by the sun, by fire, or by touching something really hot, skin burns are always painful. The seriousness of the burn depends on the depth to which it penetrates the skin. First degree burns only affect the epidermis. Second degree burns penetrate to the dermis, and third degree burns reach into the adipose tissue.

Very serious burns include those burns that cover more than 10% of the body. When burns penetrate down into the muscle, nerves, and bones, they are referred to as fourth degree burns.

Seek medical treatment right away for serious burns. To treat a first degree burn, take the following steps.

1. Cool the skin down. Get the burned area under cool water immediately. Use cold compresses if available.

2. Protect the skin. Cover with sterile, non-sticky bandages or clean clothes. Use a petroleum-based ointment, applied several times a day.

Seek medical help for any more serious burns. Or, if the burn shows signs of infec-

**Degrees of Burn**

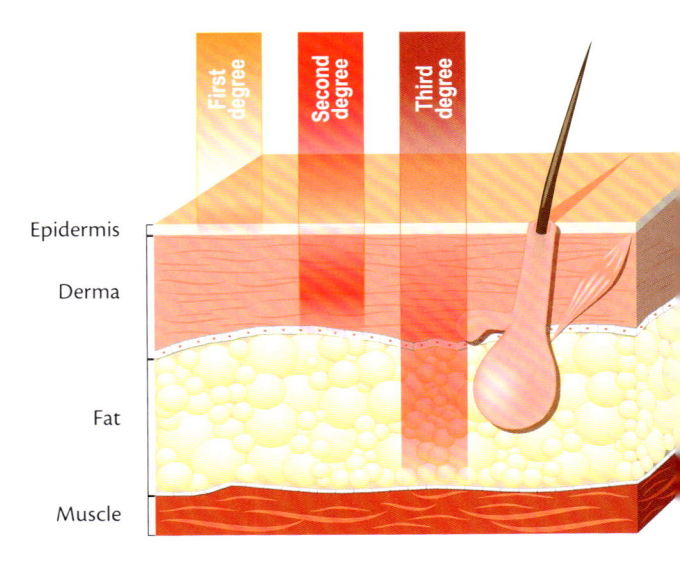

First degree
Second degree
Third degree

Epidermis
Derma
Fat
Muscle

235

GOD MADE LIFE

## Human Internal Organs

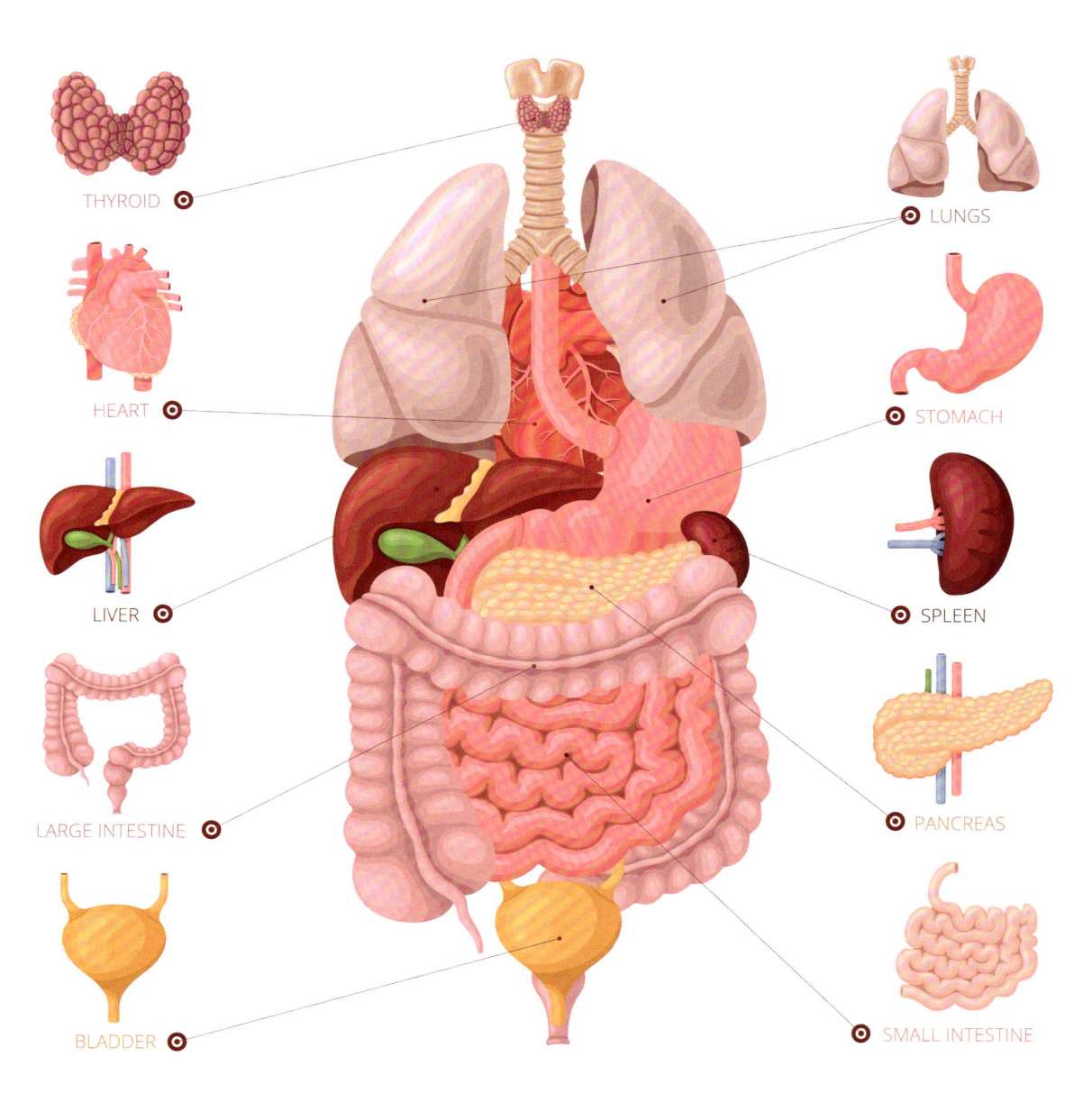

THYROID

LUNGS

HEART

STOMACH

LIVER

SPLEEN

LARGE INTESTINE

PANCREAS

BLADDER

SMALL INTESTINE

tion. That would include redness, swelling, oozing, or increased pain.

## The Wonder of the Body's Organs

To review a little of what we have already covered, the human body is made of building blocks. First, water molecules, protein molecules, and fat molecules are used to make cells. Then, these cells team up to make tissue. Finally, the tissue will form into **organs**.

An organ is a collection of tissues that God assembles into a useful, functional object for your body. Scientists have counted anywhere from 78 to 100 organs within the human body.

These organs usually play a part within a system the Creator designed for the human body. Typically, physiologists count 11 systems in the body. The tables on the following pages summarize these systems, the organs involved in each, and their functions.

For this survey, it isn't necessary to memorize all these organs and their functions. But every student should at least think about all these bodily functions. The human body is very complicated with lots of activity going on. It's at least a trillion times more complicated than an automobile. Millions of studies have been published by doctors and scientists on the human body. America spends about $180 billion each year on medical research, while American car companies spend only about $20 billion on automobile design and research each year. The human body is extremely complicated because it was designed by God to do many things.

Thank God for the functions of your body. Thank God for your nimble fingers which can snatch up small little objects. Thank God for your mouth, your teeth, and your stomach which process your food. Thank God every time you go to the bathroom. Thank God for your respiratory system. Thank God for your heart pumping blood through your body. God designed these amazing systems for human life.

For You are my hope, O Lord GOD;
You are my trust from my youth.
By You I have been upheld from birth;
You are He who took me out of my mother's womb.
My praise shall be continually of You.
(Psalm 71:5-6)

## The Body's Systems Work Together

Speaking the truth in love, [we] may grow up in all things into Him who is the head—Christ—from whom the whole body, joined and knit together by what every joint supplies, according to the effective working by which every part does its share, causes growth of the body for the edifying of itself in love.
(Ephesians 4:15-16)

The body's organs and systems do not function independently of each other. Without lungs, the heart could not function. Without

## Organ Systems in the Human Body

| System | Organ | Function |
|---|---|---|
| **Skeletal** | Bones | Give support to the body and produces blood cells |
| | Joints | Help the body move |
| **Muscular** | Muscles | Help the body move |
| **Cardiovascular** | Heart | Pumps blood |
| | Blood Vessels | Distribute blood, oxygen, and nutrition Remove carbon dioxide |
| **Respiratory** | Nose | Aids in smelling, and filters and channels air |
| | Pharynx | Channels air, food, and fluids from nose and mouth |
| | Larynx | Produces vocal sounds and channels air |
| | Trachea | Warms and moistens air as it passes into lungs |
| | Lungs | Remove waste gasses and transfers oxygen to blood |
| **Nervous** | Brain | Enables the ability to think and process information |
| | Spinal Cord | Controls body movements and organ functions |
| | Nerves | Convey information at 300 miles per hour (482 km/hr) |
| | All five senses | Receive information from the eyes, ears, touch, smell, etc., and conveys it to the brain |
| **Urinary** | Kidneys | Filter wastes out of the blood and return vitamins and good substances to the blood |
| | Ureters | Carry urine from kidneys to bladder |
| | Bladder | Stores urine and releases it at the right time |
| **Immune** | Lymph nodes | Protect against disease |

| System | Organ | Function |
|---|---|---|
| **Digestive** | Mouth | Receives air and food |
| | Esophagus | Propels food into the stomach |
| | Stomach | Stores and digests food |
| | Small Intestine | Digests food and transfers nutrients to blood |
| | Large Intestine | Fights germs, regulates water levels, and expels wastes |
| | Liver | Regulates chemicals, processes blood, etc. This organ has 500 functions. |
| | Gall Bladder | Stores bile made in the liver with which to break down fats |
| | Pancreas | Feeds digestive juices into small intestines |
| **Integumentary** | Skin | Protects the body and regulates temperature |
| | Nails | Protect fingers |
| | Hair | Warms body |
| **Endocrine** | Pituitary Gland | Secretes hormones, controls metabolism, growth, sexual maturation, reproduction, and blood pressure |
| | Hypothalamus | Keeps body in balance for temperature, thirst, emotions, sleep, etc. |
| | Thyroid Gland | Regulates metabolism, weight gain/loss, and heartbeat |
| | Parathyroid Gland | Controls calcium levels in the blood |
| | Pancreas | Produces hormones (insulin) that regulates sugars |
| | Adrenal Glands | Regulate metabolism, immune system, blood pressure, and response to stresses |
| **Reproductive** | Male organs | Produce male cells for reproduction |
| | Female organs | Produce female cells for reproduction, nurture babies in the womb, and provide milk for nursing babies |

## GOD MADE LIFE

the muscles, the intestines and the heart would never work. The skin provides vitamin D for the bones. The kidneys would never function without hormones supplied by the endocrine system. The whole body has to function as a unit or there would be no life.

The body is a very good picture of the church which Jesus has built. All of the body parts need to work if the whole body is going to function. If the lungs or the kidneys aren't functioning, the body would not survive. As we encourage each other and pray for each other, the body is healthy, and it grows. Every person plays a part in the church body, and that is how we grow up into Christ.

---

For as the body is one and has many members, but all the members of that one body, being many, are one body, so also is Christ. For by one Spirit we were all baptized into one body—whether Jews or Greeks, whether slaves or free—and have all been made to drink into one Spirit. For in fact the body is not one member but many. If the foot should say, "Because I am not a hand, I am not of the body," is it therefore not of the body? And if the ear should say, "Because I am not an eye, I am not of the body," is it therefore not of the body? If the whole body were an eye, where would be the hearing? If the whole were hearing, where would be the smelling? But now God has set the members, each one of them, in the body just as He pleased. And if they were all one member, where would the body be? But now indeed there are many members, yet one body. And the eye cannot say to the hand, "I have no need of you"; nor again the head to the feet, "I have no need of you." No, much rather, those members of the body which seem to be weaker are necessary. And those members of the body which we think to be less honorable, on these we bestow greater honor; and our unpresentable parts have greater modesty, but our presentable parts have no need. But God composed the body, having given greater honor to that part which lacks it, that there should be no schism in the body, but that the members should have the same care for one another. And if one member suffers, all the members suffer with it; or if one member is honored, all the members rejoice with it.
Now you are the body of Christ, and members individually.
(2 Corinthians 12:12-27)

---

## The Wonder of the Human Eye

Praise the LORD!
Praise the LORD, O my soul!
While I live I will praise the LORD;
I will sing praises to my God while I have my being. . . .
The LORD opens the eyes of the blind;
The LORD raises those who are bowed down;
The LORD loves the righteous.
(Psalm 146:1-2,8)

CHAPTER 8: GOD MADE MAN

Even Charles Darwin had to admit that the evolutionary theory seemed absurd when it came to the design of the human eye. Here's what he wrote: "To suppose that the eye with all its inimitable contrivances for adjusting the focus to different distances, for admitting different amounts of light, and for the correction of spherical and chromatic aberration, could have been formed by natural selection, seems, I freely confess, absurd in the highest degree."[22] Later, he said thinking about the human eye gave him "a cold shudder."

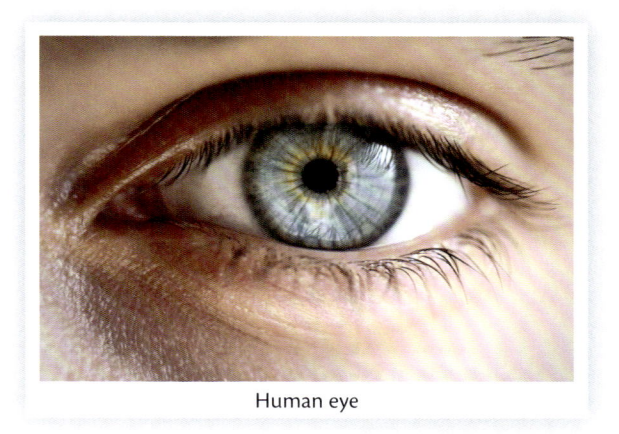

Human eye

The table on the next page lists what the parts of the human eye do, along with the similar parts found in a camera.

God made the eye to capture images, while humans have made cameras to take still pictures. Compared to a camera, the human eye is a million times more complex. Of course, the reason for this is that God is infinitely more knowledgeable than man. Here are some differences between a man-made camera and God's eye:

- A camera lens is a rigid glass, and the lens has to move mechanically to keep things in focus. The eye's lens is flexible, and muscles around the eye quickly adjust the lens so everything stays in focus all the time.

Charles Darwin

The human eye is made up of 100 million parts. As light enters the eye, there are a million nerve fibers that communicate the light signals to the brain. The eye is the second most complex organ in the human body.

- Assuming optimum vision (as God created it), the quality of the image seen by the human eye is about the same as that taken with a 52 megapixel camera. One picture taken with a 52 megapix-

241

# GOD MADE LIFE

| Part of the Eye | Function | Similar Camera Part |
|---|---|---|
| Sclera (white part) | Keeps the eye in a rounded shape | |
| Cornea | Controls and focuses light coming in | Front lens |
| Iris (colored part) | Regulates size of pupil | Aperture |
| Pupil (dark spot) | Opening that allows light to enter | Aperture |
| Lens | Focuses the light on the retina | Back lens |
| Retina | Forms the image using neurons | Film |

el camera would take about 15 MB of memory. Now, some eyes can be trained to pick up 255 frames per second. That would amount to about 4 GB per second of information transferred to the brain. Just two minutes of eyesight would fill up your hard drive on a laptop computer. That's how much information your eyes are processing all day long!

- The retina can also adjust to varying amounts of light, whereas the camera's film can't.

The eye is also built for night vision. God has equipped the human eye with two types of photoreceptors or neurons. He has placed 100 million photo-receptors called **rods** on the retina so that you can see in black and white at night. For daytime vision, He installed 6 million of a second kind of photoreceptor called **cones** capable of making out images in color. There are blue-type, red-type, and green-type cones capable of picking up these wavelengths.

There are quite a few other accesso-

**Pupil Dilation Adjusted by Exposure to Light and Dark Conditions**

Bright light    Low light

242

ries God provided the eye. For instance, the eye adjusts easily for quite a few gradations of light and dark. When you walk into a dark room, you can't see very well at first. That's because the rods aren't working yet. It takes a while for the pupils to get bigger, which lets more light into the eye. Then, some chemicals signal to the brain that the rods need to start doing their job.

**Human Eye**

Sclera
Iris
Cornea
Pupil
Lens
Ciliary body and muscle
Conjuctiva
Retina
Optic nerve
Macula
Retinal blood vessels
Vitreous body

God also supplied the eye with a wonderful fluid called the **Aqueous humor** inside the eye. This fluid cleans the cornea, provides energy to the cells working hard on the surface of the cornea, and keeps the pressure on the cornea within safe limits.

Also, the Creator supplied another gel-like fluid called the **Vitreous humor** which fills the cavity between the lens and the retina. This fluid maintains the shape of the eye.

To clean the outside of the eyes, our Creator made **tear ducts.** The tears coming from the ducts prevent the eyes from getting too dry. This salty liquid provides some oxygen and nutrients to the eye. They wash away germs or unhealthy chemicals and have a healing component. They maintain a smooth surface so you can see clearly. As the eyelids blink, they help to keep the eyes clean and bright. What a tremendous system God made for us in the human eye!

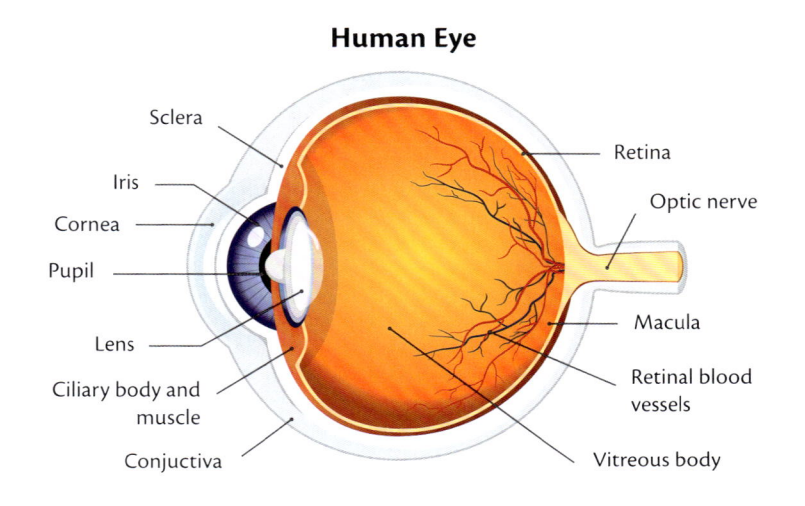

Then Job answered the LORD and said:
"I know that You can do everything,
And that no purpose of Yours can be withheld from You. . . .
I have heard of You by the hearing of the ear,
But now my eye sees You." (Job 42:1-2,5)

## The Wonder of the Nervous System

Eyes and ears are part of what is called the **peripheral nervous system**. But these sensory organs connect to the body's **central nervous system**. Indeed, the brain and the parts of the human nervous system make up the very pinnacle of God's genius seen in the physical creation. It is the most complex creation in the universe.

There is a lot of communication going on among all the people on Earth. Every day, there are 19 billion texts and about 6 billion cell phone calls made around the world. But a single body is processing about 1 trillion messages per second. There are 100 billion neurons in your body compared to 7 billion people in the world. This means your body maintains a communication system that is more complicated than the world's telecommunication system. You aren't even conscious of most of the messages. Your brain is taking care of lots of things without you being aware of it.

These neurons are interconnected through 120 trillion "connection boxes." Think of a system containing billions of telephone wires that connect all the homes in the world to each other. About 90% of the neurons in the body are contained in the brain. The neurons communicate by nerve impulses, all running at about 300 mph (482 km/hr) throughout the brain and around the body. That's a very busy communication system!

There are about 4 million "reporting stations" embedded within your skin across the surface of your body. Each station is equipped with neurons that send messages to the brain. These are all capable of sensing pain on every square inch of your body. About 500,000 of them were designed to detect surface conditions by determining whether they are soft, hard, rough, or silky. Another 200,000 of these neurons are sensitive to temperature. All of these stations are constantly reporting to the brain what is going on around you. Praise God for the body's amazing senses that give us a very acute awareness of the surrounding environment!

The nervous system picks up sensory information. It directs your muscles to move and helps you keep your balance. When you touch something hot or slimy, cold or rough, the brain lets you know. If you've tripped on something and you're about to fall, the brain quickly sends messages to your muscles to rescue yourself. Your brain processes sounds and sights and lets you know what's happening all around you.

## How the Nervous System Works

The nervous system sends messages to the brain for processing. The brain also sends messages to the body, telling it to take action. Signals go in both directions.

Nerves are made up of bundles of neurons. Each neuron is equipped with 128 dendrites, which are short branches that stick out of the neuron. The gap between neurons is called the synapse. The **dendrites** communicate with other neurons across this gap using synaptic connections. Each dendrite can create 40 connections with other neurons.

When the nerves in your fingers pick up a sudden increase in heat, like a burning sensation, the signal passes through the dendrites. The **soma** (or the neuron's cell body) determines if the signal was strong enough to pass it up the chain to other neurons. If it is, the nerve fires and creates an electrical signal. The signal travels to the other end of the neuron along another branch called the axon, where it is received by **axon terminals**.

## CHAPTER 8: GOD MADE MAN

### Human Nervous System

**Parasympathetic Nerves**

- Constrict Pupils
- Stimulate Salvia
- Constrict Airways
- Slow Heartbeat
- Stimulate Activity of Stomach
- Inhibit Release of glucose / Stimulate Gallbladder
- Stimulate Activity of Intestines
- Contract Bladder

**Sympathetic Nerves**

- Dilate Pupils
- Inhibit Salivation
- Relax Airways
- Increase Heartbeat
- Inhibit Activity of Stomach
- Stimulate Release of glucose / Inhibit Gallbladder
- Inhibit Activity of Intestines
- Secrete Epinephrine and norepinephrine
- Relax Bladder

BRAIN

Cranial Nerves

Cervical Nerves: C1, C2, C3, C4, C5, C6, C7, C8

Thoracic Nerves: T1, T2, T3, T4, T5, T6, T7, T8, T9, T10, T11, T12

Lumbar Nerves: L1, L2, L3, L4, L5

Sacral Nerves: S1, S2, S3, S4, S5

Coccygeal Nerves: Co1

Spinal Cord

Sympathetic Chain

---

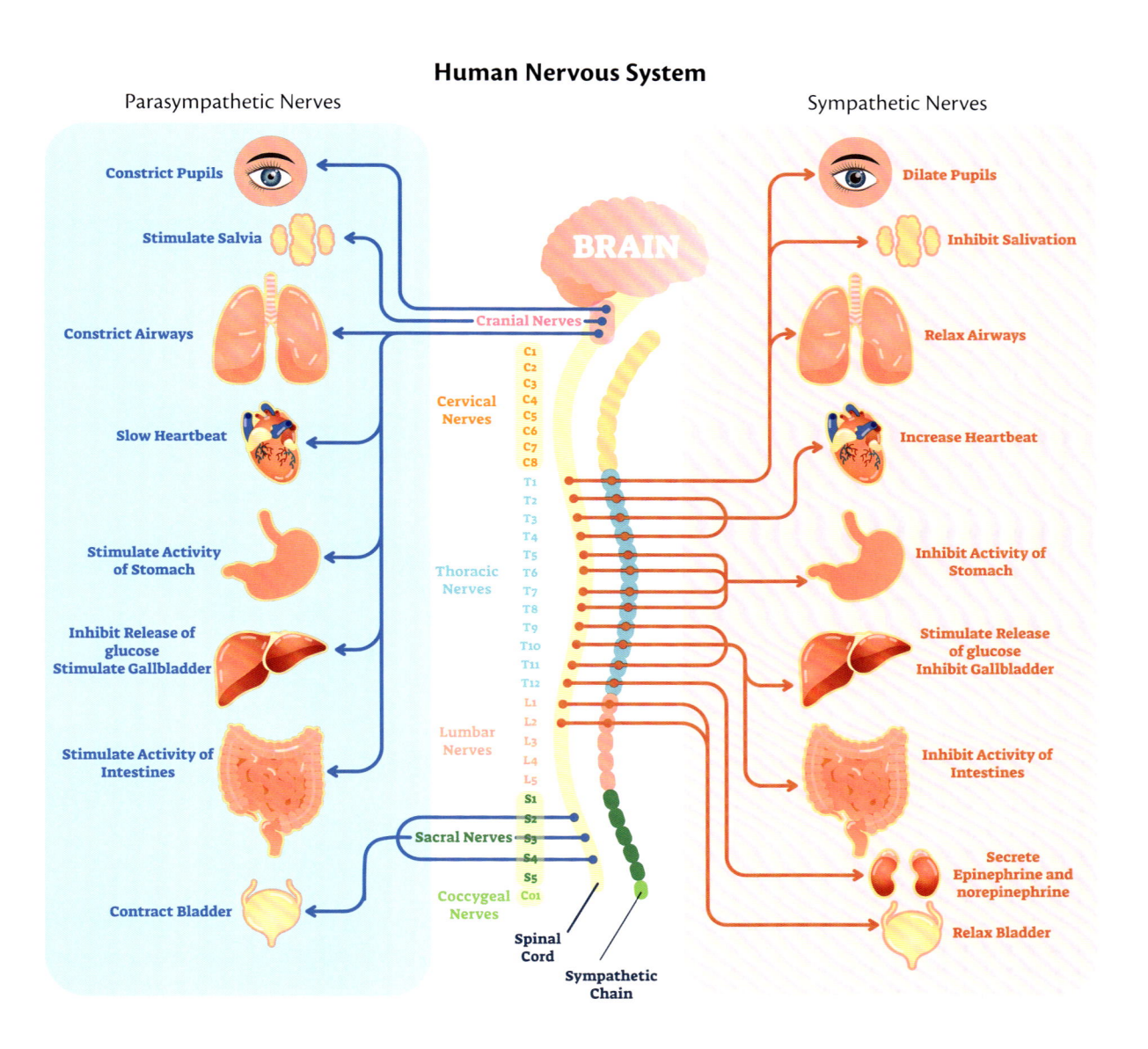

These sit close to the dendrites of another neighboring neuron that sits waiting to receive messages.

At the very tip of the axon terminals are little vesicles containing **neurotransmitters**. When the electrical signal hits these vesicles, they burst open, allowing the neurotransmitters to escape into the synapse. If enough neurotransmitters connect with the next neuron, another electrical signal will

## GOD MADE LIFE

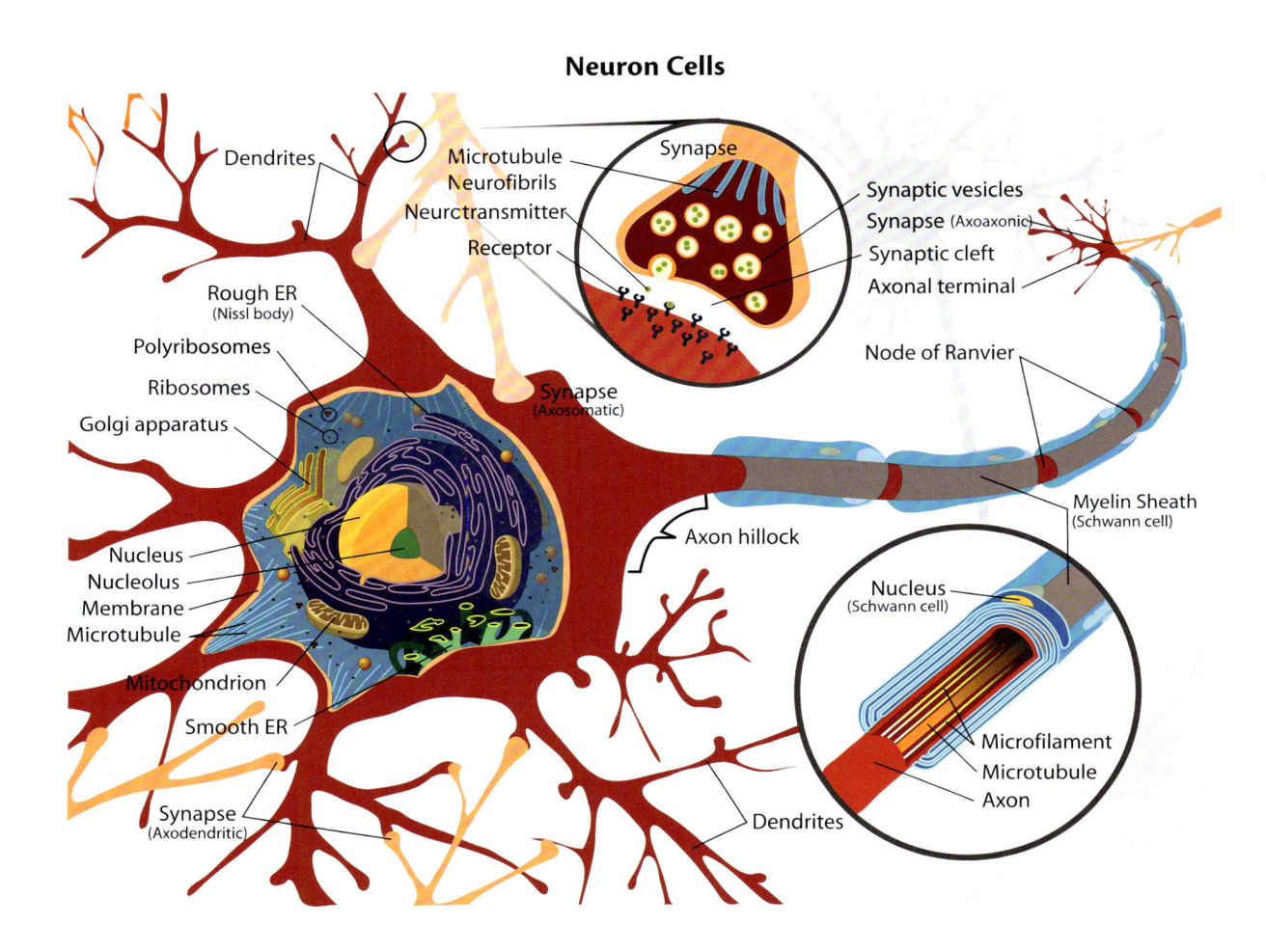

**Neuron Cells**

transmit through this neuron to the next one in the chain.

There are about 100 different kinds of neurotransmitters. These amazing chemicals control heart rate, breathing, moods, digestion, concentration, appetite, muscle movement, and sleep cycles. The better-known neurotransmitters are dopamine, adrenaline, and endorphins.

When the body picks up some sensation, the signal is passed to the **spinal cord** and then travels along to the brain. As the brain decides to act, another nerve impulse fires from one neuron to the next, traveling through the spinal cord until it reaches an **effector**. The effector could be either a muscle which immediately contracts or a gland that secretes a hormone. Two examples of these hormones are insulin and prolactin. **Insulin** is found in the pancreas and controls glucose in the body. **Prolactin** is what starts the process of milk production when a mom is nursing her baby.

Some of your body's muscular actions occur extremely quickly. They are called **reflexes**. If you are going to pick up a ball, you have to think about it first. You see the ball, and your brain tells your hand to pick it up. But reflexes act without you having to think about it. If you touch a hot stove, your hand jerks away quickly. You react more quickly because the neurons communicate the message only as far as your spinal cord; they do not carry them all the way to your brain. God made a special neuron called the **interneuron** to pass the signal on to a **motor neuron**. And the motor neurons are designed to quickly signal the muscles to contract. That's what makes your hand pull back from the stove. These signals are running through your body at 328 feet per second (360 km/hr). The time elapsing between sensing the heat and yanking your hand back is only about 0.015 seconds.

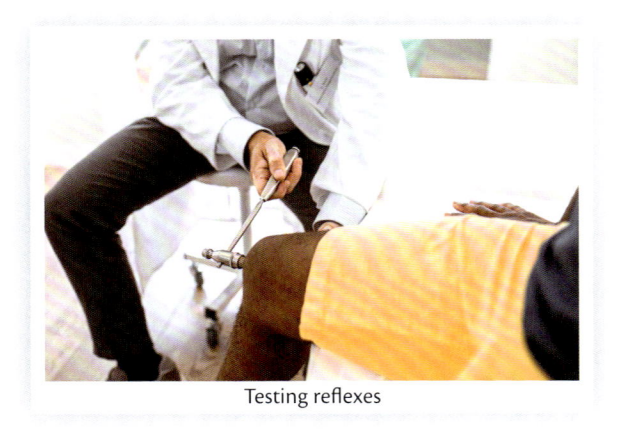

Testing reflexes

## The Wonder of the Sense of Touch

And suddenly, a woman who had a flow of blood for twelve years came from behind and touched the hem of His garment. For she said to herself, "If only I may touch His garment, I shall be made well." But Jesus turned around, and when He saw her He said, "Be of good cheer, daughter; your faith has made you well." And the woman was made well from that hour. (Matthew 9:20-23)

By faith, this woman in the story from Matthew, touched the hem of Jesus's garment. Not only did she feel the surface of His garment, but in this case, she felt the power of healing that came from the Son of God Himself. The intricate system of nerves in your skin is keeping track of the environment around you—more than you are aware. The skin is paying attention to temperature, humidity, light, radiation, wind, parasites, particles in the air, and changes in pressure. The nerves in your hand can pick up the difference between a smooth pane of glass and a 0.0004 inch (0.01016 mm) deep scratch in the glass. That's one tenth the size of what you can see with the naked eye.

Various parts of the body are calibrated to sense different levels of pressure. For example, the tough soles of your feet will feel a heavier touch at 250 milligrams per square millimeter. However, your fingertips are way more sensitive, responding to only

# GOD MADE LIFE

3 milligrams of pressure, and the tip of your tongue to 2 milligrams. But when something like a tiny eyelash touches your cornea with just 0.2 milligrams per square millimeter of pressure, your eye will respond, usually by blinking. Every square inch of your body has been designed with a special sensitivity according to the Creator's blueprints. None of this happened by chance. The human body is fearfully and wonderfully made!

Your skin can also adapt to the environment. Have you ever climbed into a hot bathtub? At first, you can barely stand the heat. But after 30 seconds or so, your skin gets used to it.

Animals and humans are made with a need to touch and to feel the touch of others. If you have witnessed dogs and cats giving birth, you know the mothers lick their puppies and kittens. Most mammals will do this. Zoologists believe this stimulates the young animals' breathing and digestion.

A high percentage of babies died in orphanage hospitals in the early 1900s. After studying the problem for awhile, doctors finally realized that the babies weren't being held enough. Infant mortality rates dropped quickly when the babies were picked up and cuddled several times a day.

*Mom holding a baby*

*Mother dog licks her puppy*

He will feed His flock like a shepherd;
He will gather the lambs with His arm,
And carry them in His bosom,
And gently lead those who are with young. (Isaiah 40:11)

## The Wonder of the Human Brain

For behold,
He who forms mountains,
And creates the wind,
Who declares to man what his thought is,
And makes the morning darkness,
Who treads the high places of the earth—
The LORD God of hosts is His name. (Amos 4:13)

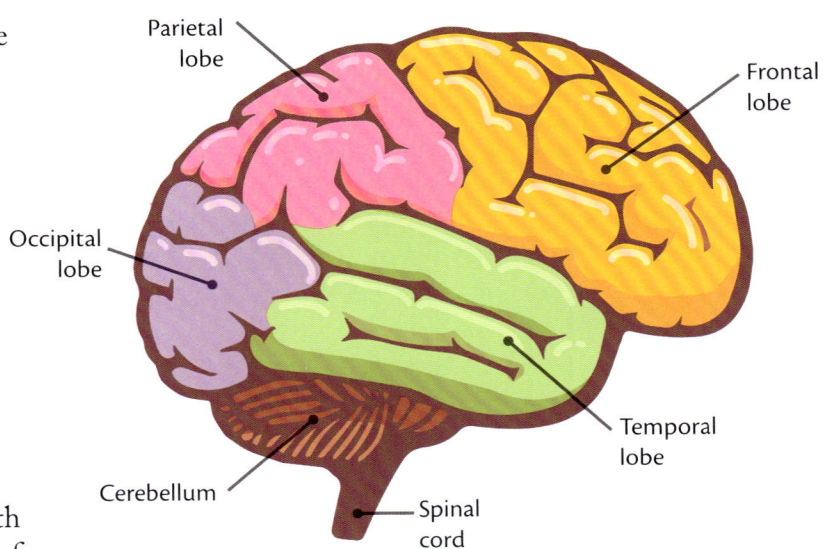

**CHAPTER 8: GOD MADE MAN**

Man was created in the image of God, capable of understanding God's revelation to him. Monkeys cannot read God's Word. If you were to tell an ape, "God tells us not to steal food from one another," he would not understand you. He would not understand the word "God." And he would continue taking food from other apes.

But God created man with a soul and a mind, capable of knowing God. Unlike monkeys, man can understand the instructions God gave to him. God spoke to man and told him to take care of the Garden of Eden. He told the man not to eat from one tree in the garden. Of course, both Adam and Eve understood what God said, but they disobeyed Him. We can't understand everything God knows. But, we can understand some of it. We can understand what He reveals to us.

**Parts of the Human Brain**

> "The secret things belong to the Lord our God, but those things which are revealed belong to us and to our children forever, that we may do all the words of this law." (Deuteronomy 29:29)

There are many things scientists do not understand. Plants are not aware of their own existence. Yet, we are conscious that we exist.

Humans are concerned about the purpose of life, the value of life, and the curse of death. Plants do not try to interpret what is going on around them. We do.

There are many things scientists do not understand. For example, they don't know why the body needs sleep. There are also many things they don't know about the brain God created. Brains can record memories similar to the way text and pictures are stored on a computer hard drive. But where are these memories stored in the brain, and how are they stored? Why do some memories fade but other memories are clear like they just happened to you? These are mysteries yet to be discovered. But there are also different kinds of memory. There is **short-term memory** where five to seven facts

# GOD MADE LIFE

or numbers can be remembered for a few seconds. There is also **long-term memory**. There is a declarative memory of names, numbers, and facts. Then, there is also non-declarative memory where we learn to do things by habitual practice. Then, there are **"flashbulb" memories** where we can remember many details of some memorable event.

How does the mind direct the constant, smooth movement required to thread a needle or to play a complicated musical piece? Robots move in a jerky motion. Nerve impulses are often quick and haphazard. How does the brain control every nerve and every muscle to result in such smooth action?

What is the imagination—this place where the brain creates experiences or objects never seen before? How does language come about? Animals have never invented one word, but man has assigned many words to objects and concepts. How does the mind conceive of God? What is the moral conscience, or where do the feelings of guilt come from when we do something sinful? Thinking involves more than chemical processes in the brain. Our thoughts can be either good thoughts or evil thoughts.

How do thoughts impact our emotions and will? The brain may recognize that one child grabbed a toy from another child. But why does the child get angry? Why does the child decide to punch the other child in the nose? Scientists will never answer these questions by studying the brain. There is an interaction between the brain and the invisible soul of man.

Old Agur in Proverbs 30 had to admit that there were things "too wonderful for me." There were parts of God's creation that he did not understand. The smartest scientists in the world cannot explain the human brain because God made it. Our only response to this amazing creation of God is to humble ourselves and praise Him for His infinite wisdom! Amen!

---

Oh, the depth of the riches both of the wisdom and knowledge of God! How unsearchable are His judgments and His ways past finding out!
"For who has known the mind of the LORD? Or who has become His counselor?"
"Or who has first given to Him And it shall be repaid to him?"
For of Him and through Him and to Him are all things, to whom be glory forever. Amen. (Romans 11:33-36)

---

## The Influence of God's Word on the Human Mind

We are responsible for our minds. We must control our body's functions. We cannot allow our hands to punch our brothers and sisters. Even if we feel ourselves getting angry, we must still control our emotions. God doesn't want us to strike our friends and family members in anger. In the same way, God wants us to control our thoughts.

**CHAPTER 8: GOD MADE MAN**

He does not want us to think sinful, bad thoughts.

Our thoughts are powerful. The more we think about something evil, the more likely we are to do something evil. The more we are discontented and covetous in our thoughts, the sooner we turn into an unthankful and grumpy person. But, if we are thinking of God's goodness and His mercy to us, we will sing the praises of God louder when we gather with the church on Sundays.

For as he thinks in his heart, so is he. (Proverbs 23:7)

And even as they did not like to retain God in their knowledge, God gave them over to a debased mind, to do those things which are not fitting... (Romans 1:28)

Because the carnal mind is enmity against God; for it is not subject to the law of God, nor indeed can be. (Romans 8:7)

And do not be conformed to this world, but be transformed by the renewing of your mind, that you may prove what is that good and acceptable and perfect will of God. (Romans 12:2)

Finally, brethren, whatever things are true, whatever things are noble, whatever things are just, whatever things are pure, whatever things are lovely, whatever things are of good report, if there is any virtue and if there is anything praiseworthy—meditate on these things. (Philippians 4:8)

Meditating on God's Word

## Pray

- Thank the Lord for the 100 organs in your body and the organ systems that work together in concert. What a beautiful, coordinated design God has provided for us!
- Praise God for the miracle of the human eye!
- Praise God for the 100 billion neurons and its communication system in your body! It is more complicated than the whole telecommunication and phone systems in the world. Praise the Lord for His wisdom that is beyond the capability of the human mind to understand!
- Praise God for the brain—a design that baffles scientists more than anything else in the physical creation!
- Thank the Lord for providing us with a mind so that we would have the ability to come to know God and realize His love for us.

## Sing

Having seen the marvelous work of God in the creation of man (all of us), an appropriate response is always worship and praise. If the student is unfamiliar with the hymn or psalm, some version of it is available on the internet and may be accessed (with supervision) for singing along.

*Now Thank We All Our God*

Now thank we all our God
With heart and hands and voices,
Who wondrous things has done,
In whom His world rejoices;
Who from our mothers' arms,
Has blessed us on our way
With countless gifts of love,
And still is ours today.

O may this bounteous God
Through all our life be near us,
With ever joyful hearts
And blessed peace to cheer us;
To keep us in His grace,
And guide us when perplexed,
And free us from all ills
Of this world in the next.

All praise and thanks to God,
The Father, now be given,
The Son and Spirit blest,
Who reign in highest heaven,
The One eternal God,
Whom heaven and earth adore;
For thus it was, is now,
And shall be evermore.

## Do

Choose at least one of the following activities and apply the lessons you have learned in this chapter.

1. Consider the amazing capacity of your brain to remember things. What are your earliest memories? How long ago did these things happen? What details can you remember about what happened? Take a moment and praise God for you brain which remembers these things that happened so long ago.

2. Take good care of your eyes.

    a. Here are some things you can do to keep your eyes healthy and working well. Which of these pointers would be helpful for you?

    - Eat well. Green leafy vegetables like spinach and kale, citrus fruits, nuts, beans, salmon, and other oily fish provide omega-3 fatty acids, zinc, and vitamins C and E. Carrots are also beneficial for your eyes.

    - Wear sunglasses when facing the sun. Do your sunglasses block out 99-100% of Ultraviolet A and B rays? Wraparound lenses are even better.

    - Use safety glasses when working on carpentry projects or playing sports like ice hockey or racquetball.

    - Look away from the computer screen at least once every 10-20 minutes. Look at something at least 20 feet away for at least 20 seconds.

    b. Doctors recommend a regular eye screening every 1-2 years. Conduct a simple test for far-sightedness and near-sightedness using eye charts available on the internet. Check one eye at a time.

3. Make a list of the ways in which God's creation of the human body is more complicated than man's creations. Explain how this disproves the evolutionary theory that the world came about by chance. Take a moment and give God the glory for His brilliant work in creating your body.

4. One way which our minds are renewed is by continual exposure to and meditation upon the Word of God. Memorizing Scripture is a great way fill your mind with the truths of the Lord. It will also equip you to prayerfully answer temptations and lies as they cross your path throughout your life. Your mind will never be able to remember things as well or as

# GOD MADE LIFE

easily as it can while you are young. To take advantage of this time, find a memorization method that works for you and set a goal to memorize a few verses, a chapter, or even an entire book of the Bible that God is using in your life currently. Therapists who work with memory care patients encourage them to repeat whatever they are trying to learn for 21 consecutive days; this creates new paths in their minds, helping them to permanently remember what they are learning. Thus, as you memorize, commit to repeating these passages for at least 21 days, and revisit them after a time to refresh your memory. By God's grace, you will find that they will become embedded in your mind and heart over time!

## Watch

To watch the recommended videos for this chapter, go to **generations.org/GodMadeLife** and scroll down until you find the video links for Chapter 8. Our editors have been careful to avoid films with references to evolution; however, we would still encourage parents or teachers to provide oversight for all internet usage. These videos may not give God the glory for His amazing creative work, so the student and parent/teacher should respond to these insights with prayer and praise.

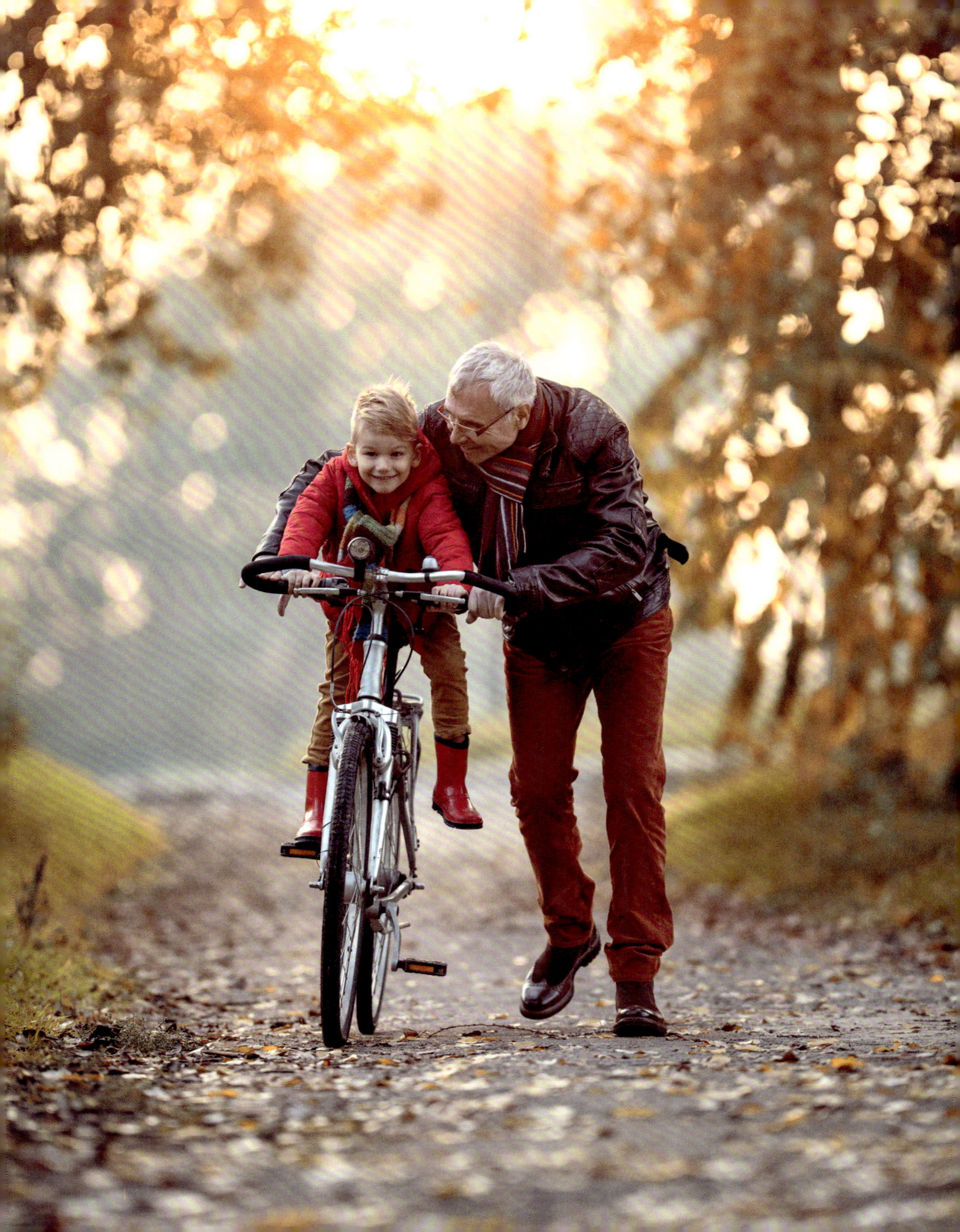

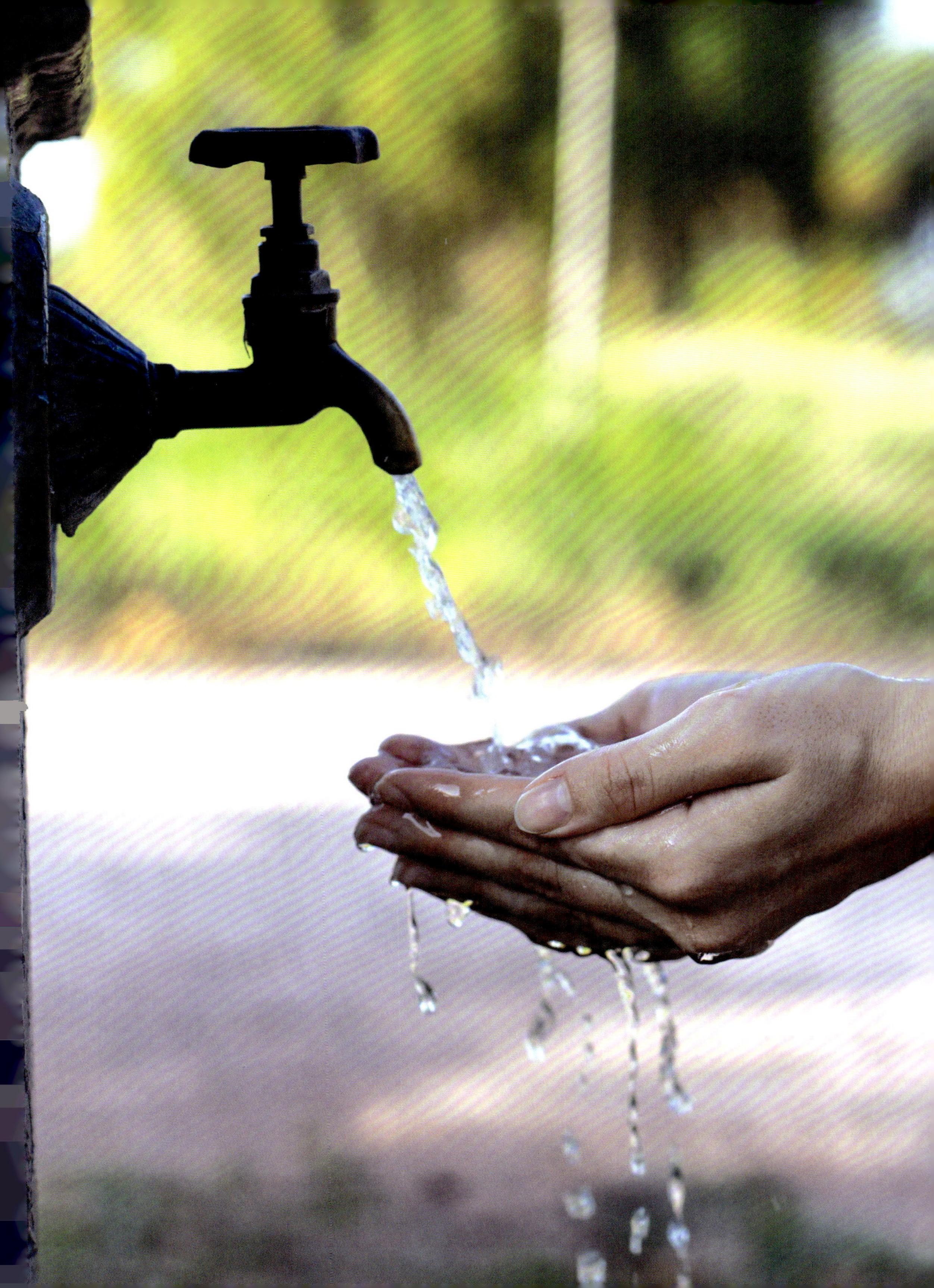

## CHAPTER 9
# GOD SUSTAINS HUMAN LIFE

[Christ, the Son of God] is the image of the invisible God, the firstborn over all creation. For by Him all things were created that are in heaven and that are on earth, visible and invisible, whether thrones or dominions or principalities or powers. All things were created through Him and for Him. And He is before all things, and in Him all things consist. (Colossians 1:15-17)

God made life, and He sustains life. He made us, and He sustains us. He provided functions in the body which help to sustain our lives day to day. At the very least, the body needs water, food, and breath to live. All of these come into the body by way of the mouth and nose. The Lord designed the mouth to be quite versatile, so it can eat, chew, taste, speak, breathe, cough, sneeze, and blow. He provided two inlets for breath so you can eat and breathe at the same time. The nose, mouth, and throat are designed to coordinate all these activities without you choking to death.

## The Wonder of The Human Nose

I will praise You, for I am
fearfully and wonderfully made;
Marvelous are Your works,
And that my soul knows very well.
(Psalm 139:14)

The **nose** has three very important and unique features. First, it filters out germs that make it into the nasal cavity. Often, it is better to breathe through your nose to take advantage of this benefit. Bacteria and viruses breathed into the nose get stuck in a

# GOD MADE LIFE

thin layer of sticky mucus. Then, tiny hair-like cilia sweep the mucus with all the germs down into your throat. When you swallow it, the junk goes down into your stomach, and the bad stuff is eliminated. There are also nasal hairs that filter out dirt, bacteria, and junk you don't want in your body.

When you get a cold, the mucous glands and sinuses will produce extra mucus, causing your nose to run. That's supposed to flush the bad viruses and bacteria out of your system. Most of the time, your body produces about ¼ gallon (1 L) of that mucus every day, most of which is swallowed. If the mucus is turning green or yellow, there is a strong possibility you have a bacterial infection.

The second way your nose benefits your bodily health is by treating the air as it enters your lungs. There are blood vessels running very close to the surface of your nose which are designed to warm up the air as it enters the nasal cavity. Since they are so close to the surface, you can have nose bleeds if your nose is too dry. Your lungs appreciate the warm, moist air coming in through your nose rather than the cold, dry air coming through your mouth when you're outdoors in cold weather. Your lungs need to stay moist, so excessively dry air can be a detriment.

The third benefit of the nose is the sneeze. When the nerves along the internal lining of the nose sense an unusual invasion of foreign materials like dust, smoke, and microbes, a signal is sent to the brain. Somehow, the brain realizes the nose needs to get this stuff out. . . now! Your whole body braces itself for the sneeze. Your eyes shut, your tongue moves to the top of your mouth, and kaboom! You sneeze. All water, mucus, and microbes explode into the air. You can expel 100,000 germs across a distance of 27 feet (8.25 m) with one strong sneeze.

## The Wonder of the Human Throat

Have you ever wondered how you can eat and breathe without choking? Sometimes food goes down the wrong tube, and you choke. But generally, your **throat** can swallow food and inhale air at the same time.

Your throat, or **pharynx**, connects your nose and mouth to the rest of your respiratory and digestive systems. In respiration, air enters your throat through your nose or mouth and passes from your pharynx through your **larynx** (or voice box), which guides the air into your lungs. The larynx is one of two safety measures God has put into your throat. It can close quickly by reflex to keep food and liquid from going into your lungs.

The second safety measure is the **epiglottis**, a flap of cartilage at the root of the **tongue**. In addition to air, your food and

# CHAPTER 9: GOD SUSTAINS HUMAN LIFE

liquid must also pass through the pharynx. However, when you swallow, your body recognizes that you are taking in something other than air, and the epiglottis flap closes over the **glottis**, which is the small opening of the larynx. This closes the pathway to the lungs and directs the food or liquid into your **esophagus** and down to your stomach. If food or liquid makes it into your lungs, it's called **aspiration** and can result in pneumonia.

What if a little food does pass into the larynx? In that case, the folds in the lining of the larynx wrap around the food, and the body starts to cough in order to remove the food from the air's pathway. If the food does slip into the **windpipe** (**trachea**) under the larynx, it could block the air's pathway into the lungs. This is when things get very dangerous. At this point, the food has made it past both the epiglottis flap and the larynx. If you can't breathe, oxygen is cut off from your body and your brain. Permanent brain damage could happen in four minutes, and you could very well be dead in eight minutes. Here are the signs that somebody is choking and may be unable to breathe at all:

- He or she is unable to talk
- He or she makes squeaky sounds when trying to breathe

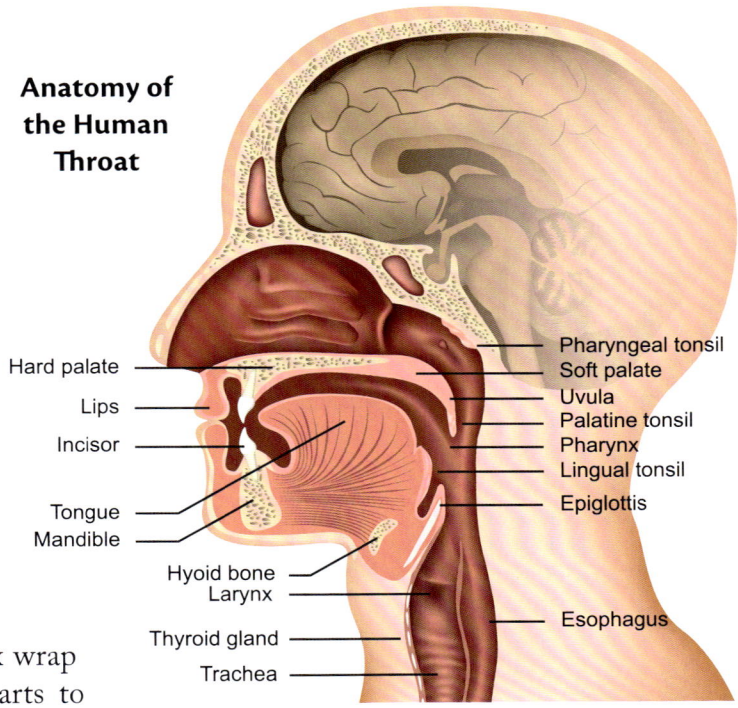

**Anatomy of the Human Throat**

Hard palate
Lips
Incisor
Tongue
Mandible
Hyoid bone
Larynx
Thyroid gland
Trachea

Pharyngeal tonsil
Soft palate
Uvula
Palatine tonsil
Pharynx
Lingual tonsil
Epiglottis
Esophagus

- He or she tries to cough, but it is very weak
- His or her skin, lips, and nails are turning blue
- He or she loses consciousness

## How to Save a Life and Help Somebody Who is Choking

Deliver those who are drawn toward death,
And hold back those stumbling to the slaughter.
If you say, "Surely we did not know this,"

259

## GOD MADE LIFE

*Does not He who weighs the hearts consider it?*
*He who keeps your soul, does He not know it?*
*And will He not render to each man according to his deeds?*
*(Proverbs 24:11-12)*

---

To be ready to help in case somebody is choking, a certified first aid course is recommended. The American Red Cross recommends a "five-and-five" approach for somebody choking. But always remember to pray first as you attempt to save somebody's life.

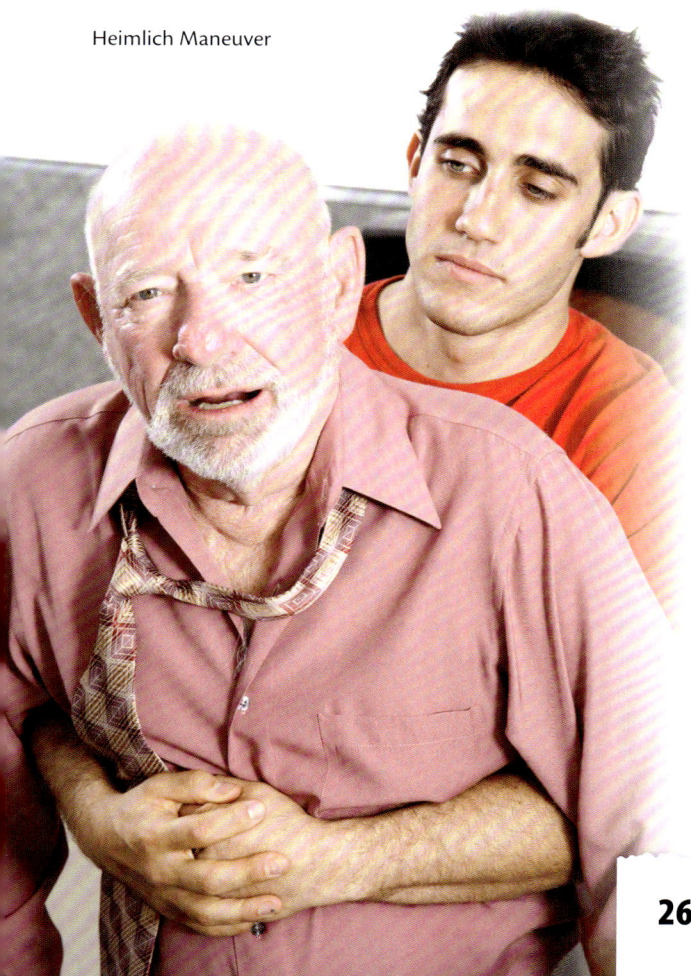

1. Pray.
2. Apply five back blows.
   - Stand just behind a choking adult, and a little to the side. In the case of a child, kneel behind him or her.
   - Place one arm around the victim's chest.
   - Have the person bend over at the waist so the upper body is parallel to the ground.
   - With the palm of your hand, give five firm blows to the victim's back right between their shoulder blades.
3. Apply five abdominal (stomach) thrusts. This is called the **Heimlich Maneuver**. This is more technical, so stick with the back blow technique if you are unfamiliar with this.
   - Stand behind the victim, and brace yourself by placing one foot in front of the other.
   - Wrap your arms around the victim's waist.
   - Tip the victim forward a little bit.
   - Make a fist with one hand, and press in just above the victim's navel.
   - Grasp your fist with the other hand, and press hard into the victim's stomach with a quick upward thrust.

Heimlich Maneuver

## CHAPTER 9: GOD SUSTAINS HUMAN LIFE

Repeat the five-and-five technique until you dislodge the food from the victim's windpipe. Preferably, you should take a first aid course before having to perform these functions. In a worst-case scenario, a doctor or medic in the area may be able to apply a **tracheotomy**. The skilled professional cuts a hole in the victim's throat and inserts a straw, enabling the victim to breathe air directly through the straw.

## The Wonder of the Human Voice

Oh, sing to the LORD a new song!
Sing to the LORD, all the earth.
Sing to the LORD, bless His name;
Proclaim the good news of His salvation from day to day.
Declare His glory among the nations,
His wonders among all peoples.
(Psalm 96:1-3)

Perhaps the most phenomenal part of the throat is the **vocal cords** surrounding the glottis. The vocal cords are two bands of smooth muscle tissue that can produce sounds. Air flows from the larynx into the lungs through the windpipe, then returns back out of the lungs, up through the windpipe, and through the larynx again. When you breathe, the glottis is wide open. But when you talk, the glottis opens only a little bit, making the vocal cords vibrate. The high and low pitches of your voice are adjusted by whether the vocal cords are vibrating very quickly or very slowly. The human voice is extremely versatile.

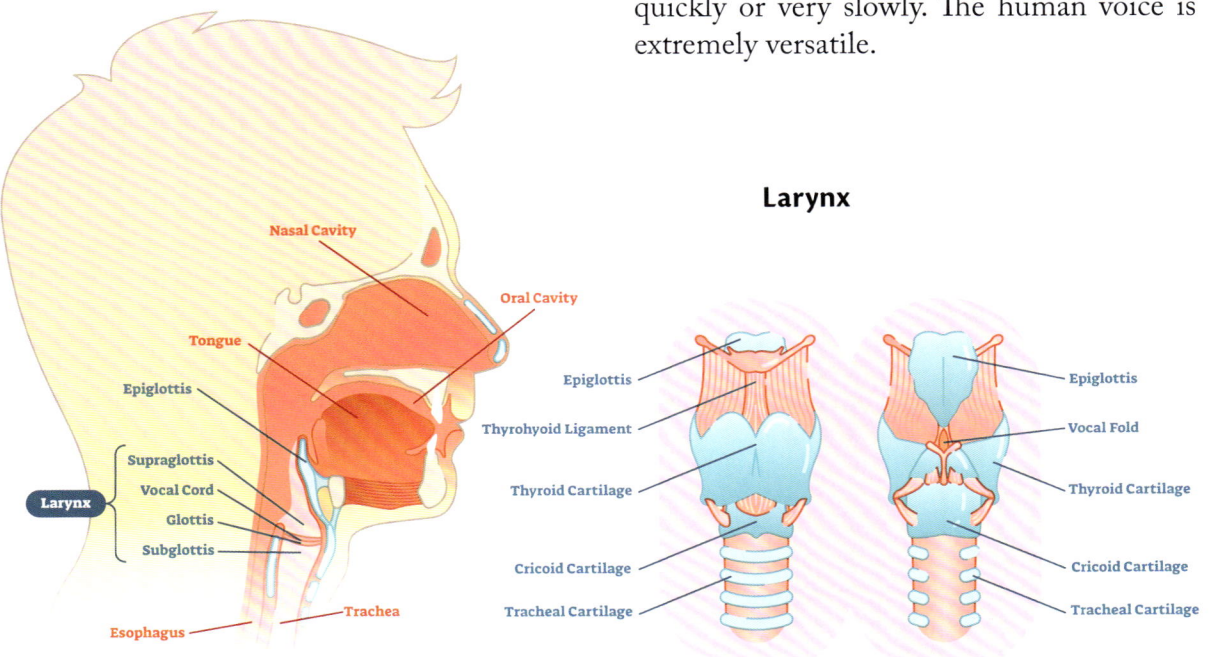

**Larynx**

# GOD MADE LIFE

The human voice is amazing! Human languages use up to 4,000 single-syllable sounds, which can be used to make millions of words. And God has designed the tongue and lips to make difficult sounds like "b," "s," "t," and "f." Now think about how the Creator has made all of us different. Humans are capable of producing a huge range of vocal sounds by adjusting airflow through the mouth and nose. Singers learn how to position their mouths, control air using their **diaphragm muscle**, and adjust their vocal cords to produce amazing, beautiful performances. Let us use this amazing design to return glory and praise to the Creator!

The penguin's voice is even more impressive when it comes to individual vocal qualities. How does a penguin find its mate in a crowd of 300,000 birds? In His infinite wisdom, God provided each with a unique voice. Penguins have a split vocal organ, and they can make two sounds at the same time!

## The Wonder of the Human Lungs

*"God, who made the world and everything in it, since He is Lord of heaven and earth, does not dwell in temples made with hands. Nor is He worshiped with men's hands, as though He needed anything, since He gives to all life, breath, and all things." (Acts 17:24-25)*

Following the air's flow from the mouth

Choir singing

CHAPTER 9: GOD SUSTAINS HUMAN LIFE

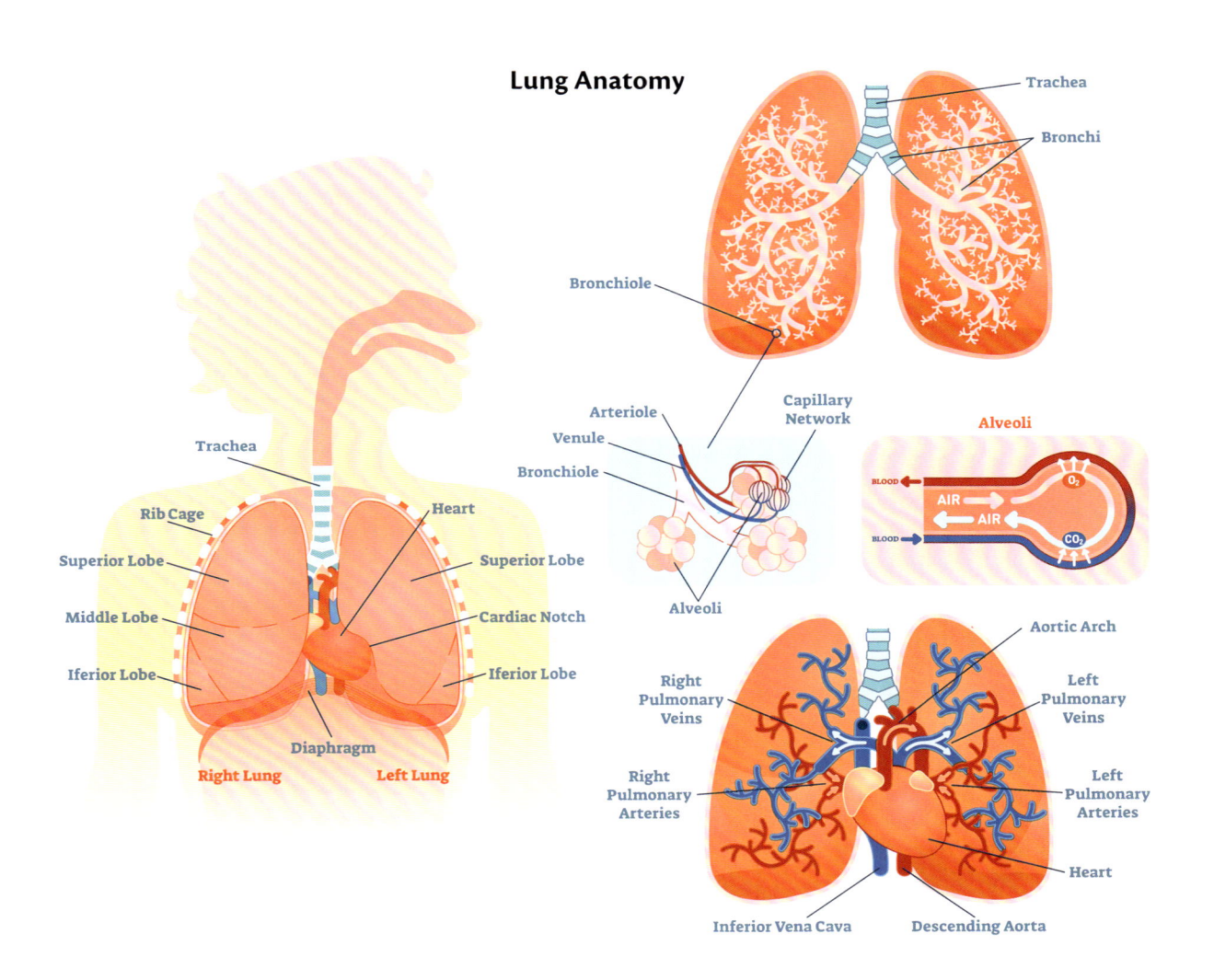

**Lung Anatomy**

down through the windpipe, two branches form, connecting into each lung. These are called **bronchi**. Inside the lungs, these bronchi branch out like a tree into smaller and smaller branches called **bronchioles**. There are about 30,000 of these tiny branches for each lung. As air passes through these tiny branches, they finally end up in the **alveoli** sacs. Here is where the respiratory system meets up with the circulatory system, which will be discussed later. The oxygen from the air is passed into the blood stream in the alveoli sacs. At the same time, the blood returning from the rest of your body is carrying unneeded carbon dioxide. This carbon dioxide is transferred back into the

263

# GOD MADE LIFE

air in the alveoli. If you were to compare the air you breathe into your lungs with the air you breathe out, you would find a pretty big difference in the oxygen and carbon dioxide content in each. You can see from the table below that your body is not using all the oxygen you are breathing in.

|  | Air Breathed In | Air Breathed Out |
| --- | --- | --- |
| Oxygen | 21% | 16.4% |
| Carbon Dioxide | 0.04% | 4.4% |

A grown person can hold about 1.45 gallons (5.5 liters) of air in both lungs. However, you don't use your full lung capacity unless you are exercising hard or holding your breath under water. Most of the time, you inhale and exhale about 1 pint (0.5 L) of air.

When some people get scared or panicky, they will start breathing very fast. This can cause **hyperventilation**. The problem is that they begin to exhale more air than they inhale. This reduces the amount of carbon dioxide in the blood. Changing this delicate balance can cause blood vessels to narrow. As a result, the blood has a hard time getting to the brain. This makes the person feel light-headed, and they might even pass out. If someone appears to be hyperventilating over a long period of time (more than 20 minutes), you should seek help from a doctor.

Praise the LORD!
Praise God in His sanctuary;
Praise Him in His mighty firmament!
Praise Him for His mighty acts;
Praise Him according to His excellent greatness! . . .
Let everything that has breath praise the LORD.
Praise the LORD! (Psalm 150:1-2,6)

## The Wonder of Blood

"But you shall not eat flesh with its life, that is, its blood. Surely for your lifeblood I will demand a reckoning; from the hand of every beast I will require it, and from the hand of man. From the hand of every man's brother I will require the life of man." (Genesis 9:4-5)

God's revelation makes it very clear to us that the physical life of man and animal is carried by the **blood**. Genesis 9 puts it clearly: "The life . . . is its blood." The blood sustains man's life, particularly because it carries oxygen to every living part of the body. Without oxygen carried by the blood, you would be dead within 10 minutes.

Oxygen keeps your body going all the time. When a car runs out of gas, it will stall on the side of the road. When key parts of your body don't get enough energy to run, the body shuts down. Remember, there are billions of activities going on even when

## CHAPTER 9: GOD SUSTAINS HUMAN LIFE

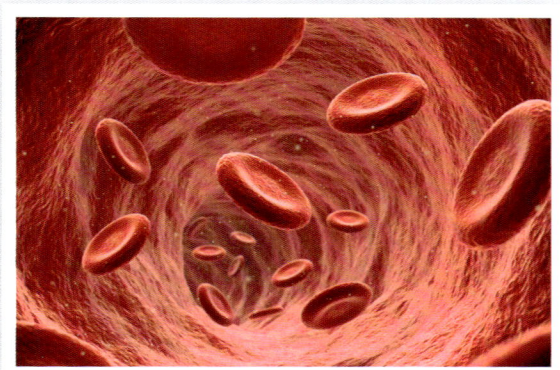

Red blood cells in the blood stream

you are asleep. It takes energy to breathe. It takes energy for your heart to beat. Oxygen is a fuel that helps turn food into energy in every corner of your body. Your brain uses about 20% of the total oxygen intake. The busy brain cells start dying within four to five minutes of losing access to oxygen.

Since breathing is all about getting oxygen into the blood, let's take a look at blood.

An adult's body contains about a gallon of blood (4-5 liters). Scientists can separate blood into three different layers of substances by spinning a sample in a **centrifuge**. The yellowish top layer is the blood's plasma. A very thin middle layer contains white blood cells and platelets. Then, the bottom layer contains red blood cells.

**Blood plasma** is made of 90% water mixed with other substances. God put proteins into the plasma. These proteins are assigned to keep the blood at the right consistency, which prevents the blood from clotting. Also, the blood plasma transports food, wastes, minerals, and hormones to the right locations around the body.

God made the **white blood cells** (or **leukocytes**) to hunt and destroy bad germs. When you skin up your knee, these little guys charge out to the injured area. If the skin is penetrated, the body knows that bad bacteria could get in and cause an infection. So, the white blood cells destroy these organisms and clean up the messed up cell fragments in the wound.

Because white blood cells normally only live for 1-12 days, the bone marrow, the thymus gland, the tonsils, and the spleen are assigned to make new ones.

**Platelets** help the blood to clot when

### Blood Cells

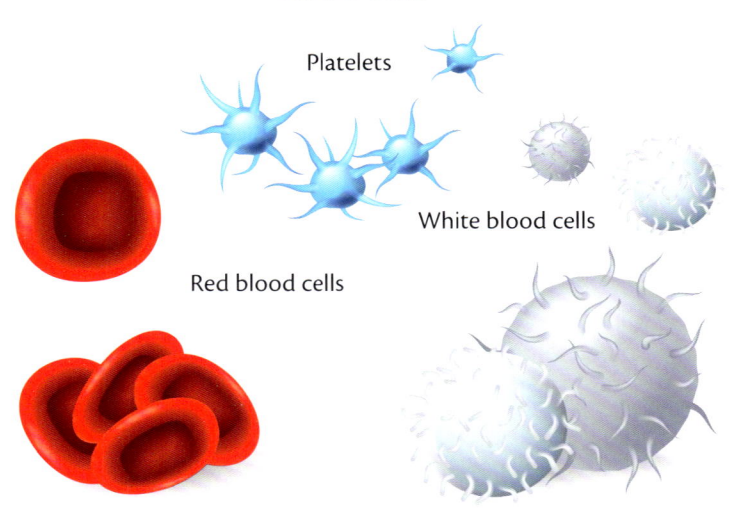

Platelets

White blood cells

Red blood cells

265

# GOD MADE LIFE

Centrifuge separates blood components

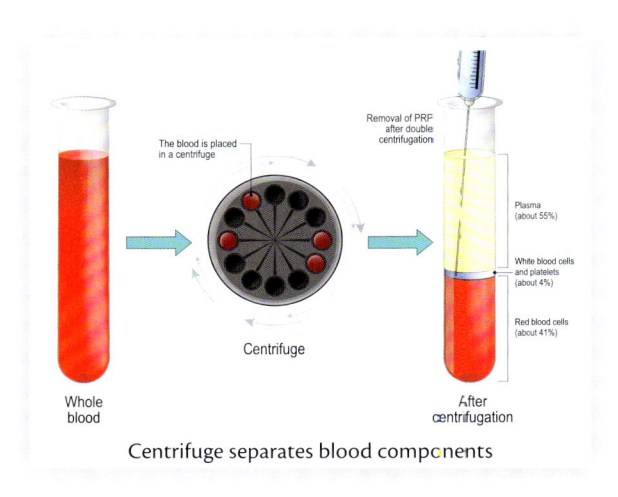

you need it. These little guys will clump together and plug up cuts or abrasions in the blood vessels. You wouldn't want your arteries or veins to develop leaks, or you could bleed to death. Doctors get concerned when a patient has a low platelet count in their blood. That means the patient could be at risk of excessive bleeding.

The **red blood cells** (or **erythrocytes**) don't contain a nucleus, so they aren't a complete cell. It's the bone marrow that produces these cells—about a billion of them every day. The iron-rich **hemoglobin** makes them look red. These red blood cells are little trucks designed to carry oxygen from the lungs to every part of the body. Since every living cell in your body needs oxygen, these tiny little trucks are fitted to deliver the oxygen everywhere.

## Blood Types

The blood contains certain **antigens**, which are proteins or sugars coating the outside of the red blood cells. Antigens are trained to stimulate an immune response from the white blood cells. Different antigens produce different immune responses. So, God designed people to have different blood types. The red blood cells in each blood type are decked out with different antigens. This is probably to make sure that everybody is not equally susceptible to the same germs.

There are eight different blood types which are divided up according to their types of antigens. There are the A, the B, and the O antigen types. The AB blood type includes a combination of the A and B antigens. Then, there are negative and positive antigens. The reason there are no AO or BO blood types is because O does not contain an antigen. AB-negative is the rarest blood type, while O-positive is the most common in the United States.

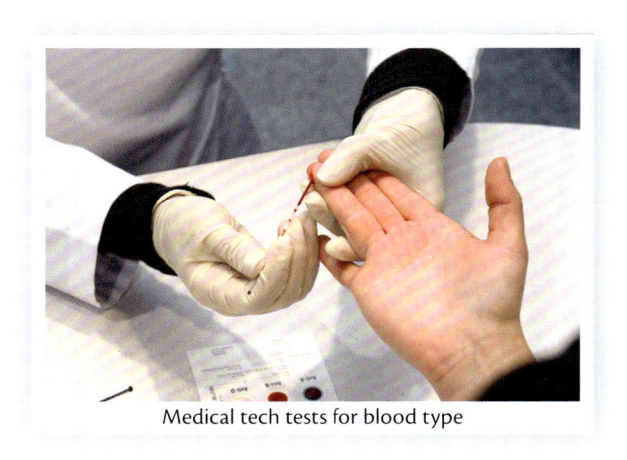

Medical tech tests for blood type

# CHAPTER 9: GOD SUSTAINS HUMAN LIFE

| Blood Type | Percent of U.S. Population with This Type |
|---|---|
| AB-negative | 0.6% |
| B-negative | 1.5% |
| AB-positive | 3.4% |
| A-negative | 6.3% |
| O-negative | 6.6% |
| B-positive | 8.5% |
| A-positive | 35.7% |
| O-positive | 37.4% |

## Blood Transfusions

And the people rushed on the spoil, and took sheep, oxen, and calves, and slaughtered them on the ground; and the people ate them with the blood. Then they told Saul, saying, "Look, the people are sinning against the LORD by eating with the blood!"

So he said, "You have dealt treacherously; roll a large stone to me this day." Then Saul said, "Disperse yourselves among the people, and say to them, 'Bring me here every man's ox and every man's sheep, slaughter them here, and eat; and do not sin against the LORD by eating with the blood.'" So every one of the people brought his ox with him that night, and slaughtered it there. Then Saul built an altar to the LORD. This was the first altar that he built to the LORD.
(1 Samuel 14:32-35)

If you were to lose 40% of your blood, you would die since your life is in your blood. Quite a few people have died from blood loss. This is especially true on the battlefield where soldiers get shot.

In 1818, a British doctor named James Blundell figured out how to transfer somebody else's blood into a patient who had suffered blood loss. Now, about 100 million people from all around the world receive blood transfusions each year. Many lives are saved because of this doctor's discovery.

When transfusing blood, the doctor must make sure that the blood type he uses as replacement blood matches the patient's blood type. However, O-negative blood can be used for anybody. Transfusing the wrong blood type results in the body forming antibodies that attack the new blood. That's because the body does not recognize the new antigen contained in the transfused blood. This can be life-threatening.

James Blundell (1790-1878)

## The Wonder of the Human Heart and the Circulatory System

The circulatory system is made up of the heart and the blood vessels, and its job is to move blood throughout the body. The system supplies tissues with oxygen and nutrients, and it helps the body get rid of waste products.

The human heart will beat about 2.5 billion times in a lifetime. That's about the same amount of movement that occurs in the engine of a car if the car runs for 350,000 miles. This is a heart-working little machine, as it just keeps pumping about 75 gallons (300 L) of blood every hour of your life. Swimming pool pumps won't last much longer than 5-10 years, but incredibly, the heart keeps pumping for 75-90 years.

The **heart** is made of muscle tissue, and it is about the size of your two hands folded together. After traveling through the body to deliver oxygen, the deoxygenated blood comes rushing into the **right atrium**. A valve opens, allowing the blood into the **right ventricle**. With a pump of the heart, that blood is sent into the lungs to pick up more oxygen from the alveoli sacs. The oxygenated blood flows back into the **left atrium** of the heart. More valves open, and the blood flows into the **left ventricle**. With another pump of the heart, the blood is pushed out through a valve into the **aorta** (the main artery) and, from there, to the rest of the body. The **arteries** are responsible for delivering the oxygen and nutrients to the rest of the body, while the **veins** carry the blood back to the heart.

But how does the heart keep pumping with a constant rhythm? The Creator provided the **sinoatrial node** which keeps time by discharging an electrical pulse 70-80 times a minute. This signals the heart muscle to contract and continue pumping at 70-80 times a minute. Should the sinoatrial node fail, God mercifully provided a back-up. There is another group of cells called the **AV node** which can discharge an electrical pulse 40-60 times a minute.

As the blood exits the heart, it circulates around the body through the arteries. Then the veins return the deoxygenated blood to

CHAPTER 9: GOD SUSTAINS HUMAN LIFE

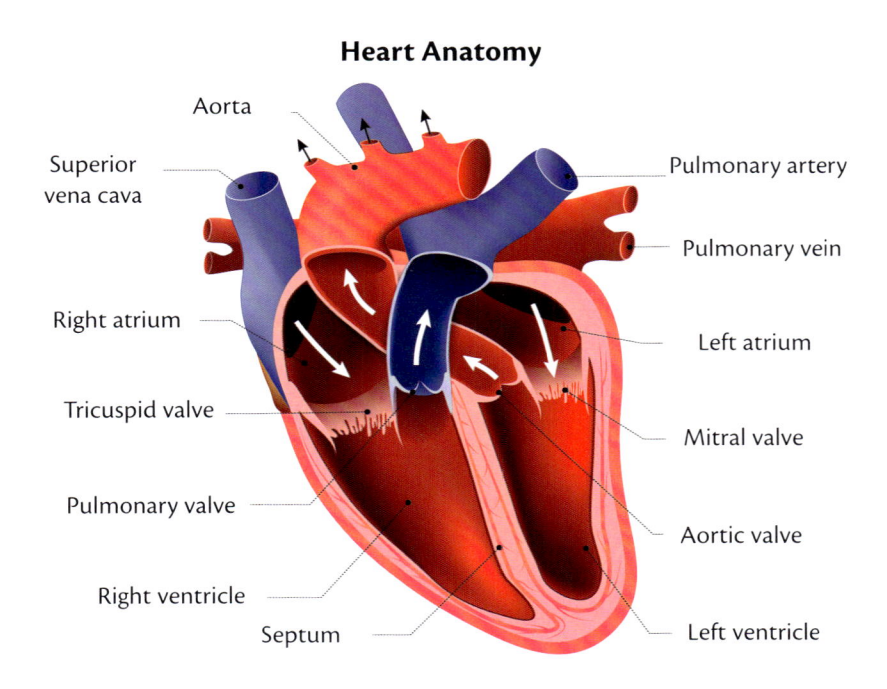

**Heart Anatomy**

- Aorta
- Superior vena cava
- Pulmonary artery
- Pulmonary vein
- Right atrium
- Left atrium
- Tricuspid valve
- Mitral valve
- Pulmonary valve
- Aortic valve
- Right ventricle
- Left ventricle
- Septum

the heart. It takes about 23 seconds for a drop of blood to take the trip through the body, back to the heart, on into the lungs, back to the heart, and off through the aorta again. These faithful little cells carry oxygen and food to the body over and over again all day long. That's 3,756 trips a day! In its 120-day lifespan, that little red blood cell takes 450,000 trips! Here is one more impressive and exceptional feature of the human body.

The arteries feed into **capillaries**. These are 10 billion little blood vessels that branch out into every tiny part in your body. If all the capillaries in your body were spread out end to end, they would stretch out over 60,000 miles (96,561 km)!

To review, the circulatory system flows according to this order:

1. The blood picks up oxygen in the lungs.
2. The blood flows into the heart.

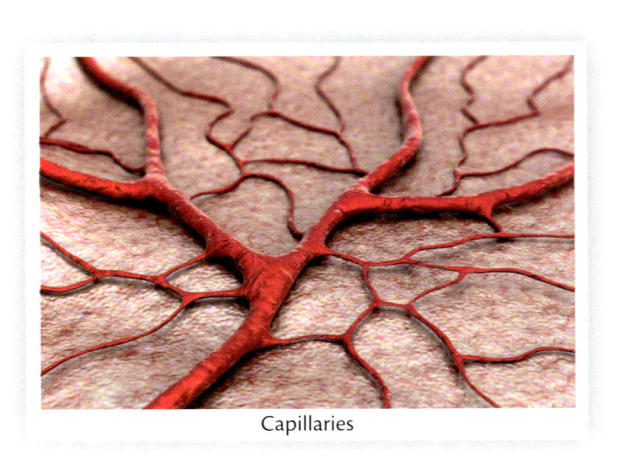

Capillaries

# GOD MADE LIFE

3. The blood flows out of the heart through the aorta.
4. The blood flows through the arteries, and branches into capillaries.
5. The blood delivers oxygen and nutrition to the cells and picks up carbon dioxide and other waste products through the capillary walls.
6. The capillaries then merge back together into veins, and the blood travels back into the heart.
7. The blood is pumped from the heart into the lungs to exchange carbon dioxide waste for more oxygen.

What a tremendously complex and carefully constructed transportation system God has created for our bodies! Let us praise Him for His wisdom, creativity, power, and care to build this amazing network to sustain our lives!

---

Daniel answered and said:
"Blessed be the name of God forever and ever,
For wisdom and might are His. . . .
He gives wisdom to the wise
And knowledge to those who have understanding.
He reveals deep and secret things;
He knows what is in the darkness,
And light dwells with Him." (Daniel 2:20-22)

---

## The World's Most Deadly Diseases

The deadliest disease in the world is **coronary heart disease**. This is where the blood vessels that carry blood and oxygen to the heart get clogged up or shrink up. This terrible disease is the reason for 15.5% of all deaths in the world. When blood vessels coming into or leaving the heart become narrowed, the heart has to work harder to keep the blood flowing. This problem is usually caused by **atherosclerosis**, which is a build-up of fatty substances like cholesterol and calcium in the arteries. Imagine a rock getting stuck in a garden hose. That would slow the water flow in the hose. The same kind of thing happens when something gets stuck in the arteries. Contributing factors to atherosclerosis include a bad diet, lack of exercise, and being overweight. If the blockage is too severe, the heart will stop beating. This is called a **heart attack**.

Sometimes an **aneurysm** will form, which is a little bulge in the wall of the artery or blood vessel. Imagine a bulge in a garden hose. This usually weakens the artery and inhibits the flow of blood. If a main artery breaks open at the bulge, the end result could be deadly.

The second deadliest disease is **stroke**, accounting for about 11% of all deaths in the world. Whereas heart disease occurs in the proximate location of the heart, a stroke occurs in the brain.

Two things could cause a stroke. Either an artery in the brain is blocked, or it leaks.

An **ischemic stroke** occurs when something like a blood clot blocks a main artery in the brain. A **hemorrhagic stroke** occurs when the artery develops a leak. Once again, this happens when the artery "hose" bulges in an aneurysm. Or, if the blood pressure coming into the brain is very high, it could burst the artery. Think about what would happen if you turned up the water pressure in a weak hose that had been run over too many times by cars.

There is also something called a **transient ischemic stroke**. This happens when the brain's blood supply is blocked for a short time (less than five minutes). People recover from these transient strokes quickly. More than likely however, they will have a more severe stroke if the conditions are not treated.

Symptoms of stroke include sudden numbness in the face, difficulty walking, sudden weakness in the limbs, blurred vision, dizziness, loss of balance, trouble speaking, confusion, and difficulty understanding speech. It is critical to get help right away when these symptoms occur. Those who get treatment within an hour or two are less likely to suffer long-term consequences. Most people don't live very long after suffering a very bad stroke—28% die within one month, 42% die within one year, and 60% died within five years.[23]

The U.S. Center for Disease Control offers the following helpful guide so you can know whether your parent or grandparent

**How Blood Flows Through the Body**

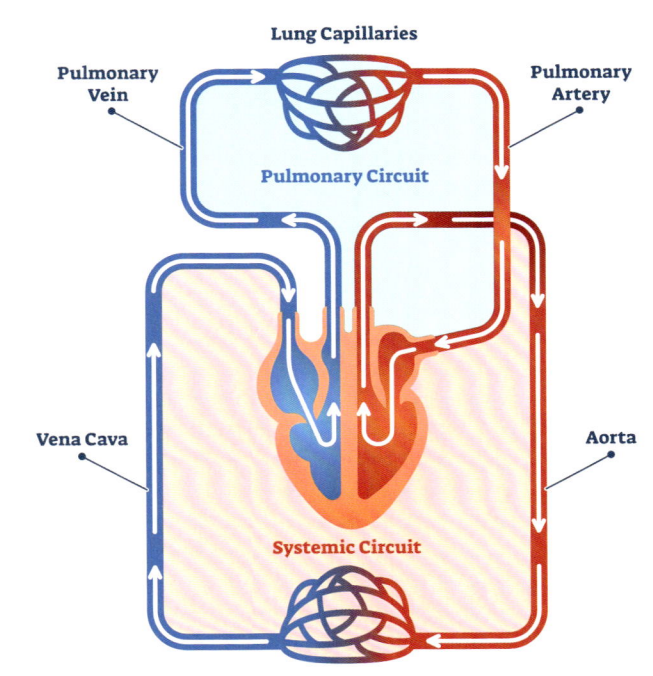

is having a stroke.[24] Try to memorize this routine. You may save a life.

Act F.A.S.T. and perform this simple test:

- **F** stands for **Face**: Ask the person to smile. Does one side of the face droop?
- **A** stands for **Arms**: Ask the person to raise both arms. Does one arm drift downward?
- **S** stands for **Speech**: Ask the person to repeat a simple phrase. Is the speech slurred or strange?
- **T** stands for **Time**: If you see any of these signs, call 9-1-1 right away.

# GOD MADE LIFE

> Then Jesus said to them, "I will ask you one thing: Is it lawful on the Sabbath to do good or to do evil, to save life or to destroy?" And when He had looked around at them all, He said to the man, "Stretch out your hand." And he did so, and his hand was restored as whole as the other. (Luke 6:9-10)

## Blood Clots

Because the blood is so critical to life, our Creator God has designed a very complex and beautiful system for keeping the blood healthy. Previously, we looked at the many proteins and enzymes He uses to keep the blood from clotting. That's because blood clots can kill you very quickly.

A **blood clot** is not like a hard golf ball in your blood stream. It's more like a small water balloon or gel-like substance that gets stuck in an artery or vein. Most clots are supposed to dissolve, but sometimes they don't. That can become a real problem.

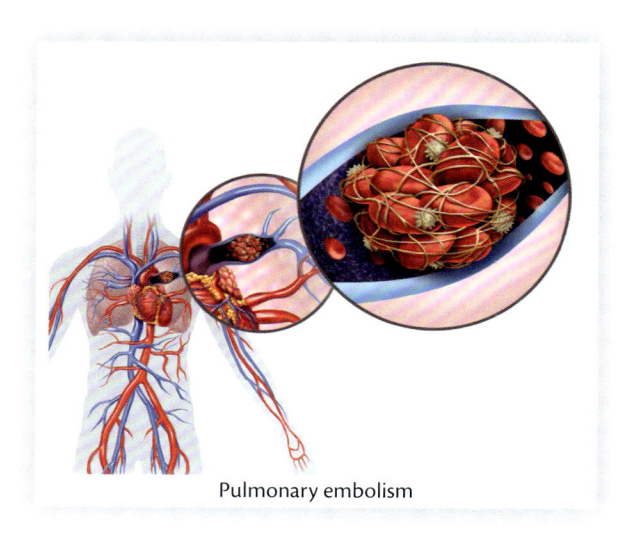

About 40% of blood clots show up in the lungs and 60% in the legs. A blood clot in the lungs is called a **pulmonary embolism**. A blood clot in the leg might give your leg a warm sensation, meaning it will feel warmer than the rest of your skin. You might also experience swelling and reddish discoloration on your skin.

Doctors treat blood clots with **anticoagulants**, or blood thinners. They won't break up the clots, but they will help the blood to keep flowing and prevent the clot from getting bigger.

What you don't want is for the blood clot to break loose in your leg and then clog up your lungs or block your heart. Some doctors will put a filter into the large vein in the abdomen to catch any blood clot that might try to get up into the lungs. Also, doctors will tell their patients who have blood clots in their legs to wear tight stockings. This will keep the blood from pooling and clotting even more.

## Blood Pressure

Most of the time, when doctors check out the condition of a patient's body, they will measure their blood pressure. They usually check both their heart rate and their blood pressure. Using a **stethoscope**, they will listen for heart murmurs, which indicate a leaky valve or a valve that won't open all the

Pulmonary embolism

CHAPTER 9: GOD SUSTAINS HUMAN LIFE

way. They also listen for **arrhythmia**, which is an abnormal rhythm. This happens when the heart skips a beat now and then or beats very quickly for a short time.

High blood pressure can lead to blood clots or aneurysms. Low blood pressure is often caused by dehydration, loss of blood, or heart disease. People with low blood pressure experience dizziness and confusion.

When you measure your blood pressure, you get a high number and a low number. The higher number is **systolic pressure**. This measures the amount of force of blood pushing against the artery walls when the heart muscle contracts to produce a heartbeat. The lower number, or the **diastolic pressure**, measures the amount of pressure in the arteries between heartbeats.

The following table lists the healthy range for a child's blood pressure.

### Healthy Blood Pressure Range

| Age | Diastolic Pressure | Systolic Pressure |
|---|---|---|
| 3-6 | 56-70 | 95-110 |
| 7-10 | 57-71 | 97-112 |
| 10-17 | 66-80 | 112-128 |

## Take Good Care of Your Heart

Your heart is the most critical organ in your body, so take good care of it! This means maintaining a healthy diet. Avoid eating too much or gaining too much weight. Don't run to food for comfort, making it an idol. Get at least 2½ hours of exercise per week—that's about half an hour a day. Limit your salt intake to about 2,300 mg of sodium per day. That's about one teaspoon per person per day. Eating too much salt can cause fluid to build up around your heart, which makes the

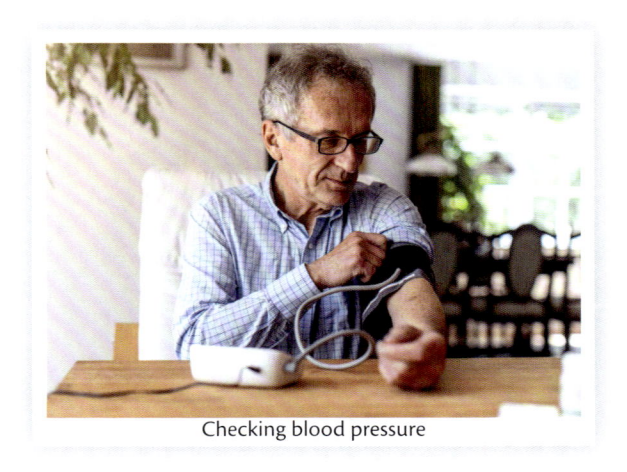

Checking blood pressure

Salt

## GOD MADE LIFE

heart work harder when pumping blood. Of course, if you tend to work hard and sweat a lot, you may need to consume a little more salt, as sweating flushes water and salt out of your body.

### Amounts of Sodium in Food

| Food | Amount of Sodium |
|---|---|
| Glass of milk | 100 mg |
| Cheese (1 oz; 0.03 L) | 400 mg |
| Salad Dressing (1 tablespoon; 0.01 L) | 100 mg |
| Apple (1) | 130 mg |
| Bread (2 slices) | 300 mg |
| Potato chips (1 oz; 0.03 L) | 170 mg |
| Baked beans (1 cup; 0.24 L) | 1000 mg |
| Canned beef stew (1 cup; 0.24 L) | 1000 mg |
| Cheeseburger (fast food) | 1000 mg |

Keep your heart with all diligence, For out of it spring the issues of life. (Proverbs 4:23)

But even more importantly, guard your invisible heart. The heart is the seat of the affections. That's the part of you that is attracted to this thing or that thing. It is the part of you that loves or does not love your brothers and sisters. It is the part of you that loves God or does not love God.

### The Provision of Energy through Food

I know that nothing is better for them than to rejoice, and to do good in their lives, and also that every man should eat and drink and enjoy the good of all his labor— it is the gift of God. (Ecclesiastes 3:12-13)

The blood functions as the transportation system for nutrients and oxygen throughout the body. So far, we have looked at how the nose and mouth feed the lungs with oxygen, and the lungs transfer the oxygen into the blood. But how does food turn into energy? That is the function of the **digestive system**—yet another glorious creation of our genius Creator!

While the nose produces a mucus to catch germs and moistens the air flowing into the lungs, the **mouth** has its own equipment. **Saliva** is produced in the **salivary glands** to moisten the food. The saliva also contains enzymes which help to break up the starches in the food. These glands produce about ¼ gallon (1 liter) of saliva a day, or about 90 gallons a year.

The two main parts to the digestive

**CHAPTER 9: GOD SUSTAINS HUMAN LIFE**

## Teeth

Teeth help to mash up food so it won't get stuck in the esophagus when you swallow. Human children get 20 teeth, and adults usually get 32 teeth. But that's nothing compared to a garden snail. The Creator decked him out with 14,000 teeth. Other snails have as many as 20,000 teeth! God made teeth very tough so they won't wear down when chewing on hard stuff. But the **limpet** sea shell creature's teeth are the hardest biological material in the world. That's because God designed these animals to scrape their food off of the surface of rocks. They can even chew through rocks. God provided squirrels and rabbits with teeth that just keep growing all the time. This prevents them from wearing their teeth out while chewing on nuts, bark, and tough leaves.

**Digestive System**

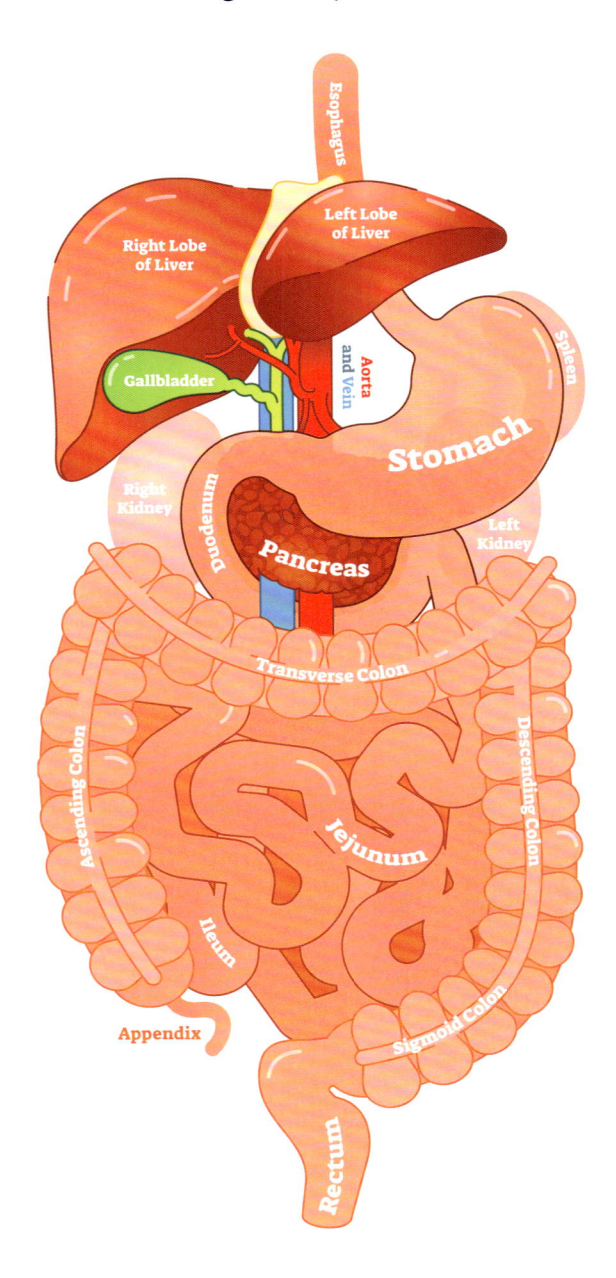

system are the **alimentary canal** and accessory organs. The alimentary canal is one long tube that processes food starting with the mouth, going down through the intestines, and exiting out the anus. It's about 30 feet (9.1 m) long, but there are a lot of twists and turns as the food passes through your system. Each part of the canal is listed in order in the following chart:

# GOD MADE LIFE

| Section of Canal | Length (for adult) |
|---|---|
| Mouth | 3 inches (8 cm) |
| Pharynx | 5 inches (13 cm) |
| Esophagus | 10 inches (25 cm) |
| Stomach | 12 inches (30.5 cm) |
| Small intestine | 22 feet (700 cm) |
| Large intestine | 5 feet (150 cm) |

As the food enters the **stomach**, it is further processed by stomach acids and enzymes. The stomach muscles get busy churning and crushing the food into a mash. The stomach produces about a gallon (4-5 Liters) of gastric juices every day. These juices contain **hydrochloric acid** which breaks down the food and kills bacteria. The stomach also makes digestive enzymes which break up the proteins in the food.

From here the food moves into the **small intestine**. The food continues to break down into tiny pieces of proteins, fat, and sugars. As they move through the 22-foot-long small intestine (700 cm), these small particles of food get transferred into the blood stream. That's how the body gets its energy from food.

## The Wonder of God's Systems for Food Processing

The two main accessory organs in the digestive systems are the liver and the pancreas. The liver is one of the most awe-inspiring organs in the human body! Scientists have counted about 500 jobs performed by this incredible gift of God.

Remember that the mouth has taken in dirt, various germs, and unusable stuff along with food. This gets processed in the stomach and then passes into the small intestine where the blood receives what the body needs. All the blood leaving the walls of the stomach and the intestines must pass through the liver before it moves on to the rest of the body. The liver provides a quality-control inspection of all the food processed. If you were eating something harmful, the liver would filter the toxins out of this blood. The harmful stuff gets carried from the liver back to the small intestine in a fluid called **bile** and becomes part of the waste in a bowel movement, which your body gets rid of every day when you use the toilet. The bile that the liver has made is God's way to help you digest fats. Once it does its job, the useful parts of bile are reabsorbed by the large intestine and sent back to the liver.

Other unusable substances in the blood are taken to the **kidneys**. As the blood runs through the kidneys, the waste products are filtered out. These waste products are processed into **urine**. The urine is then transferred into the bladder—a storage area God

## The Liver

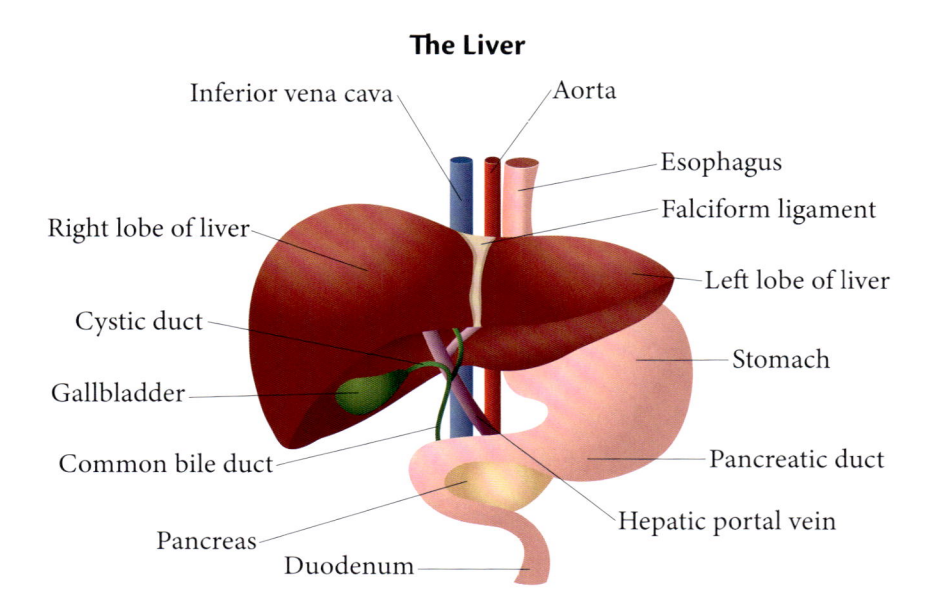

- Inferior vena cava
- Aorta
- Esophagus
- Falciform ligament
- Right lobe of liver
- Left lobe of liver
- Cystic duct
- Stomach
- Gallbladder
- Common bile duct
- Pancreatic duct
- Hepatic portal vein
- Pancreas
- Duodenum

provided for you so you wouldn't leak urine continually. Once the bladder fills up, you'll feel the urge to go to the bathroom.

The kidneys also keep the sodium and potassium content in your body well balanced. It is their job to keep your body well hydrated, too.

All of these systems are means by which God protects the body from unhealthy germs and unhelpful substances. This demonstrates God's tender care and mercy for us. Let us always give thanks for the function of the liver, the kidneys, and the intestines. Every time you go to the bathroom, remember that you are "fearfully and wonderfully made!" God has made a way to keep your body healthy.

The Lord spoke about the body's function to filter out unhealthy substances in Matthew 15:16. He knew all about this function because He created the human body.

[Jesus said], "Are you also still without understanding? Do you not yet understand that whatever enters the mouth goes into the stomach and is eliminated? But those things which proceed out of the mouth come from the heart, and they defile a man. For out of the heart proceed evil thoughts, murders, adulteries, fornications, thefts, false witness, blasphemies. These are the things which defile a man, but to eat with unwashed hands does not defile a man." (Matthew 15:16-19)

You can overwhelm your liver if you take

in too much of the bad stuff. For example, people who drink too much alcohol have destroyed their livers. Sinful activity can result in ruining the bodies God has made for us. Too much of anything can kill you. Half an ounce of caffeine would be deadly. A triple shot energy drink contains about 225 mg of caffeine, so it would take about 62 of these drinks to kill you. Consuming three cherry pits or the seeds from 18 apples could be deadly. Drinking about 1½ gallons (6 liters) of water could be deadly because your kidneys can't process that much liquid in one day.

The **pancreas** is the other major accessory in the digestive system. Its main job is to make enzymes to process food.

- The **lipase** enzyme mixes with bile from the liver to break down fat. If you're not breaking down fat and absorbing it, you may get diarrhea.

- The **protease** enzyme breaks down proteins like meat, eggs, and nuts. It also protects you from germs living in the small intestine.

- The **amylase** enzyme breaks down starches that come from foods like potatoes.

The pancreas also makes hormones like insulin, which controls the amount of sugar

**The Pancreas**

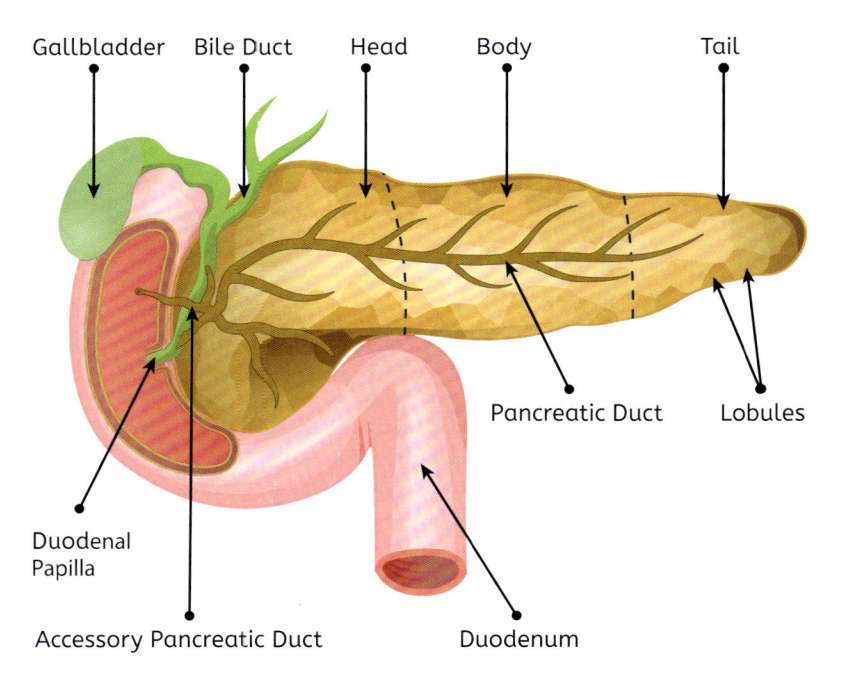

Gallbladder    Bile Duct    Head    Body    Tail

Pancreatic Duct    Lobules

Duodenal Papilla

Accessory Pancreatic Duct    Duodenum

in your body. Another hormone called **gastrin** controls the amount of gastric acids in your stomach. If you get too much gastric acid in your stomach, you may burp it up into your throat and get heartburn — a very uncomfortable thing. Isn't it wonderful that the pancreas controls the acid in your stomach through this "intelligent" hormone!

All this points to one thing: you are fearfully and wonderfully made!

## Take Good Care of Your Teeth

Tooth decay is the biggest health problem for children. The goal of oral hygiene is to protect the enamel covering of your teeth. Acids can break down the enamel. Even Fruit drinks and acidic soft drinks can break enamel down unless the fluids are washed down with water. Brushing your teeth too hard or biting hard things like pens and fingernails can also damage the enamel. Sugar is one of the biggest enemies to healthy teeth. Sugar feeds the bacteria already hanging out on your teeth where they make harmful acids. Some children get into the bad habit of grinding their teeth. This will break down the protective enamel as well.

On the other hand, calcium, phosphates, and fluoride can strengthen your teeth. Dentists recommend using fluoride as a means of making stronger teeth. But some families find there are risks involved with fluoride.

For about 50 years, governments have put fluoride into their countries' drinking water. By 2015, some governments were suggesting there was too much fluoride in the drinking water.[25] It is not the government's role to take care of our health like this. Each family should be responsible for taking care of its own health. Your family should decide how much fluoride to include in your diet and drinking water. Most tooth-

# GOD MADE LIFE

pastes contain fluoride, but it's not always healthy to swallow it. Some toothpastes have 1,000 times as much fluoride as you would find in fluoridated water.

The best thing you can do to maintain healthy teeth is to avoid consuming a lot of candy and sugary, acidic drinks. Be sure to brush your teeth, especially after drinking milk or sugary soda drinks. Also brush after eating dried fruit, hard candy, caramel, taffy, raisins, cookies, and sugary cereals. Brushing your teeth twice a day for a full two minutes at a time can help a lot.

Saliva is also good for teeth, so chewing sugar-free gum can keep the mouth moist. This bathes the teeth in the healthy minerals saliva has which can keep them strong.

## Take Good Care of Your Stomach and Intestines

People are increasingly having trouble with their "gut health." That means they suffer from upset stomach, sleeplessness, food intolerances, and autoimmune reactions.

The big lesson to take from this is that we must take good care of our digestive systems. Here are some good ways to do that:

1. Lower your stress levels. In the words of Philippians 4:6-7, "Be anxious for nothing, but in everything by prayer and supplication, with thanksgiving, let your requests be made known to God; and the peace of God, which surpasses all understanding, will guard your hearts and minds through Christ Jesus."

2. Get enough sleep. Six to 12-year-old kids should get 9-12 hours of sleep per night. And 13-18-year-old children should get anywhere from 8-10 hours of sleep per night.

3. Eat slowly.

4. Stay hydrated. Children 9-13 years of age should get at least 5 cups (1.2 L) of water per day.

5. Make sure you get healthy probiotics.

6. Be aware of the foods that give you troubles.

7. Watch your diet. Don't consume too many foods that are hard to digest, and eat plant-based foods and more fiber if possible.

Children should get 9-12 hours of sleep per night.

**CHAPTER 9: GOD SUSTAINS HUMAN LIFE**

## Pray

- Praise God for the body's amazing circulatory system. Praise Him for the thousands of miles of blood vessels serving every part of your body with oxygen and nutrition!
- Thank the Lord for guarding our bodies from unhealthy germs in so many different ways using the nose, the liver, the stomach acids, and the kidneys.
- Lift up praises to God for the amazing heart that works and works night and day without stopping.
- Praise God for the vocal cords with which we can sing and glorify Him every day.
- Praise God for the enzymes that keep the sugars and the acids balanced in the body so you don't go into sugar shock or experience acid reflux.
- Ask God for wisdom as you take care of your body with good exercise, a healthy diet, and proper hygiene.

## Sing

Having seen the goodness and the wisdom of God in the provision of life-sustaining organs, an appropriate response is always worship and praise. If the student is unfamiliar with the hymn or psalm, some version of it is available on the internet and may be accessed (with supervision) for singing along.

*All Things Bright and Beautiful*
Each little flow'r that opens,
Each little bird that sings,
He made their glowing colors,
He made their tiny wings.

Refrain:
All things bright and beautiful,
All creatures great and small,
All things wise and wonderful:
The Lord God made them all.

The purple-headed mountains,
The river running by,
The sunset and the morning
That brightens up the sky.

The cold wind in the winter,
The pleasant summer sun,
The ripe fruits in the garden,
He made them, ev'ry one.

The tall trees in the greenwood,
The meadows where we play,
The rushes by the water,
To gather every day.

He gave us eyes to see them,
And lips that we might tell
How great is God Almighty,
Who has made all things well.

## Do

Choose at least one of the following activities and apply the lessons you have learned in this chapter.

1. Be prepared to save a life. Practice the five-and-five method for choking relief with parental supervision. Watch the suggested video(s) first. Also, consider taking basic First Aid training.
2. Be prepared to save a life. Memorize the F.A.S.T. procedure for testing somebody who may be having a stroke. Review it with a parent or teacher.
3. Calculate how much sodium from salt you eat in a given day. Record everything you eat and add up the total amount of sodium in your diet. Does this exceed the recommended national average? Are you consuming too much or too little food as needed to support your lifestyle?
4. On average, children should get about 60 minutes of exercise a day, which includes walking and running. For three days a week, children should get at least 60 minutes of exercise that makes them breathe fast. How much exercise do you generally get? For a period of a week, assess how much exercise you are getting, using a log sheet like this:

### My Physical Exercise for the Week

| Day of the Week | Total Exercise | Hard-Breathing Exercise |
|---|---|---|
| Sunday | | |
| Monday | | |
| Tuesday | | |
| Wednesday | | |
| Thursday | | |
| Friday | | |

Consider adding some form of strenuous physical activity to your daily routine. Biking, jogging, swimming, or skating can all be great forms of exercise to keep your body healthy!

## Watch

To watch the recommended videos for this chapter, go to **generations.org/GodMadeLife** and scroll down until you find the video links for Chapter 9. Our editors have been careful to avoid films with references to evolution; however, we would still encourage parents or teachers to provide oversight for all internet usage. These videos may not give God the glory for His amazing creative work, so the student and parent/teacher should respond to these insights with prayer and praise.

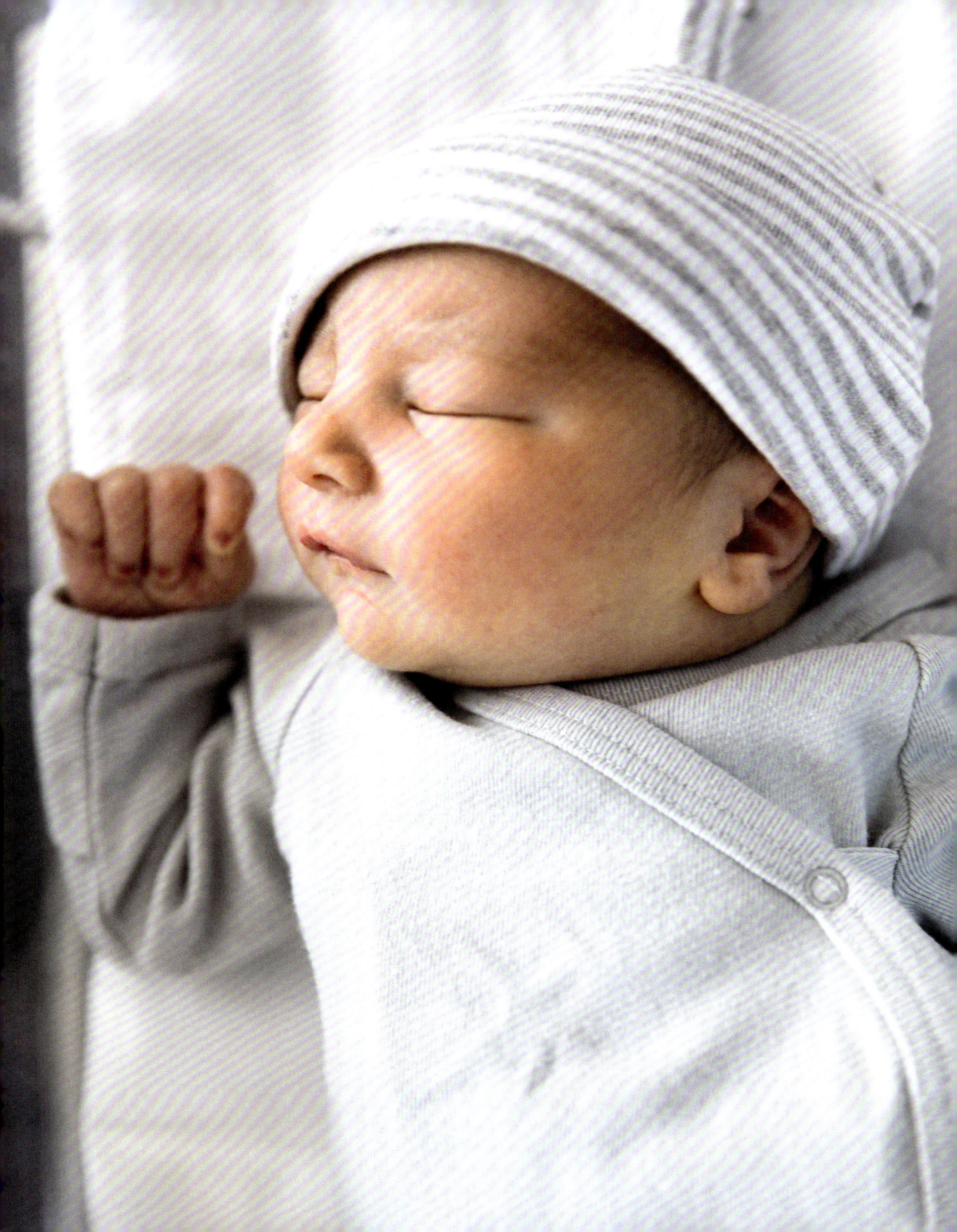

# CHAPTER 10
# GOD RESTORES AND REPRODUCES LIFE

> For You have delivered my soul from death.
> Have You not kept my feet from falling,
> That I may walk before God
> In the light of the living? (Psalm 56:13)

In the last chapter, we considered how God nourishes and maintains life for all of us, minute by minute, and day by day. Since Adam fell into sin and brought physical death upon himself and all his children, our bodies break down. Disease, germs, and wear and tear of the body is something we all experience. Because of these things, humans are ever aware of the constant threat of their impending death.

However, God does not want us to die as soon as we are born. By His mercies, He has provided the body with means of healing and restoration.

But most important of all, God has provided a means of saving our souls. That's the invisible part of us. The human soul was made to last forever. And the Lord Jesus told us that one human soul was worth more than the whole universe. That's why He came to die on the cross—to save the human soul from eternal death. When Jesus rose from the dead, He made a way for the restoration of our body and soul, and the way to eternal life!

[Jesus said], "For what will it profit a man if he gains the whole world, and loses his own soul? Or what will a man give in exchange for his soul?" (Mark 8:36-37)

GOD MADE LIFE

## Keeping the Body in Balance

I will praise You, for I am
fearfully and wonderfully made;
Marvelous are Your works,
And that my soul knows very well.
(Psalm 139:14)

Control systems are very common in our world. Heating systems in homes are controlled by thermostats. If there is no temperature control for a home's heating system, the temperature inside the home might increase to 110 °F (43 °C). That would be miserable for anybody in the home. If an air conditioner kept running without any controls, the house's temperature might decrease to 55 °F (13 °C).

Your family car won't burn gas unless there is a good fuel-to-air ratio. Fuel needs oxygen (but not too much oxygen) to burn and power the car. So, if you get too much gas or too much air in the mixture, your car will sputter and die. The fuel injection system controls how the fuel and air mix.

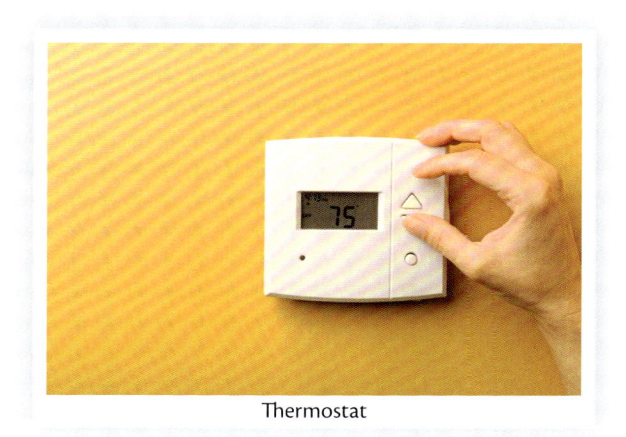

Just as humans design homes and cars with control systems, God has designed the human body with automatic controls to keep everything in balance. Most of this is done by the **endocrine system**. When everything within the body is in proper balance, we say that it is in a state called **homeostasis**. Here are some of the ways the body maintains homeostasis:

- The body must regulate the amount of water in its system. It cannot function well with too little water or too much water.

- The body's temperature must be kept between 98 °F (37 °C) and 100 °F (38 °C) for your health and strength.

- The body needs to control the amount of iron, calcium, and potassium in the blood. Iron deficiency in your body results in a condition called anemia, where the red blood cells do not have enough hemoglobin to carry oxygen around the body. When children get anemia, they feel lethargic and weak. Children should have a minimum of 12 g per dL (grams per deciliter) of iron in their blood.

- The body must maintain just the right amount of glucose in the blood. For most children ages 6-12, their blood sugar levels should stay within the range of 90 to 180 mg/dL (milligrams per deciliter). Low blood sugar is called **hypoglyce-**

Thermostat

CHAPTER 10: GOD RESTORES AND REPRODUCES LIFE

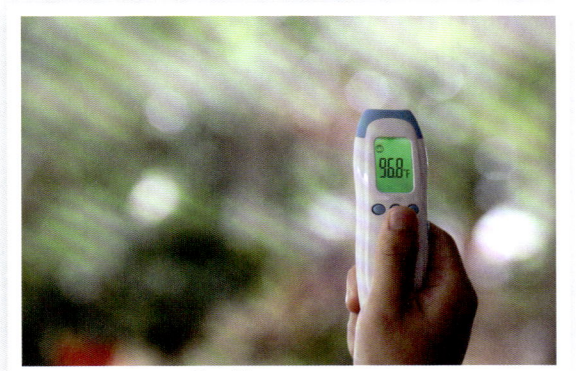

Checking temperature

mia. This will cause dizziness, lightheadedness, chills, sweating, or fainting. High blood sugar (above 180 mg/dL) is called **hyperglycemia**. Sometimes, a person with high blood sugar has **diabetes**. The symptoms for this include blurred vision, weight loss, exhaustion, increased thirst, and emptying of this excess water with frequent trips to the bathroom.

The endocrine system is made up of **glands** that keep the body in balance. Glands are little chemical factories that are always busy making **hormones**. They send out a new batch every couple of hours. Hormones are chemicals that give instructions to cells and body parts. For example, the **hypothalamus gland** produces hormones that tell the **pituitary gland** what to do. The hypothalamus also controls body temperature to a fine degree. If the body's temperature changes because of cold or heat, this gland will send signals to other glands, muscles, organs, and the nervous system. First, the nervous system lets the hypothalamus know the body is getting too cold or too warm. If the body needs to warm up only slightly, the hypothalamus releases hormones to make that happen. For example, the hormones can cause the blood vessels under your skin to shrink up, decreasing the blood flow to your skin. This allows less heat to escape from your body. However, if your body is extremely cold, its muscles will produce heat by shivering.

Next, the **thyroid gland** also receives the message that the body is too cold. So, it will release a hormone that gets the body to increase **metabolism**, or the rate at which the body burns energy. When the body

**Thyroid Hormones**

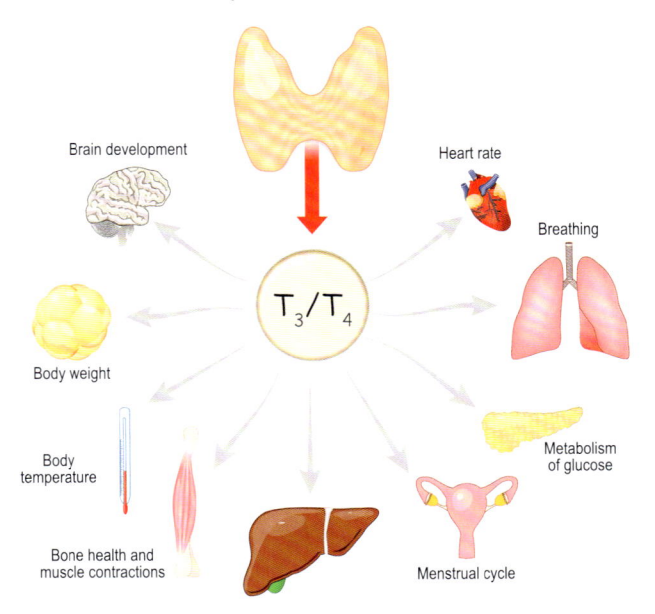

Brain development — $T_3/T_4$ — Heart rate

Body weight — Breathing

Body temperature — Metabolism of glucose

Bone health and muscle contractions — Menstrual cycle

287

# GOD MADE LIFE

burns more food, it creates more heat in the cells. But if the body gets too hot, hormones signal the blood vessels near the surface of the skin to expand, increasing blood flow. The blood carries your body heat to the surface, where the heat is convected into the surrounding air. Also, sweat glands begin to operate, forming moisture on the surface of your skin. As air moves across your body, the moisture evaporates, creating a cooling effect on your body.

Sometimes called the "master gland," the pituitary produces at least 10 different hormones. It's only the size of a marble, but this little gland regulates blood pressure, growth, milk production (for moms), and water content in the body. One of its most important functions is to produce a growth hormone. Once it leaves the pituitary, this growth hormone runs throughout the body, telling the tissues to get busy making more cells in the muscles and bones. Also, the growth hormone will break down the fat in a child's body to obtain the energy needed to make more tissue. And that's how you grow. Without these growth hormones carefully overseeing growth everywhere in your body, one of your legs might grow longer than the other one.

God provided two glands called **adrenals** to help the body deal with emergencies. Suppose a mother smells smoke in her house. Quickly, she realizes the house is on fire, and the adrenals release a shot of the hormone called **epinephrine**. As the epinephrine runs through the mother's bloodstream, her whole body goes on alert. Her liver releases a large amount of sugar into her blood to give her extra energy. The blood that is normally directed to her intestines and stomach is diverted to her muscles to give them extra strength. More blood also rushes into her brain to help her make quick decisions. So with superhuman strength, this mom quickly grabs her three little children who were taking a nap in their bedroom and carries them out of the house. This is how God has designed the body to work in times of emergency.

Hormones act as a switch on specific cells. Certain hormones attach to certain cells equipped with uniquely-fitted receptors. Once the hormone attaches, that cell will either increase or decrease specific activities in the cell. Since hormones are attaching to billions of receptors, this regulates

### Adrenal Glands

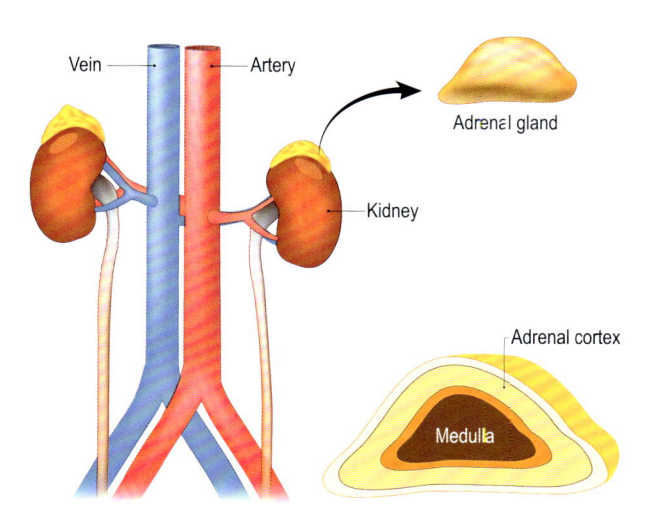

Vein — Artery — Adrenal gland — Kidney — Adrenal cortex — Medulla

## CHAPTER 10: GOD RESTORES AND REPRODUCES LIFE

entire systems in the body.

So, you can see that the whole body is carefully controlled by these glands and hormones. The Lord made about 50 different hormones to keep the body well regulated. This keeps you feeling well. And it keeps your body functioning well.

The best thing about these systems is that you don't have to think about making these adjustments. They all happen automatically. What an inconvenience it would be if you had think about adjusting your body temperature, blood pressure, water content, sugar content, and growth hormones all the time. Your body makes the adjustments all by itself! Indeed, we are fearfully and wonderfully made!

## The Provision of Healing and the Wonder of the Human Immune System

> Through one man sin entered the world, and death through sin, and thus death spread to all men, because all sinned. (Romans 5:12b)

When God created man and animals at the beginning, He declared everything to be "very good." There was no disease and no death. No tree could fall on Adam and kill him. No disease, no bacterium, no virus, and no protozoan could bring about death for man. He could live forever unless he sinned against God. After Adam ate the forbidden fruit, death fell upon all men. However, God still provided an immune system so that man could live 70-80 years on average. The body is geared up to defend itself from deadly germs. Soldier cells are employed to envelop invaders or shoot them down. Some produce weapons, some spy out the enemy, and some carry messages about the invasion. While

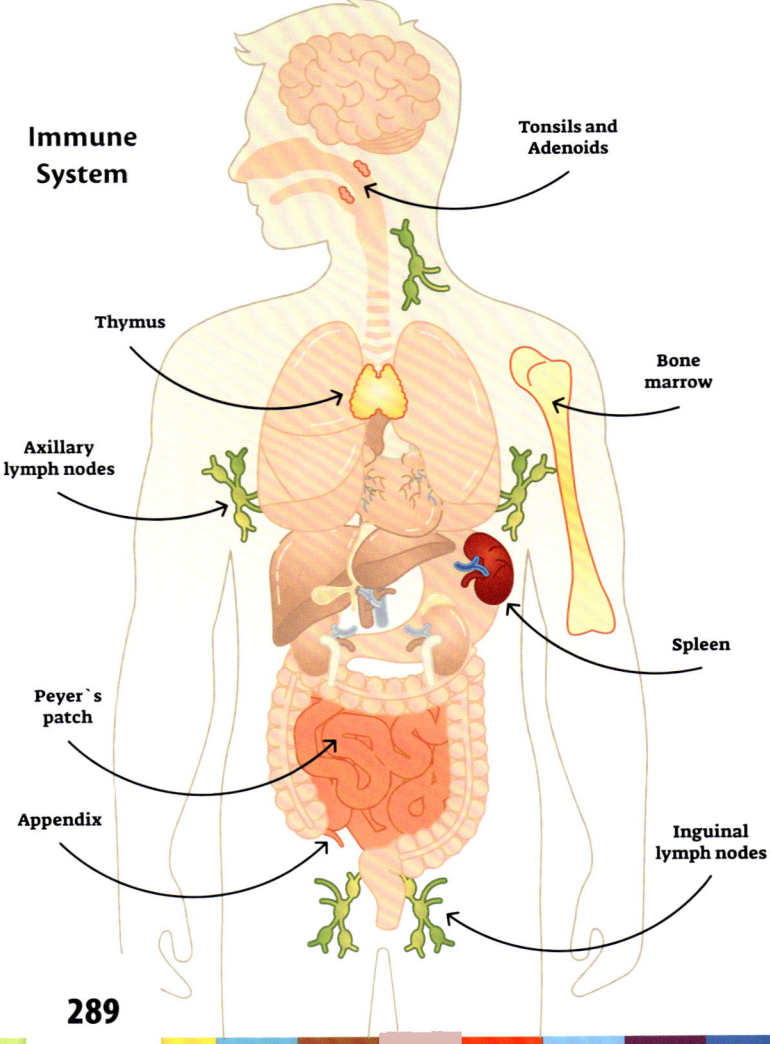

**Immune System**

- Tonsils and Adenoids
- Thymus
- Bone marrow
- Axillary lymph nodes
- Spleen
- Peyer`s patch
- Appendix
- Inguinal lymph nodes

289

# GOD MADE LIFE

each cell doesn't have a personality, these are automated systems developed by an all-wise Creator to protect your body in a dangerous world.

The **immune system** is amazing in that it can target specific enemies with a particular antibody. Also, the immune system has a remarkable memory. The body can remember a certain enemy germ for a lifetime! How does that work? Those people who get the measles once don't have to worry about getting it again. Most of the time, they will be permanently immune to it. When and if the measles virus attacks again, the body's immune system instantly recognizes it and destroys it before they get sick.

As already covered in Chapter 4, the immune system is made up antibodies, interferons, and white blood cells. An **antigen** is something in the body that stimulates an immune response. These are foreign invaders (usually a protein) that hang out on the surface of a virus or bacterium. When the body detects these foreign antigens, one variety of white blood cells—the B cells—starts producing **antibodies**. These antibody proteins are manufactured in the hundreds of **lymph nodes** scattered throughout the body. The antibodies glob on to the enemy antigen like a police dog leaping on a burglar. With the antibody hanging on to the outside of the enemy cell, it won't be able to reproduce or move around the body. This is called the **humoral immune response**.

If this is the first time the body has ever seen this particular antigen, it will take a little

## Polio

Early in the 20th century, **poliomyelitis** (or **polio**) was one of the deadliest childhood diseases. This virus would produce paralysis and even death in millions of children. Before the vaccine was developed, 2,000 children per year would die of the disease. As many as 10,000 to 50,000 children were paralyzed by the disease. But after the vaccine came out in 1955, the cases almost all disappeared. The vaccine contains a weaker version of the antigen that causes the polio virus. This weak antigen programs the body to create an antibody that will destroy the antigen. When and if a child should come upon the real polio germ later on, the body would have the antibodies ready to destroy the germ.

## CHAPTER 10: GOD RESTORES AND REPRODUCES LIFE

bit longer for the immune system to produce the new antibody. If the body has seen the antigen before, and the immune system has already produced the antibody, you may not get sick. The body will jump on the antigens and destroy the virus or bacterium.

Another function of the immune system is **cell-mediated immunity**. This is where white blood cells (**T-lymphocytes**) neutralize bacteria, fungi, cancer cells, and cells already attacked by viruses. Incredibly, these little guys shoot a protein at the germ, putting holes in its cell membrane.

There is one more very creative immune response God provided for our bodies. It's called the **complement response**. This system relies on proteins that move around the body looking for the bad-guy germs. When they find these harmful invaders, the proteins will attract **phagocytes**, whose job it is to eat the invaders. The proteins themselves will also attach themselves to the foreign cells and destroy them.

Scientists also think that the brain may be involved in defending the body from the "bad guys." When the brain is stressed, the immune system is weakened. But when the mind is at peace, that usually increases the production of antibodies and antigen-fighting proteins. Another organ called the **spleen** also increases its production of cancer-fighting cells when the brain isn't stressed.

You will keep him in perfect peace,
Whose mind is stayed on You,
Because he trusts in You.
Trust in the LORD forever,
For in YAH, the LORD, is everlasting strength. (Isaiah 26:3-4)

We would never survive without the body's immune system. The body's systems have to work as a team to fight diseases. The **thymus gland** is assigned to make white blood cells. But, if the body's immune system didn't identify the antigens correctly, the thymus would make the wrong kind of germ-fighting cells. If the cell-eating phagocytes received the wrong signal from the cells' communication systems, they would become little monsters that eat up the good cells. God made the whole body to function as a unit to keep everything healthy and strong.

Since the 1960s, **autoimmune diseases** have become

**Thymus Gland**

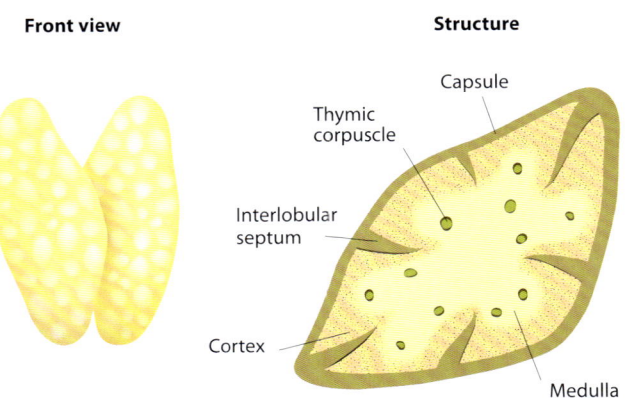

Front view

Structure
- Capsule
- Thymic corpuscle
- Interlobular septum
- Cortex
- Medulla

291

# GOD MADE LIFE

increasingly common in the modern world. Instead of attacking foreign cells, the body's immune system mistakenly begins to attack the normal cells. There are at least 80 of these autoimmune diseases active today.

## The Wonder of Tissue Regeneration

> " 'For I will restore health to you
> And heal you of your wounds,'
> says the LORD,
> 'Because they called you an outcast saying:
> "This is Zion;
> No one seeks her." ' " (Jeremiah 30:17)

As another expression of His goodness, our Creator has provided means of regenerating damaged tissue. For example, amphibians are known to grow back their limbs, jaws, or tails. A lizard will also grow back a tail after it has been severed. Such re-creations are awe-inspiring, and seem almost supernatural!

Incredibly, the human body can grow back certain parts after they are severed or badly injured. There are cases where a fingertip can grow back, or parts of the liver and bladder can also regenerate. The lining of a woman's womb grows back several times a year. But how does the body remember the original shape of the organ tissue? Scientists have yet to figure out how this works.

This **tissue regeneration** is just a small demonstration of the Creator's power to re-create and regenerate a body damaged by man's fall into sin. Most importantly, this is a picture of new life. Can God restore a resurrected body for each of us at the end of time? Let us believe this. If God has the power to grow a lizard's tail again, and if Jesus was raised with a resurrected body, then we will be raised, too.

> And God both raised up the Lord and will also raise us up by His power. (1 Corinthians 6:14)

What will our new resurrected bodies look like? That is hard for us to know. First Corinthians 15 says the new body will be like the old body, just like a watermelon seed turns into a watermelon plant once it is planted. The watermelon seed and the new plant look very different, but there is still some similarity.

> But someone will say, "How are the dead raised up? And with what body do they come?" Foolish one, what you sow is not made alive unless it dies. And what you sow, you do not sow that body that shall be, but mere grain—perhaps wheat or some other grain. But God gives it a body as He pleases, and to each seed its own body. (1 Corinthians 15:35-38)

# CHAPTER 10: GOD RESTORES AND REPRODUCES LIFE

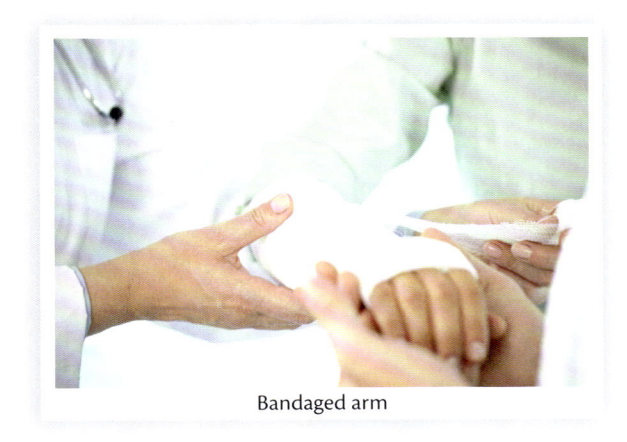

Bandaged arm

## The Wonders of Human Birth and the Reproduction of Life

For You formed my inward parts;
You covered me in my mother's womb.
I will praise You, for I am
fearfully and wonderfully made.
Marvelous are Your works,
And that my soul knows very well.
My frame was not hidden from You,
When I was made in secret,
And skillfully wrought in the lowest parts of the earth.
Your eyes saw my substance, being yet unformed.
And in Your book they all were written,
The days fashioned for me,
When as yet there were none of them. (Psalm 139:13-16)

Is there anything more wonderful in the physical world than the birth of a baby boy or girl? Where there were only two people in a home (a husband and a wife), now there are three people! A little baby joins the two parents at the dinner table. Here is another person created in the image of God. Here is another personality, another eternal soul, another child who is considered very precious by God. James 3:9 reminds us we should treat human life as valuable, for every human life is made in the likeness of God! Also, the Apostle Paul tells believing parents, "Your children are holy," which means that they are very special to God (1 Corinthians 7:14).

Where does this new life come from? The simple answer is that God creates each new life that comes into the world.

When two cells come together in one of the mother's **fallopian tubes**, a new human

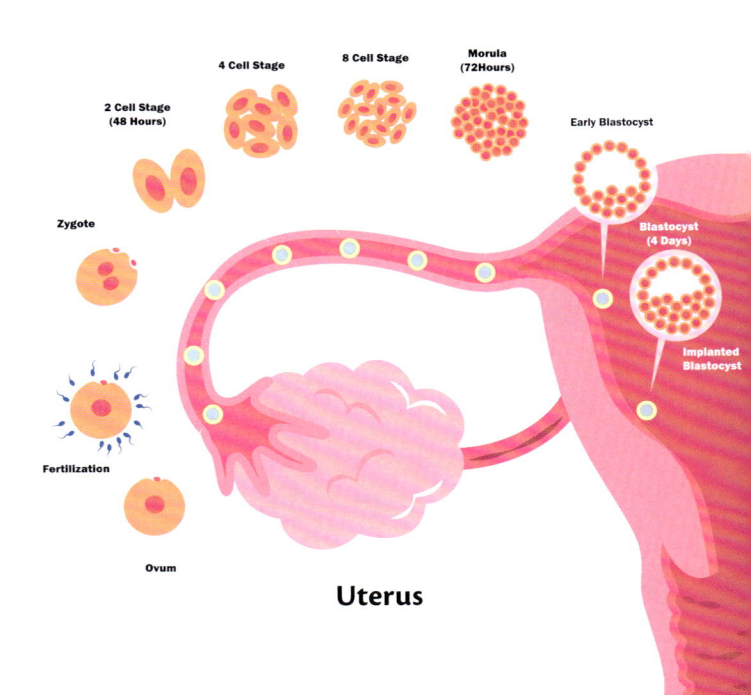

2 Cell Stage (48 Hours)
4 Cell Stage
8 Cell Stage
Morula (72 Hours)
Early Blastocyst
Blastocyst (4 Days)
Implanted Blastocyst
Zygote
Fertilization
Ovum

**Uterus**

# GOD MADE LIFE

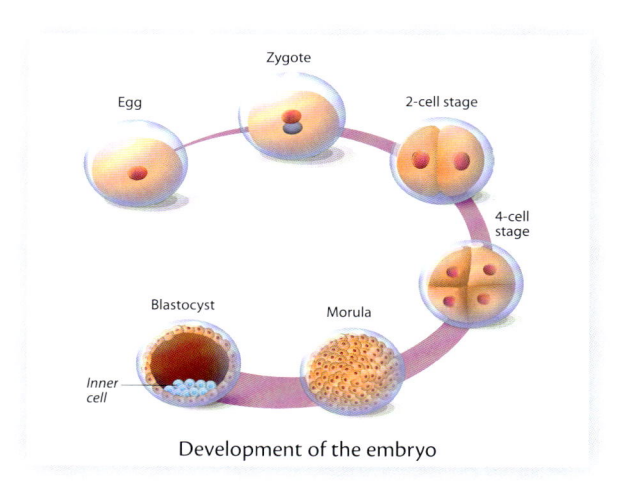

**Development of the embryo**

life is formed. The egg cell is actually the largest cell in the human body, and you can see it without a microscope. It's about the size of the width of a human hair. The male gamete is much tinier, but both the female egg cell and male **gamete** contain 23 chromosomes of DNA. As soon as these two cells come together, there are 46 chromosomes (or 23 pairs). This new cell is called the **zygote**. The nucleus of the male gamete and the nucleus of the egg combine into a single nucleus. It is the miracle of life! By the creative power of God, a new life and a separate personality has formed in this single cell.

As soon as the two cells come together to form the zygote, the outside of the egg hardens. This covering is called the **zona pellucida**, which protects the new life from interference from any other cell. The zygote divides into two cells by mitosis as we explained in Chapter 3. Within three days, the little zygote consists of 16 cells. It takes about a week for this little one to grow into 100 cells. It's still about 0.1 mm in size. That's about the size of the period at the end of this sentence.

The new life continues its journey, down one of the 4-5 inch (10-13 cm) fallopian tubes into the mother's womb. Little hair-like appendages called **fimbriae** line the fallopian tubes, coaxing the zygote along. The journey can take three to four days. Shortly thereafter, the zygote attaches itself to the womb of the mother.

## Every Tiny Baby is Either a Boy or a Girl

As soon as the baby appears as a tiny zygote, this baby is either a boy or a girl. Our Creator God assigned one of the 23 pairs of chromosomes to determine the gender of the baby. God made two kinds of chromosomes, called the **X chromosome** and the **Y chromosome**. The X chromosome is much longer and contains many more genes than the Y chromosome. The baby will automatically get one X chromosome. The baby also gets either another X chromosome or a Y chromosome to make a pair. If the pair of chromosomes are both X chromosomes, this little cell is a female. If the pair of chromosomes contains one X chromosome and one Y chromosome, this little cell is a male.

There are three big differences between a boy's body and a girl's body (besides special organs):

1. Boys get more of a hormone called **testosterone**. Girls get more of the hor-

294

mones called estrogen and progesterone.

2. Boys tend to convert their food into muscle. God made the bodies of girls to convert more food into fat, which will help them bear children.

3. Boys get more red blood cells, while girls get more white blood cells.

Remember, inside each cell in your body are all the chromosomes that make you who you are. Every cell in your body has a marker calling you out as a boy or a girl. Thirty trillion cells in your body all point to your masculinity or femininity, depending on whether you have an XX or XY chromosome. Yet, some people are unhappy with the body that God has given them. They don't like their looks. Or they are not content with the ears or the nose God has given them. Sometimes they are not content with their gender. Or they will act confused over their gender. Of course, this is rebellion against God. They don't want to accept God's plan for their lives. The Creator assigned a gender to each person and marked it out in every cell of his or her body.

> And Jesus answered and said to them, "Because of the hardness of your heart he wrote you this precept. But from the beginning of the creation, God 'made them male and female.'" (Mark 10:5-6)

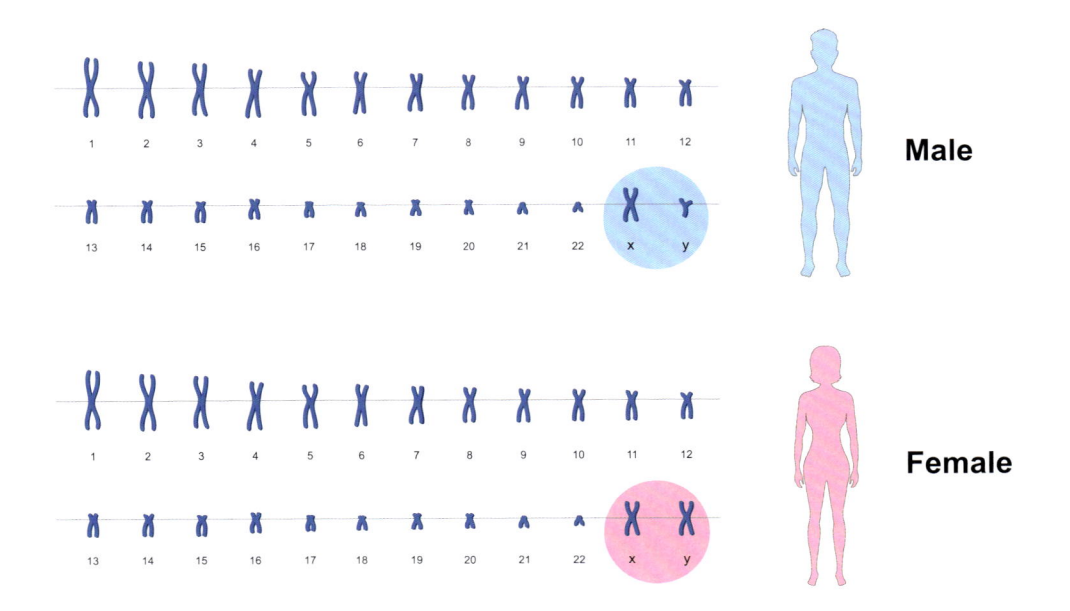

**Male**

**Female**

GOD MADE LIFE

## How God Prepares the Womb for a Baby

Thus says the LORD, your Redeemer,
And He who formed you from the womb:
"I am the LORD, who makes all things, . . ."
(Isaiah 44:24)

The mother's **womb** will be the growing baby's residence for nine months, and our wise and good Creator prepares the womb for that purpose.

First the womb needs a healthy **endometrium**, a lining rich with blood vessels. The scientific word for "womb" is **uterus**. Two hormones (estrogen and progesterone) help prepare this cushiony layer just before the baby arrives. **Estrogen** thickens the endometrium by increasing the blood supply in the uterus (or womb). **Progesterone's** job is to arrange for fluids to nourish the little zygote when it arrives in the womb. God made the uterus with the flexible capacity to stretch. This capacity increases by 500 times so it can hold an eight-pound (3.6 kg) baby at the end of nine months. The nine-month period during which the baby grows inside the womb is called the **gestational period**.

## Implantation

But You are He who took Me out of the womb;
You made Me trust while on My mother's breasts.
I was cast upon You from birth.
From My mother's womb
You have been My God. (Psalm 22:9-10)

Most babies lost in pregnancy are lost because they fail to **implant** on the wall of the uterus. This is a critical step in the child's life. To successfully implant, the new life first releases an enzyme that opens up a hole in the hardened zona pellucida's shell so the little zygote can squeeze its way out. A sticky substance is supposed to form that attaches the zygote to the wall of the mother's womb.

Then, for 11 weeks, the baby is nourished by "womb milk" secreted by glands on

**Baby in the Womb**

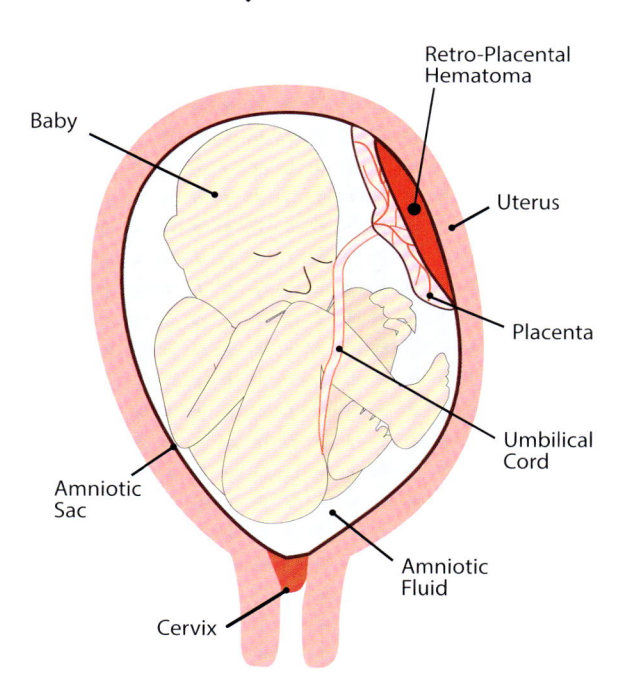

- Baby
- Retro-Placental Hematoma
- Uterus
- Placenta
- Umbilical Cord
- Amniotic Sac
- Amniotic Fluid
- Cervix

the inside surface of the womb. This is a mix of glucose (sugar) for nourishment and glycoproteins to grow new tissue. At this point, the baby is much too small to receive food through the mother's blood. However, the Lord our Creator created the placenta. This is a flattened circular organ in the uterus designed to provide nourishment transferred over from the mother's blood after the 12th week of life. During these early weeks, the placenta is much larger than the tiny little child, who is still smaller than a grape or a fig.

Usually, the placenta forms at the top of the womb. When it develops at the bottom of the womb, it is referred to as placenta previa and can cause birth complications. This wonderful organ God created not only provides oxygen and food through a supply of blood, but it also regulates the baby's temperature. It filters out germs that could cause infection as well. Since the mother's body has already produced antibodies to fight off disease, these also are shared with the baby.

The Creator forms an umbilical cord between the placenta and the baby. This 18-inch cord (46 cm) connects at the baby's belly button, and it is removed at birth. The placenta is designed to keep the baby's blood from coming into contact with the mom's blood. Since mom and baby could have different blood types, it could be fatal if they were to mix.

It's dangerous for the baby if the mother drinks alcohol or takes certain drugs while she is pregnant. These substances can pass through the placenta into the baby. Only a very thin wall of cells separates the baby from the bad stuff.

The umbilical cord contains two arteries that carry blood from the baby to the mother's body. These arteries carry waste products, which get transferred to the mother and are eliminated through the mother's system. The cord also contains one vein which brings nutrition and oxygen into the baby's body.

Amazingly, in just nine months, one cell turns into a fully functional human being, with all 78 of the baby's bodily organs working properly. The skeletal, nervous, muscular, respiratory, endocrine, immune, cardiovascular/circulatory, urinary, integumentary(skin), reproductive, and digestive systems are all ready to go. Incredibly, all of this complexity and variety comes out of one solitary cell. But that cell contained the DNA instructions needed to put together a human body.

As explained earlier in this course, DNA provides different instructions for different kinds of cells to build different body parts. But how do the cells know when to reproduce and how much they need to reproduce? How do the muscle cells form in one place, and the skin cells form in another? How do the organs form using certain kinds of cells? Somehow, cells know where they are in reference to other cells because they "talk" to each other. The genes communicate with the cells about which of the proteins need to be replicated. In the early weeks, the child's hands are like paddles with no fingers. Then, the DNA's genes tell the cells between the fingers to die.

## GOD MADE LIFE

That's what forms the fingers. According to how God has designed the genes, they will continue to send billions of messages to every part of that little child's body. That's what forms every tiny part over nine months. The toes grow. The earlobes grow out. The lungs and kidneys form. The brain develops—all according to the instructions.

To better understand the miracle of life and the development of life, scientists can look at a fertilized chicken egg. Cracking open the egg, they find the little chick. Now it is dead, but they take 10 muscle cells from the embryonic heart. These cells continue to live for a short time all by themselves. Placing these tiny cells in a saline solution, they watch as the muscles continue to beat at 350 beats a minute. Though the rest of the chick's heart has died, these muscle cells still remember the beat. After a while, something very mysterious occurs to confound all observers. First two of the cells begin to pound in unison, then three, then four, until finally all of them are keeping the same rhythm! How do these tiny cells learn this rhythm? Who is the choir director? Where do they get their directions? This is amazing! God has put the instructions for the heartbeat into every cell of the heart's muscles, and the cells somehow communicate with each other!

Legos® do not build towers by themselves without an intelligent child putting them together. Toy soldiers don't start marching together by themselves. Surely God has given intelligent directions to every one of the trillions of cells in your body through the DNA. Every cell knows what it's supposed

Brick towers do not build themselves.

CHAPTER 10: GOD RESTORES AND REPRODUCES LIFE

to do and where it belongs. Only our all-wise Creator could have created a life so wonderful, so mysterious, so complex, and so perfect. What a mind-baffling discovery that we may observe so many diverse cells working together in such graceful harmony! Think of trillions of moving parts each playing its part according to an intelligent plan. And all of these cells function in perfect, cooperative unity to produce one beautiful creation of human life. . . You!

It is good to give thanks to the LORD,
And to sing praises to Your name, O Most High; . . .
For You, LORD, have made me glad through Your work;
I will triumph in the works of Your hands.
O LORD, how great are Your works!
Your thoughts are very deep. (Psalm 92:1, 4-5)

## The Development of a Child in the Womb

"Thus says the Lord who made you
And formed you from the womb, who will help you:
'Fear not, O Jacob My servant;
And you, Jeshurun, whom I have chosen.
For I will pour water on him who is thirsty,
And floods on the dry ground;
I will pour My Spirit on your descendants,
And My blessing on your offspring;
They will spring up among the grass
Like willows by the watercourses.'
One will say, 'I am the Lord's';
Another will call himself by the name of Jacob;
Another will write with his hand,
'The Lord's,'
And name himself by the name of Israel."
(Isaiah 44:2-5)

See the chart of development on the following page of the baby in the womb. The baby is pretty well formed by the beginning of the seventh month of pregnancy. However, the child's body will continue collecting fatty deposits so he or she can stay strong and healthy after birth. Also, a **myelin** covering will form around the nerve cells in the brain, increasing the speed of the signals in the brain by a hundredfold. This helps the child to be able to think and to learn. During the

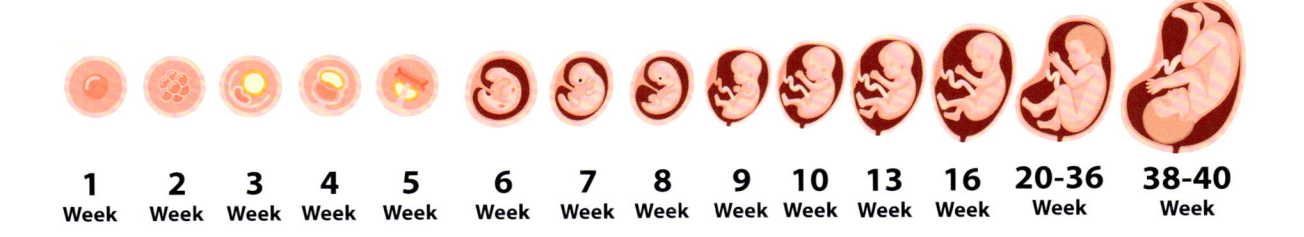

| 1 Week | 2 Week | 3 Week | 4 Week | 5 Week | 6 Week | 7 Week | 8 Week | 9 Week | 10 Week | 13 Week | 16 Week | 20-36 Week | 38-40 Week |

299

# GOD MADE LIFE

| Age of Baby in Womb | Size and Characteristics of the Child |
|---|---|
| 3 weeks | The child, sometimes called a **blastocyst**, is a ball of hundreds of cells, all multiplying very fast. The child is the size of the period at the end of this sentence. |
| 4 weeks | The child is the size of a poppy seed. |
| 5 weeks | The child is the size of a shirt button. The child's spinal cord and his little brain start to register signals. The child's heart starts to beat. Facial features develop. |
| 6 weeks | The child is the size of a lentil bean. Nose, mouth, and ears take shape. Intestines and the brain begin to form. |
| 7 weeks | The child has doubled in size in just one week. |
| 8 weeks | The child is half an inch (1.25 cm) in size. The child starts to move. The child's windpipe connects to the developing lungs. |
| 9 weeks | The child has earlobes, and all his organs are in place. The child is the size of a grape. |
| 10 weeks | Fingernails start to form. |
| 11 weeks | The child is kicking and hiccuping. The child is the size of a fig. |
| 12 weeks | The child is the size of a lime. All the basic body parts have formed. The child can open and close his fingers and curl his toes The baby can start to feel pain.[27] |
| 16 weeks | The child is 4½ inches (11 cm) long. Hard bone starts to form, replacing cartilage in the child's body. Fingerprints have developed. |
| 18 weeks | The child can hear you talking and responds to sounds. |
| 24 weeks | The child's eyes can now respond to light. The child is 12 inches (30 cm) long. |
| 32 weeks | The child is 18 inches (46 cm) long and weighs about 5 pounds (2.3 kg). |
| 36 weeks | The child can blink and weighs 5½–6½ pounds (2.5–2.9 kg). |

# CHAPTER 10: GOD RESTORES AND REPRODUCES LIFE

pregnancy, the mom has increased her blood supply by 50%. Though the child is much smaller than the mom, he still requires about half the blood she ordinarily needs.

## Childbirth

[Jesus said], "A woman, when she is in labor, has sorrow because her hour has come; but as soon as she has given birth to the child, she no longer remembers the anguish, for joy that a human being has been born into the world." (John 16:21)

When a child is born, the child's head is usually positioned downwards. First, the baby is born, and then the placenta follows. Birthing is not an easy process, and a lot of things can go wrong. When Adam and Eve fell into sin, the Lord God told the woman:

"I will greatly multiply your sorrow and your conception;
In pain you shall bring forth children."
(Genesis 3:16)

And so, while childbirth is very painful for mothers, the birth of a child is also a time of great rejoicing. There is great joy in a home and in a church community when a child is born!

For thousands of years, **infant mortality** was very high in the world. This means that many children would die in the womb or shortly after birth. In 1800, 460 out of every 1,000 children in America or England did not survive their fifth birthday. Nowadays (200 years later), only 7 out of every 1,000 children die before 5 years of age.

Improvements in childbirth and medicine came mostly from Christian doctors and scientists. By His grace, God provided special wisdom to medical researchers like Louis Pasteur, Joseph Lister, Edward Jenner, Alexander Fleming, James Young Simpson,

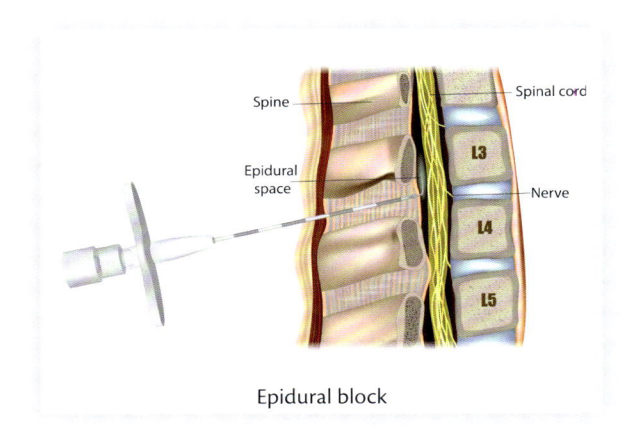

Wilhelm Roentgen, and Raymond Damadian. All these were Christians who contributed tremendously to medical science.

The most common reason why babies don't survive childbirth is that they are born too early. The second most common problem is that they don't start breathing on their own right away. Remember, the baby has been living off the mother's blood for nine months. The little one needs to take that first breath fairly quickly after he is born. About 23% of the babies who die in childbirth have contracted infections. However, all told, only 3 out of every 1,000 babies born in the United States will die at childbirth. Every life is precious, and we thank God for every one who is created in His image and for every child who survived the birthing process!

Throughout history, doctors have tried to save moms and babies using a surgery called the **cesarean section** (**c-section**). It was a dangerous surgery, and often the mother or the baby would die in the process. This procedure involves removing the baby from the mother's womb by a surgical cut.

Thankfully, the c-section is much safer today. Medical improvements have taken place since the 1840s and have helped to save a lot of lives. Cleanliness in the hospitals and sterilization of surgical tools was very important. Doctors and nurses learned that they must be very careful to wash their hands and clear all germs out of the area before doing a c-section.

A Christian doctor named James Young Simpson was the first to use anesthesia for a medical surgery in 1847. Applying **anesthesia** or **analgesia** before a c-section allows the patient to go through the surgery without feeling pain. The doctor might also use certain drugs to make the patient unconscious before surgery. Or the doctors may just numb the nerves with an analgesic drug such as the **epidural block**.

A better diet for moms is the other improvement that has helped produce healthier babies. Research has discovered that smoking and alcohol can be very bad for babies in the womb.

## The Wonder of the Human Skeleton[26]

What is this body forming inside the mother's womb? Well, 54% of the body's mass is bones and muscle. Roughly 16% of the body's mass is skin. The **skeleton** gives the body sturdiness. Over 200 bones in your body are held together by ligaments and tendons. Ligaments connect the bones to other bones, while **tendons** connect the muscles to the bones.

The ends of our bones where they connect to other bones are usually layered with **cartilage**—a soft, flexible material. If you want to know what cartilage feels like, feel the tip of your nose or the stiff parts of your ears. Remember that the baby's skeleton is made of cartilage at first. The bones form later. Cartilage is made of collagen and other proteins, and the blood does not pass through it.

**Bone** is made up of living cells, minerals like calcium and phosphorus, and blood vessels. There are also nerves in the bones. That's why it's so painful when you break one. Also, blood runs through the bones, providing ongoing oxygen and nourishment to the cells. A long narrow tube runs through the inside of some of your bones. This contains **bone marrow**, where red blood cells are produced according to the Creator's design. If you cut open the bone in a chicken leg, you'll find this marrow down the center of it.

Stop for a moment and consider how smoothly your skeleton operates. God has designed five different kinds of joints for the skeleton. Your joints work like a well-oiled machine. The **hinge joint** is what you get for your knees and elbows. Your spine uses **gliding joints**. Your hips have **ball-and-socket joints**, while your forearm uses the **pivot** joint so you can twist your arms this way and that way. The skull also has **fused joints**, which do not allow any movement.

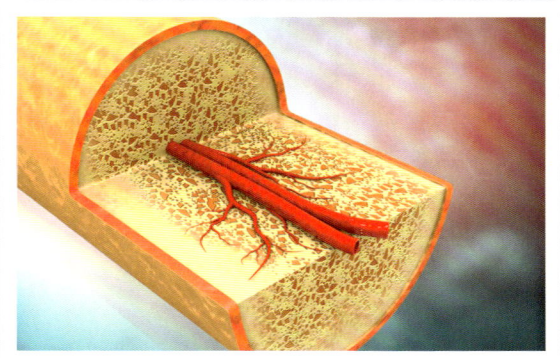

Bone marrow

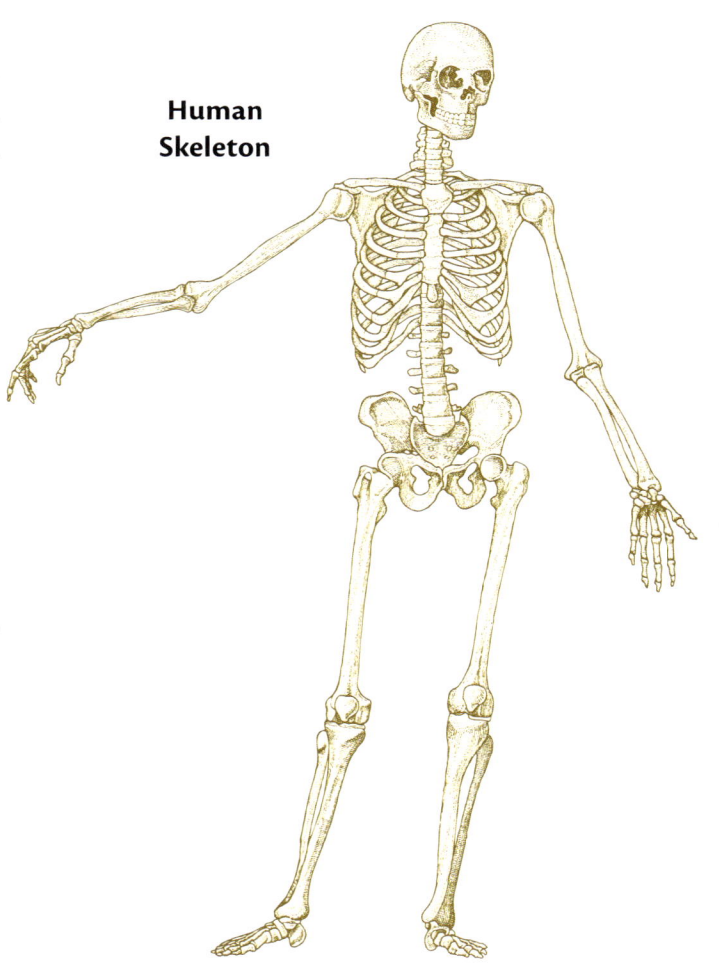

**Human Skeleton**

# GOD MADE LIFE

**Skeletal System**

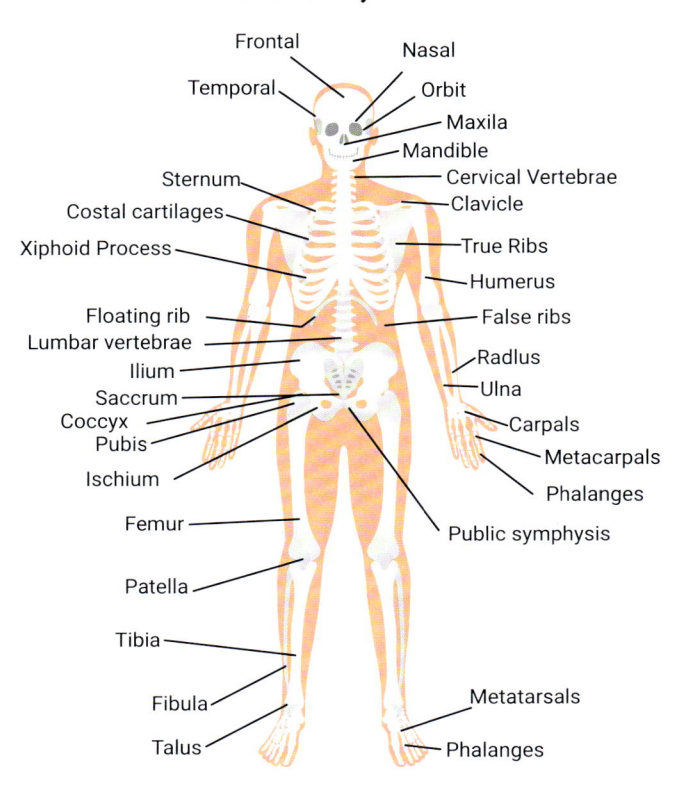

Frontal, Nasal, Temporal, Orbit, Maxila, Mandible, Sternum, Cervical Vertebrae, Costal cartilages, Clavicle, Xiphoid Process, True Ribs, Humerus, False ribs, Floating rib, Lumbar vertebrae, Radlus, Ilium, Ulna, Saccrum, Carpals, Coccyx, Metacarpals, Pubis, Phalanges, Ischium, Public symphysis, Femur, Patella, Tibia, Fibula, Metatarsals, Talus, Phalanges

## The Wonder of the Human Frame

I will praise You, for I am
fearfully and wonderfully made;
Marvelous are Your works,
And that my soul knows very well.
My frame was not hidden from You,
When I was made in secret,
And skillfully wrought in the lowest parts of the earth.
Your eyes saw my substance, being yet unformed.
And in Your book they all were written,
The days fashioned for me,
When as yet there were none of them.
(Psalm 139:14-16)

Evolutionists today think that humans descended from ape-like creatures. They have no evidence for it. But artists with vivid imaginations have created monkey-to-man charts. Yet, these charts don't highlight the big difference between monkeys and man. God created the human body to walk upright on two feet. God created man with a special dignity. There is no way that an animal walking on four legs (or a knuckle walker) could have evolved into a human walking on two feet. There are too many differences in the skeletal makeup. How could an ape-like creature turn into a human? There is no evidence to be found in fossils of apes or humans that show transitions between the species.

God gave humans super strong big toes. When you walk, your big toe rolls forward naturally, helping you to keep your balance. Apes use their thumb-like toes to grip things, but it doesn't help them to walk.

The bone structure in your **foot** is amazing. God pieced 26 uniquely-shaped bones together in each foot, forming a nice arch. All these bones fit together perfectly like a jigsaw puzzle, providing the very best design for strength. In fact, some of the bones are wedge-shaped to provide a stronger arch. The arch gives the foot the ability to carry a lot of weight, much like an arched bridge.

But the arching also provides shock absorption when running, jumping, or anytime you land hard on your feet.

The human foot is very flexible as well, so pressure can be shifted between the right foot and the left foot when walking. To help with your balance, the foot is shaped like a tripod with three points of contact—one behind the big toe, one behind the smallest toe, and one at the heel. Apes and monkeys are flat footed, which makes it harder for them to stay balanced. And they certainly cannot run on two feet.

Consider also that the Lord made human legs much longer than ape legs in proportion to the rest of their bodies. This helps humans to walk long distances without too much trouble. Although bigger than humans, lions only move a couple of miles (3 km) per day. Wolves travel up to 30 miles (48 km) a day, and elephants can lumber through 120 miles (193 km) of the African savanna in a day. Although still much smaller than elephants, some humans have been able to run 180 miles (290 km) in a single 24-hour period.

Humans can also stand on two legs for a long time because the Creator gave them a knee joint that locks in place. So, it doesn't take a toll on your muscles when you stand. The longest a person has stood motionless in one place is 30 hours and 12 minutes. This record was set in 2003. An ape would topple over in a few seconds or a few minutes.

In addition, apes are inefficient and clumsy when they run. They tend to waddle using a swaying, back-and-forth motion. That's because their femur (leg) bones are separated by quite a distance from the centerline of their torsos. Since God made man with femur bones that angle inwards, he has much better balance. This is also why you can

Humans are built to run on two legs.

Monkeys and apes walk on all fours.

# GOD MADE LIFE

stand on one leg. The world record for standing on one leg is 75 hours and 40 minutes.

While God made man with the ability to stand straight, the bone structures of apes are such that they are always bent over a bit. Humans have a pelvis bone and an upright spine, that enable them to walk with their head straight above their hips. If your head was bent over like an ape's, you would topple over. That's why apes have to catch themselves with their front hands from time to time as they walk. These are the "knuckle-walkers."

Another marvelous feature of God's design for man is found in the S-shaped human **spine**. Like a strong spring, the spine acts like a shock absorber when lifting things or running over rough ground. Also, the Lord placed a soft disk made of cartilage between the vertebrae. This cushions the back a bit. A Christian man named Paul Anderson (1932-1994) used his back to lift 6,270 pounds (2,844 kg), which is the heaviest load ever lifted by a human! That a human weighing only 360 pounds (163 kg) could hold up 20 times his weight with his back is a phenomenal testimony to God's design of the human skeleton!

## What Can Go Wrong with Your Skeletal System?

In the current state, humans are not supermen. They can fall and break their bones. There is a difference between a broken bone and a sprain. When you sprain your ankle playing soccer, you have torn or stretched your tendons and ligaments in an unusual way. In order for a sprain to heal properly, doctors usually recommend some movement but not too much. Broken bones, on the other hand, are more serious.

Once more, the Creator has provided wonderful means for healing bones. If the bone is badly broken, doctors will have to "set" the bone. This means that they will align the pieces of bones in the right place so the bones will heal themselves. Sometimes they will use screws to hold the bones in place. Sometimes they will bind the outside of the arm or leg with a cast. Usually, fractures heal in about 6 weeks (4-5 weeks for children). In what seems like a miracle, the body will send cartilage-like material into the break. This seeps into the crack like glue. Then, the body adds a little more calcium and phosphorus to it so the patch will turn into hard bone.

Just like cars, bikes, sweaters, and refrigerators, the human body does wear out. The constant movement in the joints over 70 years wears down the cartilage padding between the bones. Then, the rubbing of bone against bone and the breakdown of the cartilage causes inflammation and irritation. Many older people suffer from this painful condition called **arthritis**.

## The Wonder of the Muscular System

I will praise You, for I am fearfully and wonderfully made; Marvelous are Your works,

**CHAPTER 10: GOD RESTORES AND REPRODUCES LIFE**

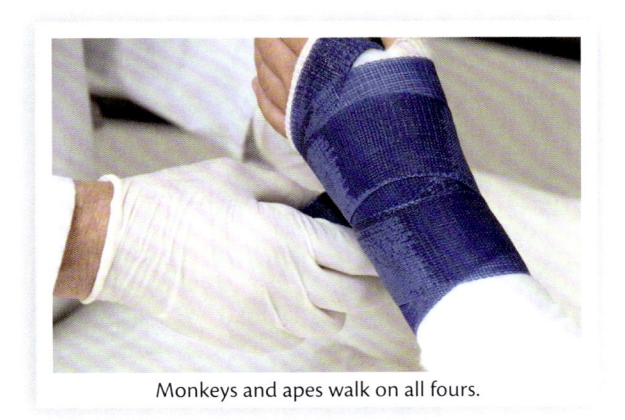

Monkeys and apes walk on all fours.

**And that my soul knows very well. (Psalm 139:14)**

There are over 600 **muscles** in the human body. The important thing to know about muscles is that they always pull, and they never push. They pull by contracting (getting shorter). Your muscles cannot work independently of your skeleton. They have to work together in perfect concert as your body moves. So God made tendons to connect muscle to bone. The whole body is working together to provide smooth movement. First, the nerves signal the muscles to contract. Next, since the muscles are attached to the bones, the whole structure will move as well.

There are two types of muscles: **voluntary muscles** and **involuntary muscles**. The voluntary muscles are skeletal muscles, which move when you want them to move. Involuntary muscles, such as those that work the movements of the heart, intestines, and stomach, will function on their own. Involuntary muscles can be divided into two kinds: **cardiac muscles** for the heart, and **smooth muscles** for the other organs and their functions.

Your skeletal muscles work in teams. When you fold up your arm, the triceps relaxes while the biceps contracts. Then when you straighten out your arm, your biceps relaxes while your triceps contracts.

The bones and muscles team up to function as a lever. When you fold up your arm, the muscles in your upper arm pull up the forearm, while the elbow works as a fulcrum.

Modern medicine has come up with **prosthetic (artificial) limbs** for people who have lost an arm or a leg. These are still a little clumsy compared to God's original design. But some of the prostheses use signals from the brain to operate. More recently, some prosthetic hands invented by scientists can

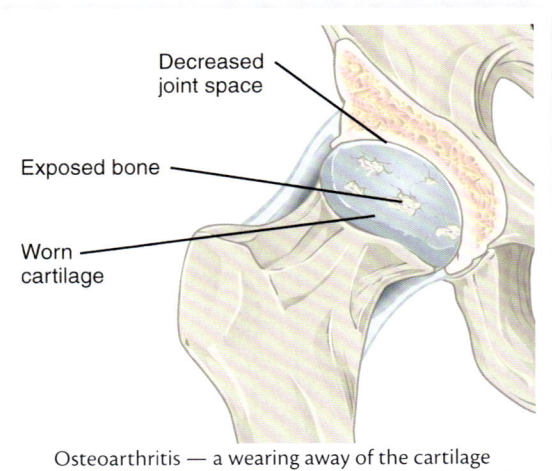

Decreased joint space

Exposed bone

Worn cartilage

Osteoarthritis — a wearing away of the cartilage

307

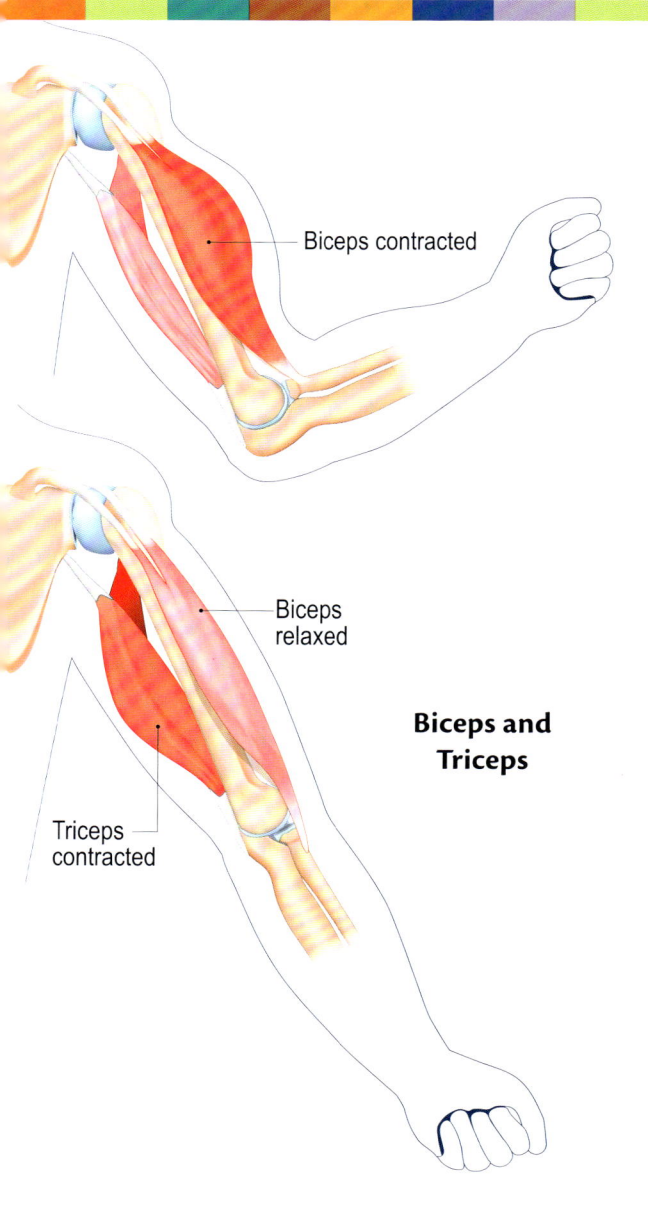

# GOD MADE LIFE

## Biceps and Triceps

- Biceps contracted
- Biceps relaxed
- Triceps contracted

## The Muscular System

- Frontalis
- Orbicularis oculi
- Sternocleidomastoid
- Deltoid
- Pectoralis major
- Rectus abdominis
- Abdominal external oblique
- Iliopsoas
- Quadriceps femoris
- Peroneus longus
- Peroneus brevis
- Temporalis
- Nasalis
- Orbicularis oris
- Rotator cuff
- Biceps brachii
- Brachialis
- Pronator teres
- Brachioradialis
- Adductor muscles
- Tibialis anterior

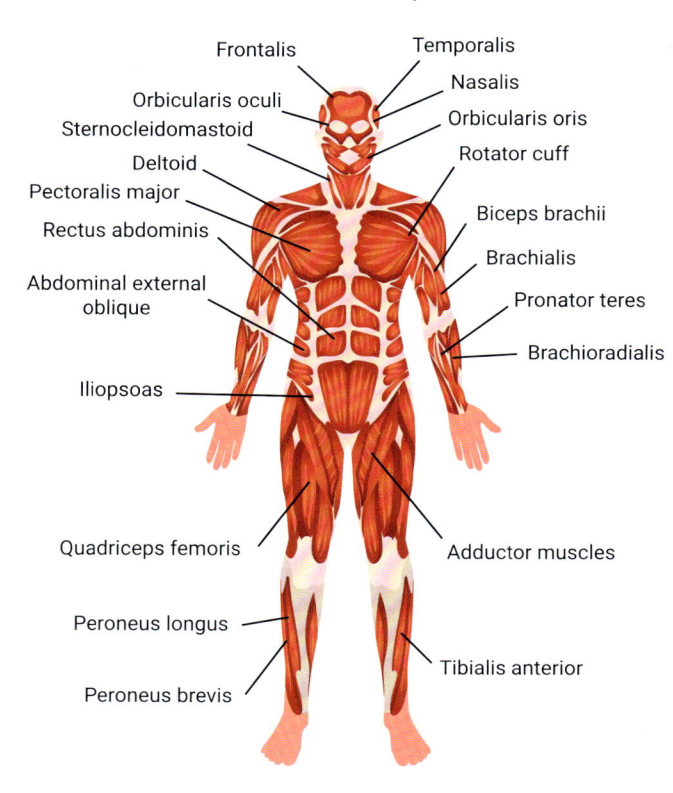

"feel" pain and texture. Pressure sensors pass signals through the nervous systems, telling the brain what the hand is "feeling." Such equipment is very complex and very expensive, but it is still not as wonderful as what God created originally for our hands.

As biologists study the human body, they have concluded that this phenomenal creation was "over designed." That means they think there is much more included in the human body than what is needed just to survive. For example, the human hand is made of almost 200 moving parts—27 bones, 70 muscles, 27 joints, and over 100 separate tendons and ligaments.

Prosthetic limb

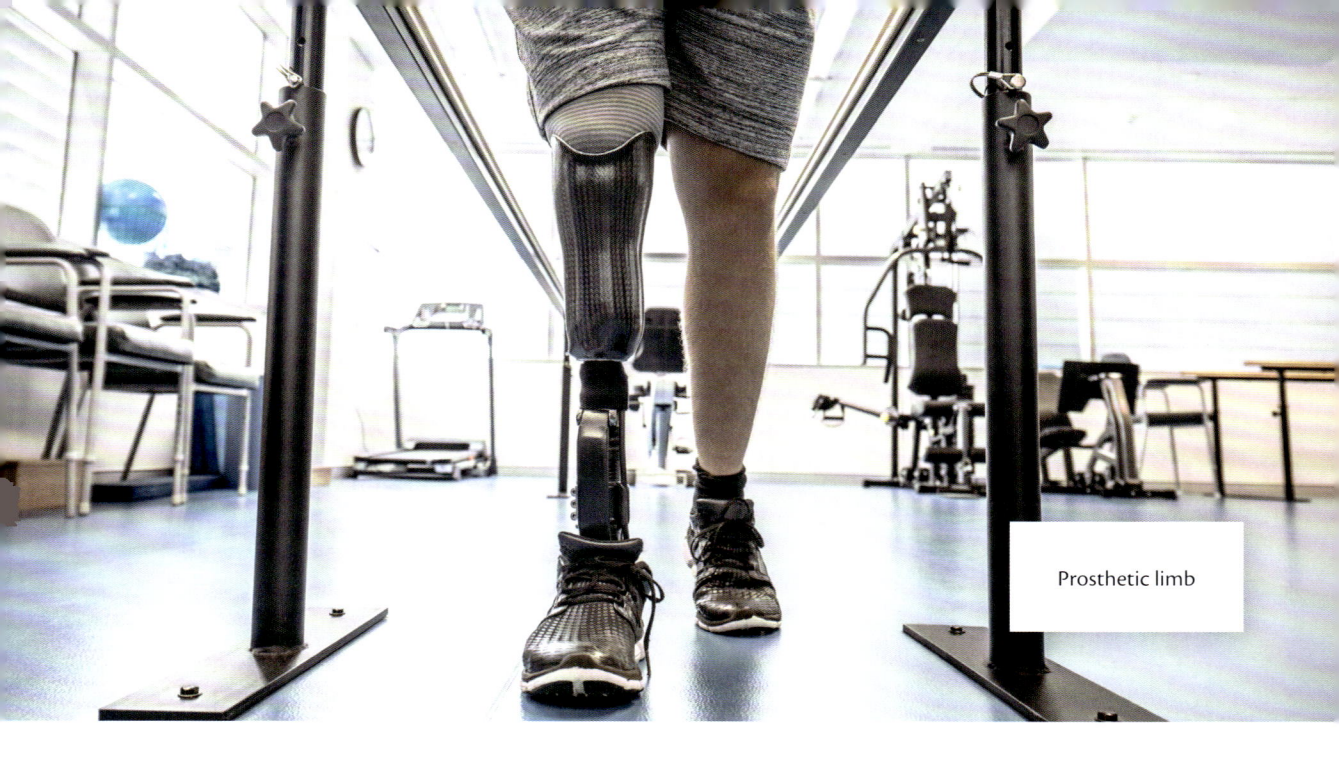

## God's Wonderful Design of the Human Hand

**If I forget you, O Jerusalem,
Let my right hand forget its skill! (Psalm 137:5)**

God made apes with clumsy hands. These guys can use a rock to crack open nuts, but they can't play the violin or juggle four balls in the air. The human is extremely skillful with his hands. These are designed for a full range of movements. Muscles consist of a bunch of fibers wrapped together. Hand muscles are made of tiny bundles designed for fine finger movement.

Animals can hunt for food and fight, but it takes special skills to write, brush your teeth, paint beautiful pictures, and perform difficult medical surgeries. An ape might be able grab a crayon and take a bite out of it. But an ape can't write with it or color a picture. Also, a child can hold a pencil using a **tripod grip**. It is a very complicated method, requiring the coordination of the thumb, the index finger, and the middle finger. Animals are too clumsy, and they could never hold a pencil that way. Of course, God intended for man to be able to write. In Deuteronomy 6:9, He commands dads and moms to write His Word on the posts of the house and the gates.

**"You shall write [these words which I command you today] on the doorposts of your house and on your gates." (Deuteronomy 6:9)**

The **motor cortex** of the human brain is in charge of the body's movements. About 25% of this part of the brain is used to control hand movement.

# GOD MADE LIFE

The brain communicates with the fingers for each movement, enabling an unimaginable sequence of carefully controlled maneuvers. For example, the fastest pianist in the world can whip out 765 strokes a minute. That's about 14 separate keys per second. The fastest typist puts out 233 words per minute, which equals about 1,500 strokes per minute. Think about how quickly the brain and the fingers must work together to coordinate such an amazing feat! From the beginning, the Lord our Creator intended for humans to be the most skilled, the most versatile, and the most advanced of all His material creatures.

## The Human Face and Its Expressions

Yet one more important difference between apes and humans appears in the construction of the face. God wanted humans to be able to relate to each other. So He created humans with the capability to speak and to write. Beyond this, He also wanted us to communicate by facial expressions. Believe it or not, the muscles in your face allow you to make 10,000 expressions! Whereas God gave humans 50 facial muscles, gorillas only have 30, none of which are for making facial expressions.

In His great wisdom, God gave some lesser creatures a few enhanced features.

Drawing on a ceramic bowl

## CHAPTER 10: GOD RESTORES AND REPRODUCES LIFE

Many of these features presented to the lesser life forms contradict the evolutionary plan. For example, we have discovered the humble trilobite possessing the most excellent eyesight. The opossum is blessed with an opposable thumb like yours. And the Creator provided the koala bear with human-like fingerprints. Bats are equipped with the most amazing system of echolocation, while some salamanders get their own unique type of red blood cells! None of these fit into the evolutionary scheme. These are just the merciful provisions made for each of these creatures—all of them according to the thoughtful purposes of our all-wise God and Creator.

*An ape makes a mess on canvas.*

## You Are the Image of God

**I will praise You, for I am fearfully and wonderfully made; Marvelous are Your works, And that my soul knows very well. (Psalm 139:14)**

Over these several chapters, we have proven the truth of that verse. Indeed, we can all say, "My soul knows it very well!" Take a look at your body. Study this work of genius. It is a technological wonder! It is a true work of art! This is the work of God. More than all of this, let us never forget that the human body and the soul are both created in the image of God.

**So God created man in His own image; in the image of God He created him; male and female He created them. (Genesis 1:27)**

That being the case, we must respect the body as created in the image of God. It is of much more value than anything within the animal kingdom. After the flood, God told Noah, "Whoever sheds man's blood, by man his blood shall be shed; for in the image of God He made man" (Genesis 9:6). Human life is precious, and God is deeply offended when we disrespect it. We ought not to be angry with our brothers and sisters or curse at them. James reminds us of this:

# GOD MADE LIFE

> But no man can tame the tongue. It is an unruly evil, full of deadly poison. With it we bless our God and Father, and with it we curse men, who have been made in the similitude of God. (James 3:8-9)

Also consider that Jesus, God's only begotten Son, took on human flesh for Himself. For nine months, Jesus was growing in Mary's womb. His body had our DNA, our proteins, our myosin, our white blood cells, and our collagen. With this same body, He died on the cross for our sins. And then He rose again with a resurrected body, with nail prints still visible upon His hands.

Psalm 22 speaks of Jesus, telling us that He was "cast upon God," trusting His Father from His mother's womb. John the Baptist was filled with the Holy Spirit when he was in his mother's womb. Every little child has a personality, and he is God's image-bearer even while he is still in the womb. Therefore, it is wrong to murder babies as well as every other person.

Sadly, many babies are killed before they are born. Some are killed by surgical **abortions**. Often, the little baby cannot make it down the fallopian tubes or attach to the womb because the mother takes certain kinds of pills that destroy his life. Some mothers use a device called an IUD. So, the babies just die before they make it into the womb. These devices result in the death of most babies in modern nations. However, let us remember that God's Word clearly forbids murder.

> "You shall not murder." (Exodus 20:13)

But God's law also forbids the harming of an unborn child, even if the harm came about by an act of foolish negligence. In Exodus 21:22-25, we find the example of men fighting with each other. In the process, they accidentally hurt the baby in the mother's womb. This tragedy is treated as a crime by God's law.

> "If men fight, and hurt a woman with child, so that she gives birth prematurely, yet no harm follows, he shall surely be punished accordingly as the woman's husband imposes on him; and he shall pay as the judges determine. But if any harm follows, then you shall give life for life, eye for eye, tooth for tooth, hand for hand, foot for foot, burn for burn, wound for wound, stripe for stripe." (Exodus 21:22-25)

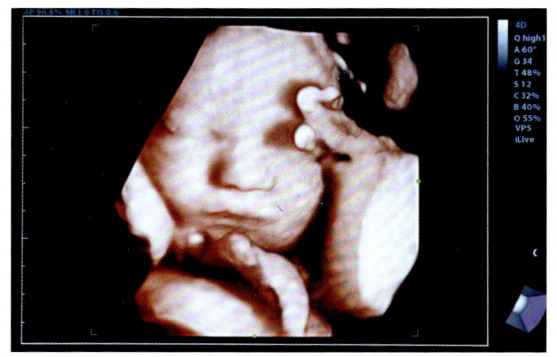

Ultrasound showing the face of the human baby in the womb

**CHAPTER 10: GOD RESTORES AND REPRODUCES LIFE**

Above all, let us respect all human life. Let us thank God for each new life. And let us be careful not to harm children created in God's image even when they are safely nestled in their mothers' wombs. Since all of us are made in the image of God, let us take good care of the elderly, the sick, and the handicapped.

## Why God Made You

But now, thus says the LORD, who created you, O Jacob,
And He who formed you, O Israel:
"Fear not, for I have redeemed you;
I have called you by your name;
You are Mine. . . .
Everyone who is called by My name,
Whom I have created for My glory;
I have formed him, yes, I have made him."
(Isaiah 43:1, 7)

Before we finish this breathtaking overview of the amazing life God created, we must firmly grasp why God made all of this. Why did God make you? He made all of creation to shine His glory! He made it so we would recognize the display of His glory. We are created to recognize beauty, wisdom, genius, power, and goodness—and we must declare it. We must declare God's glory and praise Him with every discovery of His creative genius. We must praise Him on every page of the science book, in every hour we spend in the science laboratory. We must praise Him every day we spend experiencing His vast and glorious creation! Amen and amen.

The twenty-four elders fall down before Him who sits on the throne and worship Him who lives forever and ever, and cast their crowns before the throne, saying:
"You are worthy, O Lord,
To receive glory and honor and power;
For You created all things,
And by Your will they exist and were created." (Revelation 4:10-11)

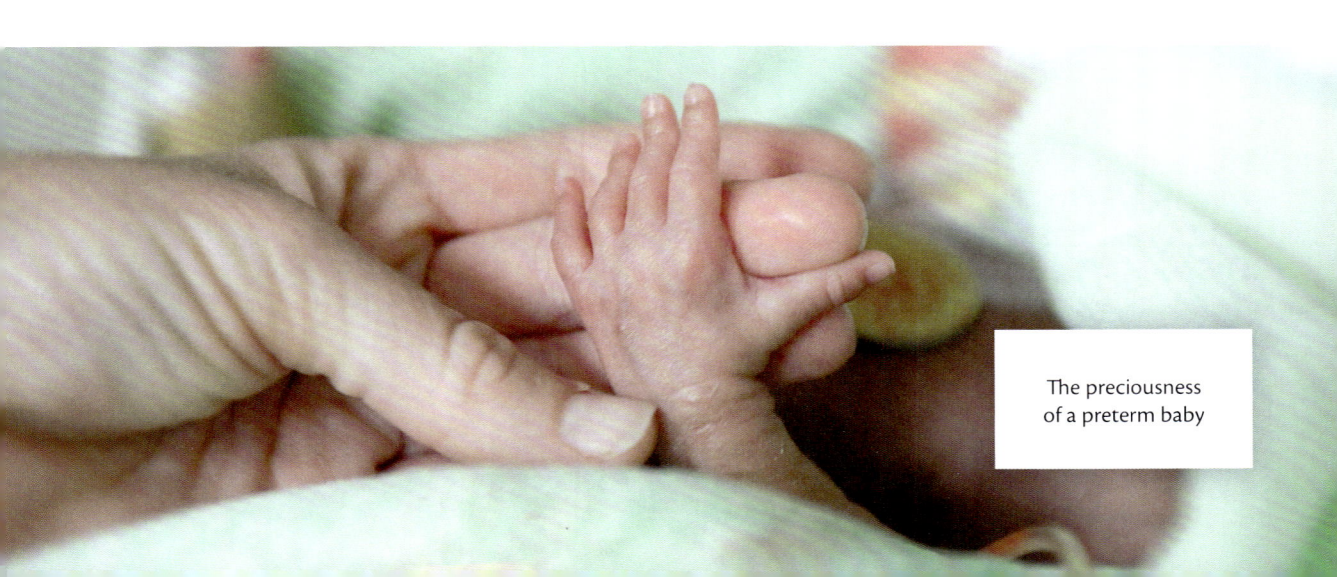

The preciousness of a preterm baby

GOD MADE LIFE

## Pray
- Take a moment to give God the glory for the wondrous creation of human life.
- Thank God for the precious births of your brothers and sisters, cousins, and friends. Thank Him for your life. Thank Him for your bodily organs, your skin, your muscles, and your bones. Thank Him for your immune system. Thank Him for your respiratory system and the air you breathe. Thank Him for your digestive system.
- Praise God for the amazing dexterity of the human hands and the facial muscles that can communicate so many expressions.
- Praise God for the ability to stand upright, to walk, and to run. Thank Him that He made us with dignity—He made us higher than the animal kingdom.
- Pray that our countries would respect human life and that dads and moms and doctors would take good care of babies in the womb.

## Sing
Now that we have seen the beauty of life development in the womb and the amazing work of God reflected in the skeletal-muscular system, the appropriate response must be worship and praise. If the student is unfamiliar with the hymn or psalm, some version of it is available on the internet and may be accessed (with supervision) for singing along.

*Hark the Herald Angels Sing*
Hark! The herald angels sing,
"Glory to the newborn King;
Peace on earth, and mercy mild,
God and sinners reconciled!"
Joyful, all ye nations rise,
Join the triumph of the skies;
With th' angelic host proclaim,
"Christ is born in Bethlehem!"

Refrain:
Hark! the herald angels sing,
"Glory to the newborn King!"

Christ, by highest Heav'n adored;
Christ, the everlasting Lord!
Late in time, behold Him come,
Offspring of a virgin's womb.
Veiled in flesh the Godhead see;
Hail th' incarnate Deity,
Pleased with us in flesh to dwell,
Jesus, our Emmanuel.

Hail the heav'nly Prince of Peace!
Hail the Sun of Righteousness!
Light and life to all He brings,
Ris'n with healing in His wings.
Mild He lays His glory by,
Born that man no more may die,
Born to raise the sons of earth,
Born to give them second birth.

Come, Desire of nations, come,
Fix in us Thy humble home;
Rise, the woman's conqu'ring Seed,
Bruise in us the serpent's head.
Now display Thy saving pow'r,
Ruined nature now restore;
Now in mystic union join
Thine to ours, and ours to Thine.

## Do

Choose at least one of the following activities and apply the lessons you have learned in this chapter.

1. Engage in a pro-life activity, encouraging friends to support life. Support a pro-life crisis pregnancy center. Write a letter to your legislator in the state government, encouraging them to support pro-life bills and oppose abortion on demand. Attend a pro-life rally with your parents at an abortion clinic or at a state capitol.

2. Your feet take a lot of abuse. After all, they carry your full body weight every day. Use the following checklist to test how you are treating your feet.

   - Foot cleanliness. Do you have athletes' foot or fungal growth? Do your shoes have adequate ventilation? Do you change your socks at least once a day? If your shoes get wet from sweat, do you air them out before putting them back on?

   - Shoe size. Are your shoes causing blisters or callouses? Are your shoes too loose, too big, or too small?

   - Shoe stability. Is your heel secured in the shoe? Do you limit the amount of time you wear crocs or flip-flops throughout the day?

   - Shoe rigidity. Does your shoe give you adequate shock absorption?

   - Flat feet. Are you flat footed? Do you need extra arch support?

## Watch

To watch the recommended videos for this chapter, go to **generations.org/GodMadeLife** and scroll down until you find the video links for Chapter 10. Our editors have been careful to avoid films with references to evolution, however, we would still encourage parents or teachers to provide oversight for all internet usage. These videos may not give God the glory for His amazing creative work, so the student and parent/teacher should respond to these insights with prayer and praise.

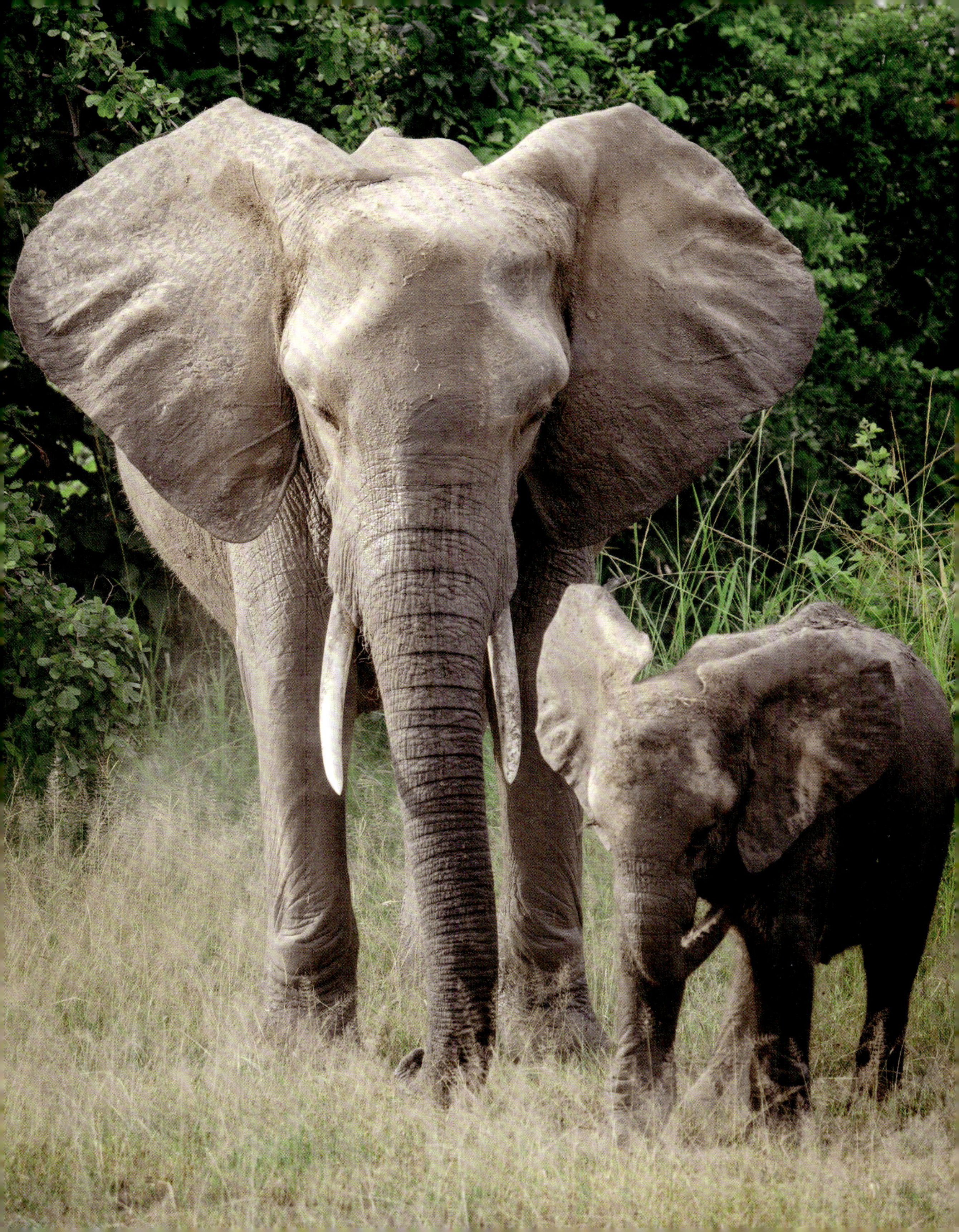

# NOTES

1. https://www.agriculturejournals.cz/publicFiles/37170.pdf
2. https://science.sciencemag.org/content/337/6102/1628.abstract
3. https://lozierinstitute.org/new-study-abortion-after-prenatal-diagnosis-of-down-syndrome-reduces-down-syndrome-community-by-thirty-percent/
4. https://www.nih.gov/news-events/nih-research-matters/common-genetic-factors-found-5-mental-disorders
5. Nolie Mumey and Edward Jenner, *Vaccination: bicentenary of the birth of Edward Jenner*, Vol. 1 (Range Press), 37.
6. John Baron, *The Life of Edward Jenner, MD*, Volv. 2 (Cambridge: Cambridge University Press, 2014), 446-447.
7. https://www.criver.com/sites/default/files/resources/MicrobialHotspotsandDiversityonCommonHouseholdSurfaces.pdf
8. https://www.sciencedaily.com/releases/2014/09/140929090354.htm
9. http://saspjournals.com/wp-content/uploads/2015/08/SJAVS-24A304-311.pdf
10. Outline taken from http://saspjournals.com/wp-content/uploads/2015/08/SJAVS-24A304-311.pdf.
11. https://www.ncbi.nlm.nih.gov/books/NBK232623/
12. https://www.ncbi.nlm.nih.gov/books/NBK544130/
13. https://www.tfah.org/wp-content/uploads/2020/09/TFAHObesityReport_20.pdf
14. https://www.webmd.com/diet/obesity/news/20181116/heres-more-evidence-obesity-can-shorten-your-life#1
15. https://www.medscape.com/answers/123702-11499/how-many-deaths-in-the-us-are-associated-with-obesity
16. https://www.ncbi.nlm.nih.gov/pmc/articles/PMC3560124/
17. https://esg.gc.cuny.edu/2020/02/18/the-opioid-crisis-in-the-new-york-area-a-first-look/
18. https://bestfoodimporters.com/mexico-food-imports-2018-main-importers-trends-and-best-strategies-for-market-entry/. The article states that Mexico is "the 7th largest food importer worldwide, despite a huge agricultural output and potential."
19. https://wiki.cancer.org.au/policy/Position_statement_-_Pesticides_and_cancer
20. Ibid.
21. https://nypost.com/2019/12/04/starbucks-holiday-themed-beverage-contains-23-teaspoons-of-sugar-study/
22. Charles Darwin, *The Origin of the Species*, Chapter 6.
23. https://www.ahajournals.org/doi/full/10.1161/hs0901.094253
24. https://www.cdc.gov/stroke/signs_symptoms.htm
25. https://www.hsph.harvard.edu/news/hsph-in-the-news/federal-government-calls-for-lowering-fluoride-levels-in-drinking-water/
26. Assistance with the outline of this section provided by Terry Mortenson, *Searching for Adam: Genesis & the Truth about Man's Origin* (Green Forest, AR: Master Books, 2016).
27. https://erlc.com/resource-library/articles/study-shows-preborn-babies-feel-pain-as-early-as-12-weeks-gestation/

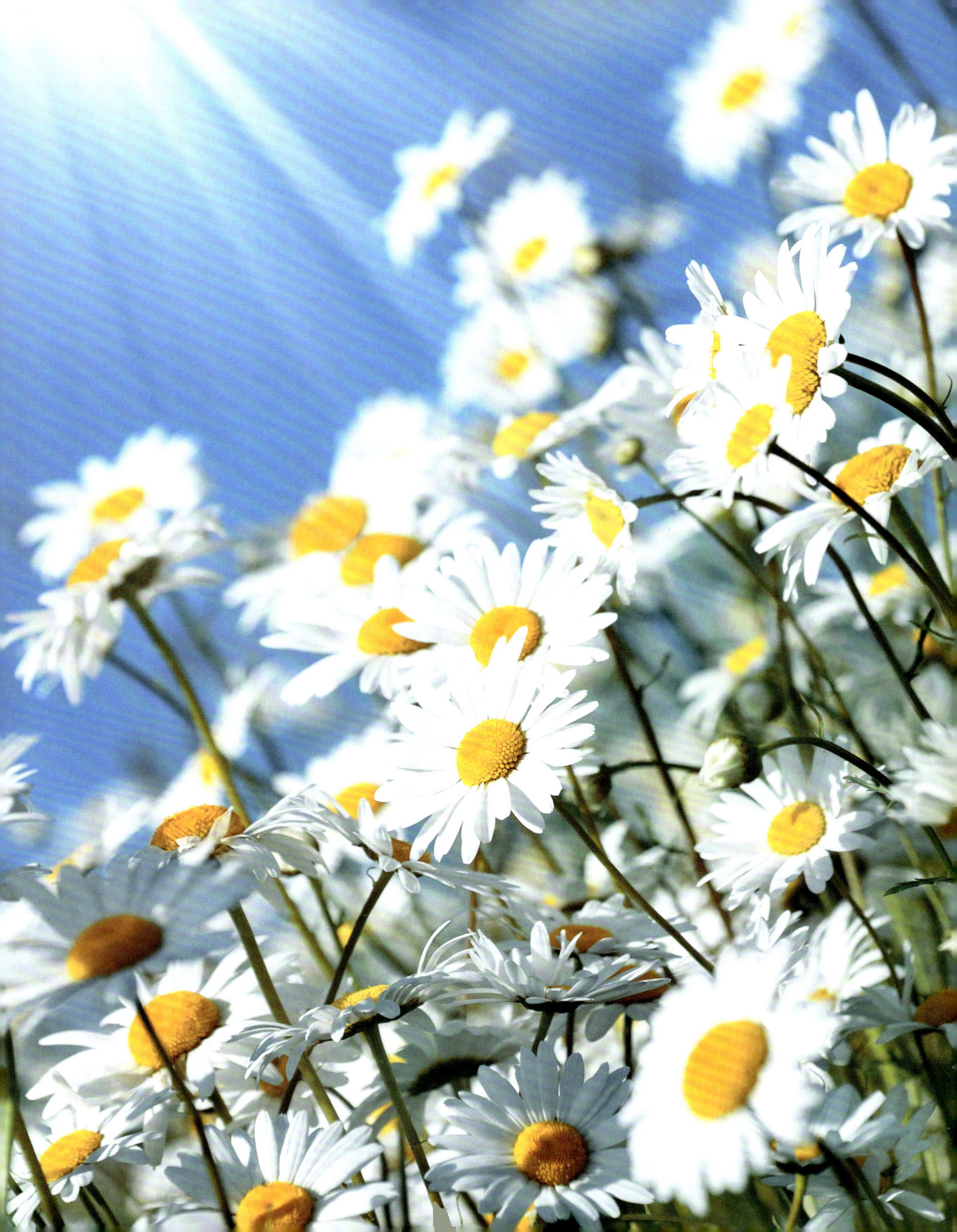

# IMAGE CREDITS

*All images from iStock.com, with the following exceptions:*

Page 17 — Alexander Fleming, Wikimedia Commons
Page 32 — Louis Pasteur, Wikimedia Commons
Page 36 — Photosynthesis, Wikimedia Commons
Page 38 — Robert Hooke, Wikimedia Commons
Page 47 — Flagellum motor, Wikimedia Commons
Page 71 — Gregor Mendel, Wikimedia Commons
Page 76 — Francis Galton, Wikimedia Commons
Page 84 — Carolus Linnaeus, Wikimedia Commons
Page 92 — Edward Jenner, Wikimedia Commons
Page 101 — Sneezes can go 27 feet, Wikimedia Commons
Page 131 — Avocado seed, Wikimedia Commons
Page 156 — Sir Richard Hawkins, Wikimedia Commons
Page 156 — James Lind, Wikimedia Commons
Page 200 — The human ear, Wikimedia Commons
Page 204 — The olfactory system in a human, Wikimedia Commons
Page 206 — Ganglia of insect, Wikimedia Commons
Page 211 — Whiptail lizard, Wikimedia Commons
Page 225 — "Lucy" skeleton, Wikimedia Commons
Page 246 — Neuron cells, Wikimedia Commons
Page 268 — James Blundell, Wikimedia Commons
Page 290 — Polio vaccine ad, Wikimedia Commons
Page 307 — Osteoarthritis, Wikimedia Commons

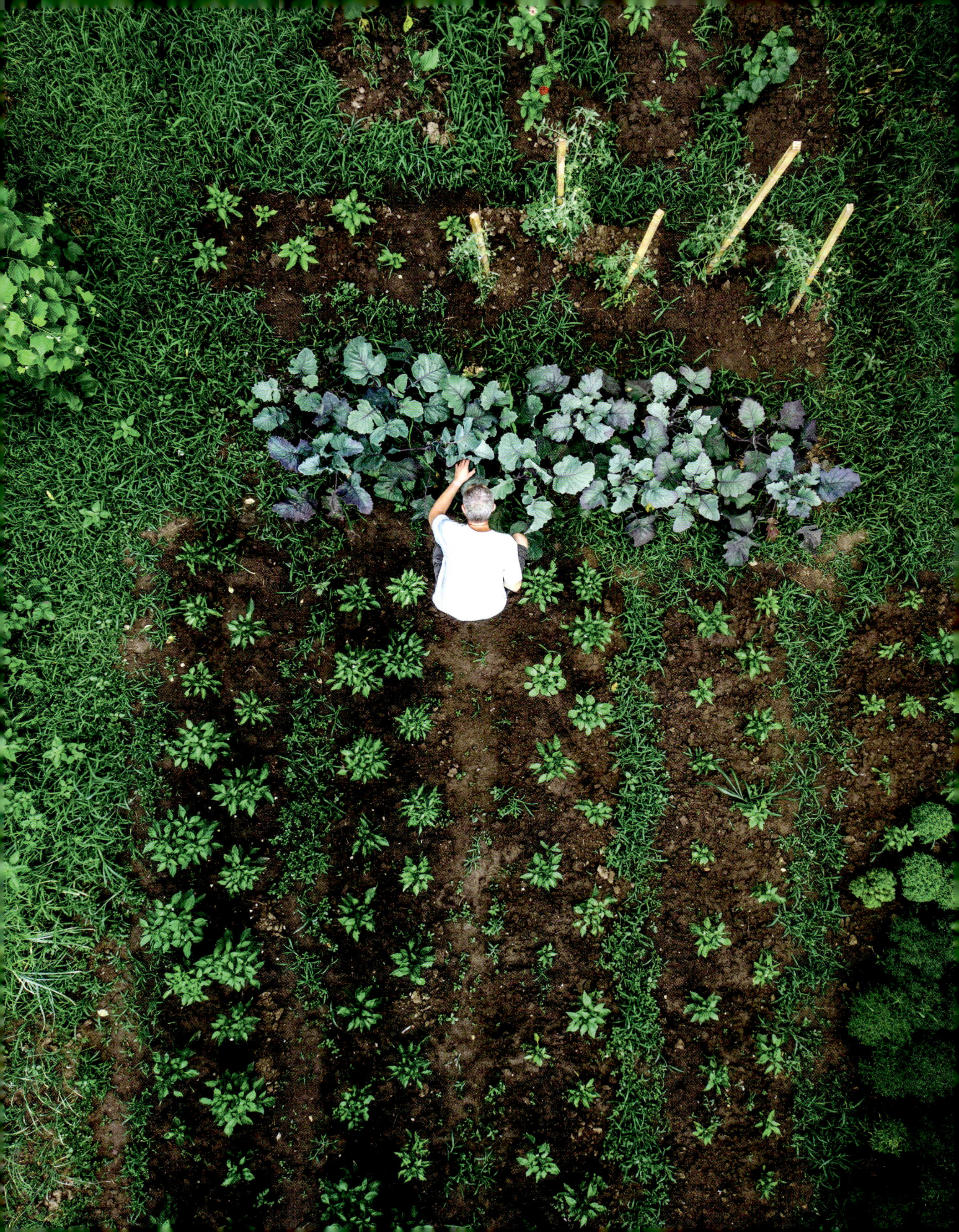

# INDEX

## A

Abiogenesis, 32
Abortion, 312
Abraham, 10–11, 73
Acids
    Amino, 14, 41, 64, 68–69
    Nucleic, 40, 42, 87, 89
Adam, 31, 60, 65–66, 73, 83, 134, 174–175, 183, 187, 223, 249, 285, 289, 301
Adenine, 67–68
Adrenals, 239, 246, 288
African Sleeping Sickness, 110
Agnatha, 221
Agur, 109, 250
AIDS, 87, 89–90, 92, 99, 110–111
Albinism, 74
Alcohol, 32, 167, 170, 172, 278, 297, 302
Algae, 108–109, 128, 192, 215
    Spirogyra, 109
Alimentary Canal, 275
Alleles, 71–72
Alveoli, 263–264, 268
Amazon Rainforest, 125
Amoebas, 34, 86, 109–110
Amylase, 278
Anabolism, 35
Anatomy, 32, 259, 263, 269
Anemia, 74–75, 77, 286
Anesthesia, 16, 302
Aneurysm, 270–271, 273
Angels, 30–31, 187, 221
Angiosperms, 127–128
Animate Creation, 191
Anthophyta, 127, 130

Antibiotics, 17, 74, 96–98, 113
Antibodies, 91, 93, 267, 290–291, 297
Anticoagulants, 272
Antigens, 266–267, 290–291
Antioxidants, 50
Anvil, 199
Aorta, 268–270
Apes, 30, 224, 249, 304–306, 309–310
Archaebacteria, 85–87, 187
Arphaxad, 73
Arrhythmia, 273
Arteries, 232, 268–273
Arthritis, 113, 168, 306
Aspiration, 259
Aspirin, 167
Atherosclerosis, 270
Atrium, 268–269
Autoimmune Diseases, 280, 291–292
Autotrophic, 86
Axon Terminals, 244

## B

Bacillus Cereus, 103
Bacteria, 17, 28, 36, 38, 45, 74, 83–87, 90, 95–99, 103–104, 107–108, 112–113, 129, 161, 187, 192
    Aerobic, 95
    Anaerobic, 95
    Bacterium, 95, 289–291
Bamboo, 141
Base Pairs, 66
Beriberi, 157–158
Beta-carotene, 50
Bile, 239, 276, 278

# GOD MADE LIFE

Binary Fission, 95, 97
Birds, 28, 39–40, 46, 94, 124, 175, 177, 188, 195, 202, 208, 214–215, 221, 262
Black Stem Rust, 113
Blastocyst, 300
Blood, 11–12, 29–30, 33, 39, 47, 63, 72, 75, 85, 87, 91, 110, 128–129, 144, 158, 160, 167–168, 189, 203, 210, 214, 216, 221, 225, 227–230, 232–233, 237–239, 243, 258, 263–274, 276, 286–291
    Clot, 62–64, 158, 160, 265, 271–273
Blundell, James, 267–268
Bombardier Beetle, 209
Bone, 37–38, 43, 49, 158–161, 190, 199, 202, 226–228, 230, 235, 238, 240, 265–266, 288, 300, 303–308
    Marrow, 265–266, 303
Botany, 32
    Botanist, 123, 127, 129
Bottom Feeders, 215
Botulinum, 107
Brain, 27, 110, 161, 163, 168, 192, 197, 199–201, 203–207, 222, 226–228, 238, 241–244, 246–250, 258–259, 264–265, 270–271, 288, 291, 298–300, 307–310
Breath, 11, 29–31, 33–34, 37, 85–87, 136, 176, 187, 189–191, 221, 224, 228, 246, 248, 257–259, 261–262, 264–265, 302, 313
Breeding, 74–75, 78, 128
Bronchi, 228, 263
Bronchioles, 263
Bubonic Plague, 96, 214
Buck, Carrie, 76
Budding, 64, 87
Burns, 168, 234–235, 244, 279

## C

Calcium, 13, 51, 160, 239, 270, 279, 286, 303, 306
Callus, 191, 232
Calories, 53, 56, 162–164, 166
Cancer, 8–9, 17, 47–50, 74, 165–166, 169, 178–179, 203, 234, 291
Capillaries, 230, 269–270
Carbohydrates, 35, 40, 43, 135, 159, 162, 172–173
Carbon, 20, 33, 36, 41, 43, 119, 122, 124, 135, 159, 238, 263–264, 270
    12, 20
    14 Dating, 20
    Dioxide, 36, 119, 122, 124, 135, 238, 263–264, 270
Carcinogens, 165
Carnivores, 36, 93, 215
Cars, 39, 44–45, 60, 125, 181, 223, 271, 286, 306
Cartilage, 258, 300, 303, 306–307
Catabolism, 35
Catalyst, 41, 209

Cats, 31, 33, 75, 93, 138, 198, 213–214, 221, 248
Cell-mediated Immunity, 291
Cesarean Section, 302
Chickenpox, 89, 99
Chihuahua, 74–75
China, 93, 96, 98, 121–122, 181
Chlorophyll, 122, 135, 145–146
Chloroplasts, 44, 55, 135, 191
Chondrichthyes, 221
Chromosome, 66–67, 294–295
Cilia, 43, 45, 109, 243, 258
Clam, 190, 193–194, 206
Classification, 84–85, 123, 127–128
Cochlea, 199–200
Codeine, 170
Codons, 67–68
Coffee, 12, 101, 203
Complement Response, 291
Cones, 242
    Pine, 59–60, 128, 142, 147–148
Conifers, 127–128, 142–143
Conscience, 31, 223, 250
Consumers, 36–37
Coronary Heart Disease, 270
Cotyledons, 133
Creation, 11, 27–33, 37, 39–40, 44, 59–60, 62, 69, 72, 78, 83–86, 89, 97, 110, 113, 123, 130–133, 140, 147–148, 170, 181, 183, 187–188, 190–191, 197, 200, 206–207, 209, 211–214, 222–223, 243, 250, 274, 299, 308, 313
    Creationists, 69
Crop (Invertebrate Digestion), 210
Crop Rotation, 179
Crustaceans, 194, 211
Cycads, 127–128, 142
Cyclosporine, 113
Cystic Fibrosis, 74–75, 77
Cytokinesis, 61
Cytoplasm, 33, 39, 43, 61, 87
Cytosine, 67–68
Cytoskeleton, 43
Cytotoxic Granules, 91

## D

Darwin, Charles, 18–19, 46, 70, 72, 76, 241
Damadian, Raymond, 17–18, 302
Death, 9–10, 22, 31, 33, 47–48, 60, 63, 73, 77, 84, 93, 96–97, 110–111, 137, 167–171, 249, 257, 266, 270, 285, 289–290, 312
Decapods, 194

# INDEX

Declaration of Independence, 18
Decomposers, 36, 112
Deepest Subcutaneous Layer, 231
Dehydration, 107, 273
Deimatic Behavior, 208
Dendrites, 244–245
Diabetes, 166–167, 203, 287
Dicots, 130–131
Digestive System, 44, 161, 210, 258, 274–276, 278, 280, 297
Diploid, 132
DNA, 38, 40–45, 47, 60–61, 63, 66–69, 73, 75–76, 86–87, 89, 95, 113, 120, 131–133, 165, 180, 191, 224, 226, 294, 297–298, 312
Doberman Pinscher, 203
Dogs, 31, 72, 138, 201, 203, 207, 213–214, 221–222, 248
Dolphins, 189, 201
Dominant, 71–72
Dominion, 31, 213–214
Dormant, 131
Down Syndrome, 75–77
Dragonfly, 195, 197, 207
Drugs, 89, 113, 167, 169–171, 297, 302–303
Dynein Arms, 109

# E

E. Coli, 101, 103
Eardrum, 199
Earthquakes, 18, 20, 202
Earthworm, 192–193, 196, 206, 210–211, 214
Echolocation, 201, 311
Ecosystem, 36–37
Ecuador, 98
Effector, 246
Eggs, 8–9, 103, 106, 120, 158, 160, 164, 173, 278
  Flower, 129–130, 133
Egypt, 96, 98
Eijkman, Christiaan, 157
Endocrine System, 239–240, 286–287
Endometrium, 296
Energy, 34–37, 39–40, 44, 47, 89, 95, 113, 135, 145–146, 158–160, 168, 172–173, 192, 200, 225–226, 230, 243, 264–265, 274, 276, 278, 287–288
Entamoeba Histolytica, 110
Enzymes, 14, 35, 41–42, 44, 62, 89, 91, 112, 161, 194, 209–210, 272, 274, 276, 278, 296
Epidermis, 231–232, 235
Epidural Block, 302
Epiglottis, 258–259
Epinephrine, 288

Erythrocytes, 225, 266
Esophagus, 206, 210, 228, 239, 259, 275–276
Estrogen, 295–296
Eubacteria, 85–87, 187
Eugenics, 76
Euglena, 109
Eukaryotic Cells, 38–39, 87, 108, 226
Evolution, 17, 19, 45–47, 69, 73–74, 76, 132, 188–189, 197–199, 207, 209, 223–225, 241, 304, 311
  Evolutionists, 45–47, 69, 73, 188–189, 197–198, 223–224, 304
Excretory Tubules, 211
Exercise, 8, 50, 102, 270, 273
Exoskeleton, 37, 190
Eye, 11, 13–14, 20, 61, 65–66, 72, 83, 100, 134, 138, 196–199, 201, 208, 229, 238, 240–243, 247–248, 258, 300, 311
  Compound, 196–197

# F

Fat, 33, 35, 40, 42–44, 54, 134, 159, 162–165, 167, 171–173, 225, 227, 230, 235, 237, 270, 276, 278, 288, 295, 297, 299
Fallopian Tubes, 294, 312
Fern, 127–130, 147
Fertilizer, 15–16, 36, 136, 177–178, 214
Fibonacci Sequence, 147–149
Fiddleheads, 130
Fimbriae, 39, 294
Flagella, 43, 45, 56, 192
Fleas, 96, 214
Fleming, Alexander, 17, 97, 113, 301
Flowers, 33, 35, 70–71, 123, 126–128, 130, 132–134, 137, 139–140, 147–148, 174, 177, 179, 196
Fly Larvae, 42
Food, 9, 12, 28, 31, 34–37, 39–40, 42, 44, 46–47, 50, 84, 86–87, 93, 95, 100–101, 103–104, 106–109, 111–113, 119–120, 122, 133–137, 139, 141, 144, 155, 158–167, 171–175, 178, 180–182, 191, 194–196, 202, 204, 207, 210–211, 213, 216, 222, 226, 228, 237–239, 249, 257–259, 261, 265, 269, 273–276, 278, 280, 288, 295, 297, 309
  Poisoning, 87, 103–105, 107
Foot, 87–88, 113, 130, 141, 166, 177, 193, 203, 206–207, 234, 240, 260, 276, 304–305
Fossils, 19–21, 130, 188, 224, 304
Frequencies, 201
Fronds, 130
Fruit, 40, 50, 73, 76, 103, 105–106, 119–120, 124, 128, 130, 134, 137–140, 155, 157–161, 164, 168–169, 173–174, 177–178, 180, 183, 195, 279–280, 289
FSF, 64
Fungi, 83–87, 112–113, 128, 177, 187, 210, 291

**323**

# GOD MADE LIFE

Fungicides, 177
Funk, Casimir, 157

## G

Galton, Francis, 76
Gamete, 132–133, 294
Gastrin, 279
Gender, 294–295
Gene, 61, 64, 66, 71–72, 75–78, 95, 109, 180–181, 294, 297–298
General Sherman, 125
Genetic Mutation, 19, 73–75
Genetics, 65–67, 70, 72, 75–76
Germ, 12, 17, 63, 87, 95, 100–101, 104, 197, 230, 239, 243, 257–258, 265–266, 274, 276–278, 285, 289–291, 302
Germination, 130–131, 147
Gestational Period, 296
Ginkgoes, 127, 142
Gizzard, 210–211
Glands, 133, 210, 228, 232, 233, 239, 258, 274, 287–289, 296
    Hypothalamus, 239, 287
    Pituitary, 239, 287
    Thymus, 239, 265, 291
    Thyroid, 239, 287
Global Flood, 18–21, 73, 97, 104, 130, 143, 188, 195, 197, 214, 221, 311
Glucose, 40, 135, 144, 172, 246, 286, 297
Glycogen Globules, 40
Glycoproteins, 297
GMO, 181
Grass, 120, 130, 141–142, 163, 174, 177, 179, 210
Grijns, Gerrit, 157
Growth, 12, 34, 39, 95, 112, 129, 158, 160, 239, 288–289
    Hormone, 288–289
Guanine, 67–68
Gymnosperms, 127–128, 143

## H

Hammer, 199
Harvard University, 66
Harvest, 15–16, 174, 177, 181–182
Hawkins, Sir Richard, 155–156
Heart, 8–10, 47, 70, 85, 111–112, 155, 160, 166–169, 182, 190, 211, 216, 221, 228–229, 237–240, 246, 265, 268–270, 272–274, 279–280, 298, 300, 307
    Attack, 63, 270
Heimlich Maneuver, 260
Hemoglobin, 75, 266, 286

Hemolymph, 29, 210–211
Hemophilia, 64, 74, 77
Herbivores, 36
Heterotrophic, 86
Histone, 66
HIV, 87, 89–90, 99, 110–111
Homeostasis, 286
Hooke, Robert, 38
Hormones, 55, 173, 239–240, 246, 265, 278–279, 287–289, 295–296
Humor
    Aqueous, 243
    Vitreous, 243
Humoral Immune Response, 290
Hydra, 60
Hydrochloric Acid, 276
Hyperion, 125
Hyperventilation, 264
Hyphae, 112
Hypoglycemia, 286

## I

Immune System, 11, 48, 50, 97, 113, 158, 160, 230, 239, 289–292
Immunotherapy, 48
Implantation, 296
Infant Mortality, 248, 301
Influenza, 89–90, 92, 99
Insects, 29, 37, 85, 136–137, 177, 179, 188–191, 194–195, 197–199, 202, 206, 209–214
Insulin, 226, 239, 246, 278
Interbred, 71, 128
Interferons, 91, 290
Interneuron, 247
Interphase, 60–61
Intestine, 210–211, 224, 239–240, 275–278, 280, 288, 300, 307
Intravenous Fluids, 107
Invertebrates, 29–30, 190–192, 197, 199, 205, 207, 210–212, 221–222
Ion
    Negative, 42
    Positive, 42
Irreducible Complexity, 45–46, 63, 199
Isopropyl Alcohol, 32

## J

Jellyfish, 60, 190–191, 193, 205–206
Jenner, Edward, 92, 301
Joint

# INDEX

Ball-and-Socket, 303
Fused, 303
Gliding, 303
Hinge, 303
Pivot, 303

## K

Kidneys, 161, 168, 228, 238, 240, 276–278, 298
Kingdoms, 84–86, 93, 95, 108, 112–113, 127–128, 187, 192, 204, 223, 311

## L

Lachrymatory-Factor Synthase, 14
Larynx, 238, 258–259, 261
Leprosy, 87, 99
Leukocytes, 265
Life
   Eternal, 31, 285
   Forms
      Higher, 29, 189
      Lower, 19, 29
   Physical, 27, 31 264
   Spiritual, 31
Light, 34, 40, 43, 95, 100, 131, 176, 191, 196–199, 207, 241–243, 247
Ligaments, 227, 302, 306, 308
Limpet, 275
Lind, James, 156
Linnaeus, Carolus, 84
Lipase, 278
Lipids, 40, 44, 55
Lister, Joseph, 301
Listeria, 103–104
Liver, 110, 215, 225, 276–278, 288, 292
   Cells, 225
Lyell, Charles, 18–19

## M

Magnesium, 51, 160
Manchineel, 138
Mantis Shrimp, 197
Masseter, 229
Mather,
   Cotton, 16, 91–92
Matter, 35–36, 40, 102, 134–135
Mechano-sensory Hairs, 196
Melanoma, 234

Memory
   Flashbulb, 250
   Long-term, 250
   Short-term, 249
Mendel, Gregor, 70–72
Mercy of God, 92–93, 168, 230, 251, 277
Mestral, George de, 28
Metabolic Rate, 159–160
Metabolism, 55, 159–160, 239, 287
Metamorphosis, 212
Metastasis, 48
Microbes, 83, 85–87, 89, 93, 101–103, 108, 112, 221, 233–234, 258
Microbiology, 16, 32, 84
Microorganisms, 32, 84–86, 113, 179
Microscopes, 14, 27, 35, 38, 83, 85, 87, 95–96, 144, 191, 196, 294
Microvilli, 226
Milk, 12, 103, 107, 121, 164, 173, 177, 214, 239, 247, 274, 280, 288, 296
Minerals, 40, 51, 107, 113, 120, 124, 129, 143–144, 158–162, 178, 265, 280, 303
Mitosis, 61, 64, 87, 95, 109, 294
Monocots, 127, 130
Morality, 31
Morphine, 170–171
Moses, 73
Mosquitoes, 28, 94, 110, 213–214
Moss, 127–129, 178
Motor Neuron, 247
Mourning Dove, 208
Mouth, 37, 100, 103, 192, 204, 206, 210, 228, 237–239, 257–258, 262, 274–276, 280, 300
MRI, 17–18
Muscle, 35, 43, 56, 96, 109, 155, 160, 173, 192–193, 202, 210, 226–228, 235, 240–241, 244, 246–247, 250, 261–262, 268, 273, 276, 287–288, 295, 297–298, 302–303, 305, 307–310
   Cardiac, 228–229, 307
   Cells, 39, 225
   Skeletal, 228–229
   Smooth, 228–229, 307
   Voluntary, 228–229, 307
Mutation, 19, 73–75, 165, 188
Mycorrhizae Fungi, 113
Myelin, 299
Myosin, 43, 312
Myriapods, 194

## N

Neanderthal Man, 224

Nerve, 39, 158, 168, 192, 199, 203–206, 226–228, 232, 235, 238, 241, 244–247, 250, 258, 299, 302–303
    Cells, 39, 226
Nervous System, 158, 160, 192, 243–244, 287, 308
    Central, 243
    Peripheral, 243
Neurotransmitters, 200, 245–246
Noah, 18, 73, 97, 130, 143, 214, 311
Node
    AV, 268
    Lymph, 290
    Sinoatrial, 268
Non-Vascular Plants, 128–129
Nose, 13–14, 35, 65, 67, 89, 101, 103, 202–204, 238, 250, 257–258, 262, 274, 295, 300, 303
Nucleotides, 42, 63, 66–68
Nucleus, 38, 43, 61, 66, 86, 95, 108, 191, 226, 266, 294
Nutrition, 39, 120, 124, 141, 144, 159, 161, 165, 175, 210, 270, 297

# O

Obesity, 50, 161, 166–167
Occipital Lobes, 199, 249
Octopus, 190, 193, 204, 206–207, 215, 222
Olfactory
    Bulbs, 203–204
    Receptor Neurons, 203
Omnivores, 36, 215
Onions, 13, 122, 138, 234
Opium, 169–170
Opossums, 208, 311
Optic Nerve, 199, 243
Organic
    Compounds, 32, 41, 181
    Farms, 178
Organs, 38, 43, 96, 130, 135, 161, 207, 215, 222, 227–228, 230–231, 237, 239, 243, 275–276, 287, 294, 297, 300, 307
    Organelles, 38–40, 43–44, 61, 87, 89, 108, 191, 226
        Storage Granule, 226
Ovary, 128, 132–134
Oxygen, 11, 75, 95, 120, 135–136, 189–190, 192, 210–211, 224–225, 238, 243, 259, 263–266, 268–270, 274, 286, 297, 303

# P

Painted Lady Butterfly, 195
Pancreas, 226, 239, 246, 276, 278–279
    Cells, 226, 279
Paramecia, 109

Parasite, 110–111, 213, 247
Parthenogenesis, 211
Pasteur, Louis, 17, 32, 301
Penicillin, 98, 113
Perennial Plants, 139, 142
Perforin, 91
Pesticides, 177–179
Petals, 132–133, 147–148
Ph Level, 129
Phagocytes, 291
Pharynx, 210, 238, 258–259, 276
Phloem, 144
Phosphate, 51, 161, 279
Photoreceptors, 199, 242
Photosynthesis, 36, 109, 135–137, 144–145, 191
Physiology, 32, 237
Phytochemicals, 50
Pigment, 74, 145–146
Pistil, 71, 133
Placenta, 297, 301
    Previa, 297
Planarian, 196, 211
Plants
    Annual, 139–140
    Biennial, 139–140
    Carnivorous, 136–137
    Cobra, 136–137
    Vascular, 128–130
Plasmin, 64
Plasmodium, 110
Platelets, 63, 265–266
Poliomyelitis, 290
Polymer, 40
Pressure
    Diastolic, 273
    Systolic, 273
Pride, 7–8, 20, 76
Probiotics, 103, 279
Producers, 36–37
Progesterone, 295–296
Prokaryotic, 38–39, 86–87, 95
Prolactin, 246–247
Prop Roots, 146
Prosthetic Limbs, 307, 309
Protease, 278
Proteins, 33, 35, 40–45, 62–64, 66–69, 87, 91, 97, 131, 159, 162, 171–173, 190, 192, 225–226, 230, 237, 265–266, 272, 276, 278, 290–291, 297, 303, 312

# INDEX

53BP1, 45
  Number Eight, 64
Protista, 85–87, 108, 110, 113
Protists, 108, 187–188
Protonema, 129
Protozoa, 45, 83, 86, 90, 108–110, 213, 289
Pulmonary Embolism, 272

## Q

Quarantine, 99
Quark, 40

## R

Rabbits, 59, 214, 275
Radiation, 49, 247
Rebuilding, 62, 64
Recessive, 71–72
Red Blood Cells, 75, 210, 225, 265–266, 269, 289, 295, 303, 311
Redwood Tree, 112, 125, 127, 135, 145
Reflexes, 247, 258
Reproduction, 39, 56, 59–60, 64, 86–87, 128–129, 132, 134, 160, 191, 211, 239, 293
  Asexual, 86
  Number, 99
  Sexual, 86
Resurrection, 31, 77, 292, 312
Retina, 199, 242–243
Rhinoceros Beetle, 194
Rhinovirus, 89
Rhizoid, 129
Rhizome, 130
Ribosomes, 38–39, 44, 47, 62, 68–69
Ringworm, 87, 113, 234
RNA, 43–44, 68–69, 75, 87
Robotic Limbs, 28
Rocks, 11, 18–20, 28, 30, 32–35, 69, 188, 197, 207, 275, 309
Rods, 242, 243
Röntgen, Wilhelm Conrad, 17, 302
Roots, 36, 113, 129–130, 137, 139–140, 142, 144–147, 175, 183, 191
  Adventitious, 146
  Aerial, 147
Rosary Pea, 138
Roundworms, 177, 196, 205, 213

## S

Saliva, 213, 274, 280

Salivary Glands, 274
Salmonella, 101, 103–104
Salt, 107, 161, 204, 233, 243, 273–274
Scabbing, 62
Science, 8–10, 12–20, 27–29, 32, 37–38, 40, 42–45, 49–50, 62–63, 66, 69–70, 75–76, 78, 83, 87, 91–92, 99–100, 109, 113, 127–128, 131–132, 157, 165, 167, 180, 188, 191, 194, 196, 197, 202, 209, 211, 223–224, 231, 233, 237, 249–250, 265, 276, 291–292, 296, 298, 301–302, 307, 313
Scurvy, 155–157
Semicircular Canals, 200
Sepals, 132–133
Shem, 73
Simpson, James Young, 301–302Siphon, 193
Sister Chromatids, 61, 66
Skeleton, 37, 43, 190, 224–225, 228, 302–303, 306–307
Skin, 38, 42, 48, 61–63, 74, 91, 103, 113, 155, 190, 203–205, 209, 227–228, 230–235, 239–240, 244, 247–248, 259, 265, 272, 287–288, 297, 302
Smoking, 48–49, 169, 302
Soil, 36–37, 104, 113, 131, 135–136, 142, 146–147, 175, 178–180, 182–183, 193, 196, 203
Soma, 244
Spinal Cord, 246–247
Spine, 303, 306
Spleen, 265, 291
Sponges, 101–102, 190–192
Spontaneous Generation, 32
Spores, 65, 87, 112, 128–131, 134
  Sporing, 65
Stamens, 132–133, 165
Staphylococcus, 95, 98
  Aureus, 98
Starch, 33, 274, 278
Starfish, 193–194
Stethoscope, 272
Stigma, 133
Stimuli, 34, 39, 44, 191–192
Stirrup, 199–200
Stolon, 112
Stomach, 42, 44, 49, 110, 161, 193–194, 204, 206, 228, 237, 258–260, 276, 279–280, 288, 307
Stomata, 135
Strengthening, 63–64
Streptococcus, 95–96
Stroke, 167, 169, 270–271
  Hemorrhagic, 271
  Ischemic, 271
  Transient Ischemic, 271

**327**

Style, 132–133
Succulents, 142
Sugar, 33, 36, 40, 44, 50, 113, 124, 135–136, 141, 144, 161–162, 166–167, 173, 203, 226, 234–235, 266, 276, 278–280, 286–289, 297
Sunlight, 34, 36–37, 74, 135, 177
Superbugs, 97–98
Sustainable Farming, 179
Sweat Glands, 232–233, 288
Symbiosis, 110

## T

T-lymphocytes, 291
Tapeworms, 190, 213
Taste, 107, 160, 195, 202, 204, 257
    Tastebuds, 204
Tear, 13–14, 146, 232, 234, 243, 285
    Ducts, 243
Tendons, 227, 302, 306–308
Testosterone, 295
Thiamine, 157
Throat, 89, 142, 204, 257–258, 261, 279
Thymine, 67
Tissue, 35, 38, 89, 96, 128, 226–230, 232, 235, 237, 261, 268, 288, 292, 297
    Connective, 230
    Epithelial, 228
    Regeneration, 292
Tobacco, 49, 169, 180
Tomato, 15–16, 50, 107, 124, 137, 158, 160, 172, 182
Tongue, 204, 248, 258, 262, 312
Tracheotomy, 261
Trait, 44, 65–66, 70–72, 112
Transpiration, 145
Trees
    Deciduous, 142
Trilobite, 197–198, 311
Tripod Grip, 309
Truth, 7–8, 10, 18, 22, 70, 73
Tsetse Fly, 110
Tuberculosis, 16, 87, 98, 110
Tympanic Membrane, 199

## U

Ukraine
Umbilical Cord, 296–297
Uracil, 67
Urine, 87, 228, 276–277

US Supreme Court, 76
Uterus, 293, 296–297

## V

Vaccine, 12, 16, 89, 92, 290
Vascular System, 144
Vegetable, 50, 76, 103, 105, 107, 120, 137, 139, 158–159, 161, 169, 171, 174, 178
Veins, 107, 128, 232, 266, 268–269, 270, 272, 297
Velcro, 28
Ventricle
    Left, 268–269
    Right, 268–269
Venus Flytrap, 37
Vertebrates, 29, 190, 210, 221–222, 224
Virus, 28, 87–93, 113, 213, 230, 234, 257–258, 289–291
Vitamin, 50, 157–162, 178, 240
    C, 50, 157–158
    E, 50, 158
Vocal Cords, 261–262
Volcanoes, 18, 20, 202

## W

Weed, 174–176, 179
White Blood Cells, 91, 139, 230, 265–266, 290–291, 295, 312
Wikipedia, 14
Windpipe (Trachea), 259, 261, 263
Wisdom, 14, 44, 62, 69, 78, 83, 93, 137, 148, 155, 183, 188, 191, 205, 233, 250, 262, 270, 301, 308, 313
Womb, 66–67, 75, 292, 294, 296–297, 301–302, 312–313
Wood, 40, 110, 120, 123–125, 134, 138, 141, 173, 178
Worldview, 77

## X

X-ray, 17, 49
Xylem, 144–145

## Y

Yeast, 64–65, 112, 233–234
Yersinia Pestis, 96

## Z

Zona Pellicida, 294, 296
Zoology, 32
Zygote, 132–133, 294, 296